STUDENT STUDY GUIDE

&

SELECTED SOLUTIONS MANUAL

FRANK L. H. WOLFS

UNIVERSITY OF ROCHESTER

FOURTH EDITION

VOLUME I

PHYSICS

for

SCIENTISTS & ENGINEERS

with Modern Physics

GIANCOLI

PEARSON

Prentice
Hall

Upper Saddle River, NJ 07458

Sponsoring Editor: Christian Botting
Assistant Managing Editor, Science: Gina Cheselka
Project Manager, Production: Kristy Mosch
Operations Specialist: Amanda Smith
Director of Operations: Barbara Kittle
Supplement Cover Manager: Paul Gourhan
Supplement Cover Designer: Victoria Colotta
Cover photographs top left clockwise: Daly & Newton/Getty Images,
Mahaux Photography/Getty Images, Inc.-Image Bank, The Microwave Sky
NASA/WMAP Science Team, Giuseppe Molesini, Istituto Nazionale di
Ottica Florence

© 2008 Pearson Education, Inc.

Pearson Prentice Hall

Pearson Education, Inc.

Upper Saddle River, NJ 07458

Printed in the United States of America

10 9 8 7 6 5 4 3 2

ISBN 13: 978-0-13-227324-4

ISBN 10:   0-13-227324-1

Pearson Education Ltd., *London*
Pearson Education Australia Pty. Ltd., *Sydney*
Pearson Education Singapore, Pte. Ltd.
Pearson Education North Asia Ltd., *Hong Kong*
Pearson Education Canada, Inc., *Toronto*
Pearson Educación de Mexico, S.A. de C.V.
Pearson Education—Japan, *Tokyo*
Pearson Education Malaysia, Pte. Ltd.

Contents

Preface

This *Student Study Guide & Selected Solutions Manual* has been prepared to accompany Volume 1 of the fourth edition of the textbook *Physics for Scientists & Engineers with Modern Physics* by Douglas C. Giancoli.

The study guide includes the following components for each chapter:
- **Chapter overview and objectives**. The list of objectives is comprehensive and your course may omit some of these objectives.
- **Summary of equations**. This list includes both the fundamental laws of physics in equation form and equations that result from application of the laws or principles to specific situations. It is important to know what each equation is stating and to what situations it can be applied.
- **Chapter summary**. The chapter summary provides a brief review of the principles covered in the chapter. Examples, illustrating these principles, are discussed.
- **Practice quiz**. A practice quiz is included at the end of each chapter. The quiz consists of several conceptual questions and several quantitative problems. The level of difficulty of both the conceptual questions and quantitative problems vary from relatively easy to difficult. The answers to the quiz questions are in the back of the study guide.
- **Question responses**. The responses to select odd end-of-chapter questions from the textbook are included in the study guide. These solutions were prepared by Katherine Whatley and Judith Beck.
- **Problem solutions**. The solutions to select odd end-of-chapter problems from the textbook are included in the study guide. These solutions were prepared by Bob Davis, J. Erik Hendrickson, and Michael Ottinger. To make the best use of these solutions, make an honest attempt to solve the problems directly from the text before looking at the solutions. Use the solutions to verify that you have correctly solved the problems or to find the step in the reasoning used to solve the problem that is giving you problems.

I have tried to assure that everything in this study guide is correct. I am solely responsible for any errors or omissions that remain in this study guide. I am happy to receive any feedback that readers send, both positive and negative, that may help me improve any future editions of this study guide.

I would like to thank everyone who assisted me directly or indirectly while preparing this study guide. I like to thank Christian Botting at Pearson for his assistance during this project. I very much appreciated being able to include select solutions to the end-of-chapter questions and problems, prepared by Katherine Whatley, Judith Beck, Bob Davis, J. Erik Hendrickson, and Michael Ottinger. And finally, I like to thank my wife Jean and my children Frank, Madeline, and Jeanie for giving me the time to work on this project and keeping me on track by continuing to ask "What chapter are you on?" Thanks Jean, Frank, Madeline, and Jeanie!!!!

Finally, I like to acknowledge three people who have had a significant influence on the career path I have chosen to follow. I like to thank my first physics teacher, Dr. Henk Klein Haneveld, who got me excited about physics. Henk, bedankt! I also like to thank my parents, who always have been very interested in what I am doing and taught me to follow my interests. Pama, bedankt!

Frank L. H. Wolfs
Department of Physics and Astronomy
University of Rochester
wolfs@pas.rochester.edu

Chapter 1: Introduction, Measurement, Estimating

Chapter Overview and Objectives

In this chapter, the nature of science is described. The terms scientific theory, model, and law are defined and the measurement process, its uncertainties, and units are described. This chapter also introduces the concept of dimensional analysis and how it is used in analyzing relationships between physical quantities.

After completing this chapter you should:
- Understand the nature of the process of science.
- Know the difference between scientific models, theories, and laws.
- Know that measurements always have some uncertainty and how these uncertainties can be reported.
- Know the SI units of length, mass, time, and their abbreviations.
- Know most of the common metric prefixes.
- Know the procedure for converting physical quantities from one set of units to another.
- Be able to make order of magnitude estimates of parameters encountered in everyday life.
- Understand how dimensional analysis can be used to analyze the relationships between physical quantities.

Summary of Equations

Percent uncertainty:
$$\text{percent uncertainty} = \frac{\text{uncertainty}}{\text{measured value}} \times 100\%$$
(Section 1-3)

Chapter Summary

Section 1-1. The Nature of Science

The goal of science is to develop an understanding of the phenomena we observe in the world around us. One of the basic components of science is making **observations**. Observations are made in an organized way by designing and performing experiments in the laboratory or other appropriate environment. The results of these observations are used to create theories that not only explain the observations, but are used to create quantitative testable predictions that predict the outcome of future observations. Since we never can observe a phenomenon under all possible conditions, it is not possible to prove that a theory is absolutely correct. However, a single observation that disagrees with a theory can be used to prove a theory wrong.

Section 1-2. Models, Theories, and Laws

Scientists create **models** of particular phenomena to help describe and understand their observations in terms of more familiar phenomena. Models may lead to additional experiments that can result in a deeper understanding of the phenomenon being studied. A good example of a model is the description of light waves in terms of the properties of water waves (which we can easily visualize).

The words model and theory are often used as synonyms, but these terms have quite different meanings in science. A **theory** is a general rule that applies to a variety of phenomena and is able to explain the observations with quantitative accuracy and precision. A good example of a theory is the electromagnetic theory of light that describes light in terms of electric and magnetic fields, which can be measured directly with the appropriate equipment.

A **law** is a very general concise rule that appears to describe the behavior of nature in general. A good example of a law is Newton's third law that states that forces come in pairs that are equal in magnitude and directed in opposite directions. This law applies under a wide range of conditions; it applies to the forces between charged particles, to the forces between the planets and the sun in our solar system, etc.

It is important to remember that the statements made by scientists concerning the behavior of the universe are descriptive statements. The theories and laws are general statements about the way the universe seems to behave.

Section 1-3. Measurement and Uncertainty; Significant Figures

Observations in physics involve the measurement of physical quantities. The measurement process usually includes a comparison to a standard (e.g., measuring the length of an object using a calibrated ruler). The result of a measurement usually has a limited precision. For a measurement to be useful, its estimated uncertainty due to limited precision and accuracy must be known. That precision can be stated explicitly, as in $d = 34.6 \pm 0.1$ cm. This notation implies that the quantity d is somewhere between 34.5 cm and 34.7 cm. Sometimes the **percent uncertainty** is the quantity that is reported. It is defined as the ratio of the uncertainty to the measured value, multiplied by 100. The percentage uncertainty of d is equal to

$$\text{percent uncertainty} = \frac{\text{uncertainty}}{\text{value}} \times 100\% = \frac{\pm 0.1 \text{ cm}}{34.6 \text{ cm}} \times 100\% = \pm 0.3\%$$

Often, particularly in textbooks, the uncertainty in a number is given implicitly by using **significant figures**. A physical quantity reported by a number such as 78.5 means 78.5 ± 0.05. The implied uncertainty is one half of the least significant figure reported.

When significant figures are used to represent the uncertainty in quantities, it is important to follow some rules regarding the number of significant figures in the results of calculations using these quantities. The result of a multiplication or division has as many significant figures as the number with the least number of significant figures used in the calculation (e.g., $1/3 = 0.3$ and not 0.333). The result of an addition or subtraction is no more precise than the least precise number used in the calculation (e.g., $4.8 - 0.62 = 4.2$; $4.80 - 0.62 = 4.18$). Intermediate results during calculations should keep an additional digit beyond the most significant figure to minimize cumulative round-off errors in the final result.

Numbers such as 3900 create an ambiguity in determining the number of significant figures. Are the two zero digits significant figures or are they only place holders to get the 3 and the 9 into the proper powers of ten? To avoid this ambiguity, numbers can be written in **scientific notation**. Scientific notation expresses numbers as a decimal number between 1 and 10, multiplied by a power of ten. If this were applied to the number 3900 above, it would be written as 3.9×10^2 if only the 3 and the 9 were significant figures. It would be written as 3.900×10^2 if all the digits were significant figures.

The uncertainty in the quantities we measure include contributions from both the **precision** of the measurement, which is a reflection of the reproducibility of the measurement, and the **accuracy** of the tools used, which is a reflection of how close a measurement made with these tools is to the true value.

Section 1-4. Units, Standards, and the SI System

Most measurements involve comparisons to a standard (e.g., a calibrated ruler). The value of the measurement is only meaningful if the standard is specified. Writing all quantities with **units** appended to the number representing the quantity provides this information.

The most common system of units used in scientific work is the **Systeme International** or **SI** system. In the SI system the standard of length is the meter, the standard of time is the second, and the standard of mass is the kilogram. These standards are defined as follows

- The standard unit of length (the **meter,** abbreviated as **m**) is defined as the distance light travels in a time of 1/299,792,458 of a second.
- The standard unit of time (the **second,** abbreviated as **s**) is defined as the time for 9,192,631,770 periods of the radiation emitted by a cesium atom when it transitions between two particular atomic states.
- The standard unit of mass (the **kilogram**, abbreviated as **kg**) is defined as the mass of a platinum-iridium cylinder kept in Paris, France.

In the metric system, larger and smaller units are built from the standard units by adding prefixes that represent powers of ten. The table on page 3 gives the prefix name, its abbreviation, and the power of ten it represents.

A second metric-based system is the **cgs** system of units. In the cgs system the centimeter is the standard of length, the second is the standard of time, and the gram is the standard of mass. A third system of units in use today is the **British engineering system**. In the British engineering system the standard of length is the foot, the standard of time is the second, and the standard of force is the pound.

Prefix	Abbreviation	Power
exa	E	10^{18}
peta	P	10^{15}
tera	T	10^{12}
giga	G	10^{9}
mega	M	10^{6}
kilo	K	10^{3}

Prefix	Abbreviation	Power
centi	c	10^{-2}
milli	m	10^{-3}
micro	μ	10^{-6}
nano	n	10^{-9}
pico	p	10^{-12}
femto	f	10^{-15}

Section 1-5. Converting Units

Quantities are often given in one type of units, but need to be converted to a different type of units. The process of converting units relies on the fact that multiplying a certain quantity by one does not change the quantity. In order to convert units we use **conversion factors**. A selection of frequently used conversion factors can be found in the front of the textbook.

Example 1-5-A: Unit conversions. What is the area of a one-foot by one-foot floor tile in square meters?

Approach: Use the conversion factors for length, shown in the front of the textbook, to convert feet to meters: 1 ft = 0.305 m.

Solution: The area of a one-foot by one-foot floor tile is one square foot. It can be converted to square meters by using the conversion factor 1 ft = 0.305 m:

$$1\,\mathrm{ft}^2 = 1\,\mathrm{ft}^2 \times \frac{1\,\mathrm{ft}}{1\,\mathrm{ft}} \times \frac{1\,\mathrm{ft}}{1\,\mathrm{ft}} = 1\,\mathrm{ft}^2 \times \frac{0.305\,\mathrm{m}}{1\,\mathrm{ft}} \times \frac{0.305\,\mathrm{m}}{1\,\mathrm{ft}} = 0.09\,\mathrm{m}^2$$

Note that the ratios are written as 0.305 m/ft instead of 1 ft/0.305 m in order to ensure that the units of ft cancel. If the book provided the conversion factor to convert meters to feet, 1 m = 3.28 ft, we would have solved the problem in the following manner:

$$1\,\mathrm{ft}^2 = 1\,\mathrm{ft}^2 \times \frac{1\,\mathrm{m}}{1\,\mathrm{m}} \times \frac{1\,\mathrm{m}}{1\,\mathrm{m}} = 1\,\mathrm{ft}^2 \times \frac{1\,\mathrm{m}}{3.28\,\mathrm{ft}} \times \frac{1\,\mathrm{m}}{3.28\,\mathrm{ft}} = 0.09\,\mathrm{m}^2$$

The answer is of course the same.

Section 1-6. Order of Magnitude: Rapid Estimating

There are times at which we want to make rough estimates of a quantity based on our intuition and experience. We may want to do this to avoid the time it takes for a detailed calculation or possibly we may not really know how to make a detailed calculation. At other times we might want to check a detailed calculation by making a rough estimate, frequently called an **order-of-magnitude estimate**. An order-of-magnitude estimate is usually accurate to within a factor of 10.

Example 1-6-A: Estimation. Estimate the number of grains of sand needed to make a beach volleyball court.

Approach: To solve this problem we need to estimate the volume of sand required and the volume of a grain of sand.

Solution: A volleyball court is approximately 10 m × 20 m. The sand must be about 0.2 m deep. This means the volume of sand needed is about $V = 10\,\mathrm{m} \times 20\,\mathrm{m} \times 0.2\,\mathrm{m} = 40\,\mathrm{m}^3$. The volume of a single grain of sand depends on how fine the sand is. Assuming each sand grain is a millimeter (10^{-3} m) on a side, we estimate that the volume of a single grain of sand is equal to $V_{grain} = 10^{-9}\,\mathrm{m}^3$. Dividing the total volume of sand by the volume of one grain gives us an estimate of the total number of grains:

$$N = \frac{V}{V_{grain}} = \frac{40}{10^{-9}} = 4 \times 10^{10}$$

Try counting the grains of sand on a court sometime and see how close this estimate is!

Section 1-7. Dimensions and Dimensional Analysis

The **dimension** of a quantity provides us with information about the base units or combination of base units that are used to quantify it. We should not confuse the dimensions of a quantity with the units of that quantity. For example, the width of a table can be quantified using units such as the meter, inch, and foot, but the dimension of the width will be length, [L]. The dimensions of mass and time will be [M] and [T], respectively.

When we derive an equation involving physical quantities, the dimensions on each side of the equation must be identical. A quantity with dimensions of time on one side of the equation cannot be equal to a quantity with dimensions of length divided by time; it does not make sense to add a quantity with a dimension of length to a quantity with a dimension of time. Applying dimensional analysis to an equation can be used to identify incorrect relationships, but it cannot be used to show that a relationship is correct.

Example 1-7-A: Dimensional analysis. Determine which of the following relationships cannot possibly be correct expressions for the volume of a cone with a regular hexagonal base with side s and height h:

a) $\pi s h^2$ b) $\frac{3}{2}\sqrt{3}s^2 h$ c) $\sqrt{3}sh/2$ d) s^3/h^2 e) $\pi s^2 h \cos(s/h)$

Approach: The dimension of volume is $[L^3]$. The correct expression for the volume of the cone, in terms of s and h, must thus have $[L^3]$ as its dimension.

Solution: The length of a side of the base, s, and the height of the cone, h, have a dimension [L]. Expressions of the form $s^i h^j$ have a dimension $[L^i][L^j] = [L^{i+j}]$. Any expression for the volume of the cone of form $s^i h^j$ can only be correct if $i+j = 3$. The dimensions of the relationships listed in the problem are

a) $[L^3]$ b) $[L^3]$ c) $[L^2]$ d) $[L^1]$ e) $[L^3]$

Only expressions a, b, and e have the correct dimensions. Expressions c and d have the wrong dimensions and thus cannot represent expressions for the volume of the cone. This problem illustrates that using dimensional analysis you can identify incorrect expressions, but there will be an infinite number of expressions that have the correct dimensions. Dimensional analysis is a quick way to check that your answer is consistent with the answer sought, but the consistency of the dimensional analysis does not prove that your solution is correct.

Example 1-7-B: The Atwood machine. In Chapter 4 you will examine the Atwood machine, which consists of two objects, with masses m_1 and m_2, connected via a string that is routed over a pulley. When the two objects have equal mass, they will remain at rest, but when their masses differ, the lighter object will move up while the heavier object will move down. We will see in Chapter 4 that the acceleration a of the objects is expected to be proportional to the gravitational acceleration g. Use dimensional analysis to obtain possible expressions for the acceleration a.

Approach: We are looking for an expression of the type $a = f(m_1, m_2)\, g$. Since the acceleration a and the gravitational acceleration g have the same dimensions, the function f must be dimensionless.

Solutions: If the masses of the two objects are equal, $m_1 = m_2$, the system will be at rest and $a = 0$ m/s^2. This immediately implies that the function f must be proportional to some positive power of $(m_1 - m_2)$. However, since f must be dimensionless, $(m_1 - m_2)$ must be divided by a quantity that has dimension [M]. This quantity most likely will be a function of m_1 and m_2, and a reasonable guess would be $(m_1 + m_2)$. Possible expressions for the acceleration a would be

$$a = \frac{m_1 - m_2}{m_1 + m_2}g \quad \text{or} \quad a = \left(\frac{m_1 - m_2}{m_1 + m_2}\right)^2 g \quad \text{or} \quad a = \sqrt{\frac{m_1 - m_2}{m_1 + m_2}}\,g \quad \text{or} \ldots\ldots$$

Although expressions such as

$$a = \frac{m_1 - m_2}{m_1}g \quad \text{and} \quad a = \frac{m_1 - m_2}{m_2}g$$

have the correct dimensions, they appear to be less likely since they do not have symmetry under the exchange of m_1 and m_2, which is expected to change the sign of the acceleration, but not its magnitude.

Practice Quiz

1. Estimate the number of revolutions a car tire must make for a car to travel one mile.
    a) 10
    b) 100
    c) 1,000
    d) 10,000

2. Given a measurement of the length of a stick to be 0.3 m, what would be an appropriate expression for the length in feet?
    a) 1 ft
    b) 0.98 ft
    c) 0.984 ft
    d) 0.1 ft

3. What is the conversion factor between cubic light-years (ly^3) and cubic meters (m^3)?
    a) $8.5 \times 10^{47} \ m^3/ly^3$
    b) $9.5 \times 10^{15} \ m^3/ly^3$
    c) $1.2 \times 10^{-48} \ m^3/ly^3$
    d) $9.0 \times 10^{31} \ m^3/ly^3$

4. An expression for the position of a certain car as a function of time is given by At^3/d where t is time and d has dimensions of length. What are the dimensions of A?
    a) $[L/T^2]$
    b) $[L^2/T^2]$
    c) $[L^2/T^3]$
    d) $[L^3/T^4]$

5. Estimate the number of crayons sold in the United States each year.
    a) 10^5
    b) 10^7
    c) 10^9
    d) 10^{11}

6. If the fractional uncertainty in x is 1%, what is the approximate fractional uncertainty in x^2?
    a) 10%
    b) 2%
    c) 1%
    d) 0.01%

7. If the fractional uncertainty in x is 1%, what is the approximate fractional uncertainty in x^n?
    a) $(1 + n)$%
    b) $(n \times 1)$%
    c) 1%
    d) $(1/n)$%

8. Estimate the number of word entries in a typical college dictionary of length 500 pages.
    a) 1,000
    b) 10,000
    c) 100,000
    d) 1,000,000

9. One year is often approximated as $\pi \times 10^7$ s. What fractional accuracy associated with using this approximation?
    a) 0.005%
    b) 0.5%
    c) 1.4%
    d) 1.4×10^5%

10. What would your hourly wages in dollars be if you earned a megadollar/gigasecond?
    a) $3.60
    b) $0.001
    c) $1.00
    d) $3600

11. Divide 1.546 by 0.05673 and report the result with the appropriate number of significant figures.

12. What is the percent uncertainty in the area of a rug that is a rectangle with dimensions 3.44 m by 4.67 m?

13. What percentage error is being made if a liter container is mistaken for a quart container?

14. Estimate the number of baseballs used in major league baseball during a season.

15. Quantity A has dimensions $[ML/T^3]$, quantity B has dimensions $[M/T]$, quantity C has dimensions $[L/T]$ and quantity D has dimensions $[M^2]$. Determine a possible expression for quantity A in terms of quantities B, C, and D.

Responses to Select End-of-Chapter Questions

1. (*a*) A particular person's foot. Merits: reproducible. Drawbacks: not accessible to the general public; not invariable (could change size with age, time of day, *etc.*); not indestructible.
   (*b*) Any person's foot. Merits: accessible. Drawbacks: not reproducible (different people have different size feet); not invariable (could change size with age, time of day, *etc.*); not indestructible.
   Neither of these options would make a good standard.

7. You should report a result of 8.32 cm. Your measurement had three significant figures. When you multiply by 2, you are really multiplying by the integer 2, which is exact. The number of significant figures is determined by your measurement.

Solutions to Select End-of-Chapter Problems

1. (*a*) 14 billion years = $\boxed{1.4 \times 10^{10} \text{ years}}$

   (*b*) $\left(1.4 \times 10^{10} \text{ y}\right)\left(3.156 \times 10^7 \text{s}/1 \text{ y}\right) = \boxed{4.4 \times 10^{17} \text{s}}$

7. To add values with significant figures, adjust all values to be added so that their exponents are all the same.

$$\left(9.2 \times 10^3 \text{s}\right) + \left(8.3 \times 10^4 \text{s}\right) + \left(0.008 \times 10^6 \text{s}\right) = \left(9.2 \times 10^3 \text{s}\right) + \left(83 \times 10^3 \text{s}\right) + \left(8 \times 10^3 \text{s}\right) = 100.2 \times 10^3 \text{s} = \boxed{1.00 \times 10^5 \text{s}}$$

   When adding, keep the least accurate value, and so keep to the "ones" place in the last set of parentheses.

13. Assuming a height of 5 feet 10 inches, then 5'10" = (70 in)(1 m/39.37 in) = $\boxed{1.8 \text{ m}}$. Assuming a weight of 165 lbs, then (165 lbs)(0.456 kg/1 lb) = $\boxed{75.2 \text{ kg}}$. Technically, pounds and mass measure two separate properties. To make this conversion, we have to assume that we are at a location where the acceleration due to gravity is 9.80 m/s².

19. (*a*) $\left(1 \text{km/h}\right)\left(\dfrac{0.621 \text{ mi}}{1 \text{ km}}\right) = 0.621 \text{mi/h}$, and so the conversion factor is $\boxed{\dfrac{0.621 \text{mi/h}}{1 \text{km/h}}}$.

   (*b*) $\left(1 \text{m/s}\right)\left(\dfrac{3.28 \text{ ft}}{1 \text{ m}}\right) = 3.28 \text{ ft/s}$, and so the conversion factor is $\boxed{\dfrac{3.28 \text{ ft/s}}{1 \text{ m/s}}}$.

   (*c*) $\left(1 \text{km/h}\right)\left(\dfrac{1000 \text{ m}}{1 \text{ km}}\right)\left(\dfrac{1 \text{ h}}{3600 \text{ s}}\right) = 0.278 \text{m/s}$, and so the conversion factor is $\boxed{\dfrac{0.278 \text{m/s}}{1 \text{km/h}}}$.

25. The textbook is approximately 25 cm deep and 5 cm wide. With books on both sides of a shelf, the shelf would need to be about 50 cm deep. If the aisle is 1.5 meter wide, then about 1/4 of the floor space is covered by shelving. The number of books on a single shelf level is then

$$\tfrac{1}{4}\left(3500 \, \text{m}^2\right)\left(\frac{1 \, \text{book}}{\left(0.25 \, \text{m}\right)\left(0.05 \, \text{m}\right)}\right) = 7.0 \times 10^4 \, \text{books} .$$

With 8 shelves of books, the total number of books stored is equal to

$$\left(7.0 \times 10^4 \, \frac{\text{books}}{\text{shelf level}}\right)\left(8 \, \text{shelves}\right) \approx \boxed{6 \times 10^5 \, \text{books}} .$$

31. Consider the diagram shown (not to scale). The balloon is a distance h above the surface of the Earth, and the tangent line from the balloon height to the surface of the earth indicates the location of the horizon, a distance d away from the balloon. Use the Pythagorean theorem.

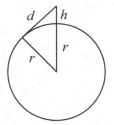

$$\left(r+h\right)^2 = r^2 + d^2 \;\rightarrow\; r^2 + 2rh + h^2 = r^2 + d^2$$

$$2rh + h^2 = d^2 \;\rightarrow\; d = \sqrt{2rh + h^2}$$

$$d = \sqrt{2\left(6.4 \times 10^6 \, \text{m}\right)\left(200 \, \text{m}\right) + \left(200 \, \text{m}\right)^2} = 5.1 \times 10^4 \, \text{m} \approx \boxed{5 \times 10^4 \, \text{m}} \left(\approx 80 \, \text{mi}\right)$$

37. (a) The quantity vt^2 has units of $(m/s)(s^2) = ms$, which do not match with the units of meters for x. The quantity $2at$ has units $(m/s^2)(s) = m/s$ which also do not match with the units of meters for x. Thus this equation $\boxed{\text{cannot be correct}}$.

    (b) The quantity $v_0 t$ has units of $(m/s)(s) = m$ and $(1/2)at^2$ has units of $(m/s^2)(s^2) = m$. Thus, since each term has units of meters, this equation $\boxed{\text{can be correct}}$.

    (c) The quantity $v_0 t$ has units of $(m/s)(s) = m$ and $2at^2$ has units of $(m/s^2)(s^2) = m$. Thus, since each term has units of meters, this equation $\boxed{\text{can be correct}}$.

43. Assume that the alveoli are spherical, and that the volume of a typical human lung is about 2 liters, which is .002 m^3. The diameter can be found from the volume of a sphere:

$$\tfrac{4}{3}\pi r^3 = \tfrac{4}{3}\pi\left(d/2\right)^3 = \frac{\pi d^3}{6}$$

$$\left(3 \times 10^8\right)\pi \frac{d^3}{6} = 2 \times 10^{-3} \, \text{m}^3 \;\rightarrow\; d = \left[\frac{6\left(2 \times 10^{-3}\right)}{3 \times 10^8 \, \pi} \, \text{m}^3\right]^{1/3} = \boxed{2 \times 10^{-4} \, \text{m}}$$

49. Make the estimate that each person has 1.5 loads of laundry per week, and that there are 300 million people in the United States.

$$\left(300 \times 10^6 \, \text{people}\right) \times \frac{1.5 \, \text{loads/week}}{1 \, \text{person}} \times \frac{52 \, \text{weeks}}{1 \, \text{year}} \times \frac{0.1 \, \text{kg}}{1 \, \text{load}} = 2.34 \times 10^9 \, \frac{\text{kg}}{\text{year}} \approx \boxed{2 \times 10^9 \, \frac{\text{kg}}{\text{year}}}$$

55. The person walks 4 km/h, 10 hours each day. The radius of the Earth is about 6380 km, and the distance around the Earth at the equator is the circumference, $2\pi R_{\text{Earth}}$. We assume that the person can "walk on water," and so ignore the existence of the oceans.

$$2\pi\left(6380 \, \text{km}\right)\left(\frac{1 \, \text{h}}{4 \, \text{km}}\right)\left(\frac{1 \, \text{d}}{10 \, \text{h}}\right) = \boxed{1 \times 10^3 \, \text{d}}$$

61. (*a*) Note that $\sin(15.0°) = 0.259$ and $\sin(15.5°) = 0.267$ and so $\Delta\sin\theta = 0.267 - 0.259 = 0.008$.

$$\left(\frac{\Delta\theta}{\theta}\right)100 = \left(\frac{0.5°}{15.0°}\right)100 = \boxed{3\%} \qquad\qquad \left(\frac{\Delta\sin\theta}{\sin\theta}\right)100 = \left(\frac{8\times10^{-3}}{0.259}\right)100 = \boxed{3\%}$$

(*b*) Note that $\sin(75.0°) = 0.966$ and $\sin(75.5°) = 0.968$ and so $\Delta\sin\theta = 0.968 - 0.966 = 0.002$.

$$\left(\frac{\Delta\theta}{\theta}\right)100 = \left(\frac{0.5°}{75.0°}\right)100 = \boxed{0.7\%} \qquad\qquad \left(\frac{\Delta\sin\theta}{\sin\theta}\right)100 = \left(\frac{2\times10^{-3}}{0.966}\right)100 = \boxed{0.2\%}$$

A consequence of this result is that when using a protractor, and you have a fixed uncertainty in the angle ($\pm0.5°$ in this case), you should measure the angles from a reference line that gives a large angle measurement rather than a small one. Note above that the angles around 75° had only a 0.2% error in $\sin\theta$, while the angles around 15° had a 3% error in $\sin\theta$.

67. (*a*) $\dfrac{SA_{Earth}}{SA_{Moon}} = \dfrac{4\pi R_{Earth}^2}{4\pi R_{Moon}^2} = \dfrac{R_{Earth}^2}{R_{Moon}^2} = \dfrac{\left(6.38\times10^3\,km\right)^2}{\left(1.74\times10^3\,km\right)^2} = \boxed{13.4}$

(*b*) $\dfrac{V_{Earth}}{V_{Moon}} = \dfrac{\frac{4}{3}\pi R_{Earth}^3}{\frac{4}{3}\pi R_{Moon}^3} = \dfrac{R_{Earth}^3}{R_{Moon}^3} = \dfrac{\left(6.38\times10^3\,km\right)^3}{\left(1.74\times10^3\,km\right)^3} = \boxed{49.3}$

Chapter 2: Describing Motion: Kinematics in One Dimension

Chapter Overview and Objectives

In this chapter, the quantities that are used to describe one-dimensional motion of objects are defined. These quantities include position, displacement, velocity, and acceleration. The relationships between these quantities are discussed and special cases, such as constant acceleration, are studied.

After completing this chapter you should:
- Know and understand the definitions of the kinematic variables used to describe one-dimensional motion.
- Know the equations of motion at constant acceleration.
- Know the integral relationships between the different kinematic variables.
- Know the SI units of kinematic variables.
- Be able to apply the definitions of kinematic variables and the equations of motion to solve problems about motion in one dimension.

Summary of Equations

Definition of average velocity: $\qquad \bar{v} = \dfrac{\Delta x}{\Delta t}$ $\qquad\qquad$ (Section 2-2)

Definition of instantaneous velocity: $\qquad v = \dfrac{dx}{dt}$ $\qquad\qquad$ (Section 2-3)

Definition of average acceleration: $\qquad \bar{a} = \dfrac{\Delta v}{\Delta t}$ $\qquad\qquad$ (Section 2-4)

Definition of instantaneous acceleration: $\qquad a = \dfrac{dv}{dt}$ $\qquad\qquad$ (Section 2-4)

Motion with constant acceleration: $\qquad v(t) = v_0 + at$ $\qquad\qquad$ (Section 2-5)

$$x(t) = x_0 + v_0 t + \frac{1}{2} a t^2 \qquad\qquad \text{(Section 2-5)}$$

Integral relationships between kinematic variables: $\quad v(t_2) = v(t_1) + \displaystyle\int_{t_1}^{t_2} dv = v(t_1) + \int_{t_1}^{t_2} a\, dt$ $\qquad$ (Section 2-8)

$$x(t_2) = x(t_1) + \int_{x_1}^{x_2} dx = x(t_1) + \int_{t_1}^{t_2} v\, dt \qquad\qquad \text{(Section 2-8)}$$

Chapter Summary

Section 2-1. Reference Frames and Displacement

Measurements of position, velocity, speed, and acceleration depend on the **frame of reference** of the observer. A coordinate system, defined by a set of coordinate axes, is used to describe positions within the frame of reference. The change in position from one location to another is defined as the **displacement**. Displacement is a vector quantity, a quantity that has magnitude (or size) and direction.

To describe motion in one dimension we do not yet need to introduce vectors. To specify a position in one dimension we specify an x coordinate. The magnitude of the x coordinate is the distance between the location and the origin of the coordinate system. The sign of the x coordinate indicates whether the position is located to the right of the origin ($x > 0$) or to the left of the origin ($x < 0$).

Section 2-2. Average Velocity

There are two closely related physical quantities used to describe motion, speed, and velocity. In everyday language, the names of those two quantities are often used synonymously, but in physics they have distinct meanings. The **average speed** is defined as the distance traveled by an object along its path divided by the amount of time it takes to travel that distance:

$$\text{average speed} = \frac{\text{distance traveled}}{\text{time elapsed}}$$

Since the distance traveled is always positive, the average speed is always positive.

The **average velocity** is defined as the displacement of an object divided by the time it takes to make that displacement:

$$\text{average velocity} = \bar{v} = \frac{\Delta x}{\Delta t} = \frac{x_2 - x_1}{t_2 - t_1}$$

The average velocity is positive if $x_2 > x_1$ (motion to the right) and negative if $x_2 < x_1$ (motion to the left). The unit of velocity is the m/s.

Example 2-2-A: Average speed and average velocity. A car travels along a road that runs east-west. It first travels east at 24.2 m/s for 562 s. It then turns around and travels back to the place it started in a time of 634 s. What is the average velocity for the entire trip? What is the average speed for the entire trip? What is the average velocity for the return trip?

Approach: In order to determine the average speed and the average velocity we have to determine the total distance traveled and the displacement.

Solution: At the end of its trip, the car returns to its starting position. The total displacement for the entire trip is thus zero and the average velocity for this trip, which is the displacement divided by the elapsed time, is zero. To determine the average speed for the trip, we need to know the total path length traveled by the car during the trip. We need to use the information about the eastward part of the trip to determine how far the car traveled east.

$$\Delta x = v\Delta t = 24.2 \times 562 = 1.36 \times 10^4 \text{ m}$$

The total path length is twice this distance because the car travels this distance again when returning to its starting position. The time the trip takes is 562 s + 634 s = 1196 s. The average speed for the entire trip is

$$\text{average speed} = \frac{\text{distance traveled}}{\text{time elapsed}} = \frac{2 \times (1.36 \times 10^4)}{(562 + 634)} = \frac{2.72 \times 10^4}{1196} = 22.7 \text{ m/s}$$

The average velocity during the second leg of the trip is the displacement during the second leg of the trip, 1.36×10^4 m west, divided by the time that leg takes, 634 s:

$$\bar{v} = \frac{\Delta x}{\Delta t} = \frac{1.36 \times 10^4}{634} \text{ west} = 21.5 \text{ m/s west}$$

Section 2-3. Instantaneous Velocity

The **instantaneous velocity** is defined as the average velocity over an infinitesimal small time interval. The instantaneous velocity is thus the time derivative of the position:

$$\text{instantaneous velocity} = v = \lim_{\Delta t \to 0} \frac{\Delta x}{\Delta t} = \frac{dx}{dt}$$

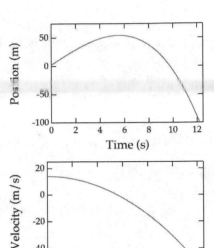

The **instantaneous speed** is the magnitude of the instantaneous velocity.

Example 2-3-A: Calculating the instantaneous velocity. The position of an object is given as $x(t) = 2.3 + 14t - 0.15t^3$. What is the velocity of the object as a function of time?

Approach: The problem provides us with the time dependence of the position of the object. The instantaneous velocity is obtained by differentiating the position with respect to time.

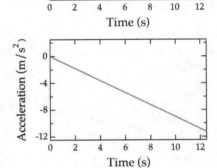

Solution:

$$v = \frac{dx}{dt} = \frac{d}{dt}\left(2.3 + 14t - 0.15t^3\right) = 14 - 0.45t^2$$

The Figure on the right shows the position and velocity as function of time between $t = 0$ s and $t = 12.5$ s. We note that the position has a maximum value at time $t = 5.6$ s. The velocity at this time is 0 m/s; it is positive at times $t < 5.6$ s and negative at times $t > 5.6$ s.

Section 2-4. Acceleration

Acceleration is a measure of the rate of change of velocity. The **average acceleration** is the change in velocity divided by the amount of time it takes to make the change:

$$\text{average acceleration} = \frac{\text{change in velocity}}{\text{time elapsed}} = \frac{v_2 - v_1}{t_2 - t_1} = \frac{\Delta v}{\Delta t}$$

The **instantaneous acceleration** is the limit of the average velocity over an infinitesimal small time interval, or the time derivative of the velocity:

$$\text{instantaneous acceleration} = \lim_{\Delta t \to 0} \frac{\Delta v}{\Delta t} = \frac{dv}{dt} = \frac{d^2 x}{dt^2}$$

The unit of acceleration is the m/s^2.

Example 2-4-A: Calculating the instantaneous acceleration. What is the instantaneous acceleration of the object discussed in Example 2-3-A?

Approach: In Problem 2-3-A we used the known time dependence of the position of the object to determine its velocity. Using the definition of the instantaneous acceleration in terms of the instantaneous velocity we can use the solution of Problem 2-3-A to calculate the instantaneous acceleration.

Solution:

$$a = \frac{dv}{dt} = \frac{d}{dt}\left(14 - 0.45t^2\right) = -0.9t$$

The Figure above shows the instantaneous acceleration as function of time. We see that the acceleration becomes more negative as time increases. Comparing the graph of the acceleration and the graph of the velocity we note that a negative acceleration does not require the velocity to be negative; it only requires the slope of the velocity versus time graph to be negative.

Example 2-4-B: Calculating the instantaneous acceleration. An object has a position x that depends on time t as

$$x(t) = 1.24t^4$$

What is the acceleration of the object at time t?

Approach: The problem provides us with the time dependence of the position of the object. The instantaneous acceleration is obtained by differentiating the position twice with respect to time.

Solution:

$$a = \frac{d^2x}{dt^2} = \frac{d}{dt}\frac{dx}{dt} = \frac{d}{dt}\frac{d}{dt}\left\{1.24t^4\right\} = \frac{d}{dt}\left\{4.96t^3\right\} = 14.9t^2$$

Section 2-5. Motion at Constant Acceleration

A common situation of accelerated motion is the case where the acceleration is constant (or approximated by a constant). In the case of constant acceleration, the relationships between position, velocity, acceleration, and time take the following form:

$$x(t) = x_0 + v_0 t + \frac{1}{2}at^2$$

$$v(t) = v_0 + at$$

Be careful to make sure that you only use these equations when the acceleration is constant.

Example 2-5-A: Bringing a car to a stop. An automobile is traveling with a speed of $v_0 = 28.5$ m/s when the driver sees a stop sign in front of him, a distance $d = 93.7$ m away. a) What constant acceleration must the automobile have so that the car comes to rest at the stop sign? b) How long does it take the automobile to reach the stop sign with this acceleration?

Approach: In this problem we are dealing with an object that moves with a constant acceleration. The initial and final positions and velocities are provided. Using this information we can determine the acceleration and the time required to come to a stop.

Solution: We will define time $t = 0$ s to correspond to the time that the driver sees the stop sign. We will choose our coordinate system such that at this time the driver is located at the origin of the coordinate system: $x(0) = 0$ m. If the car has come to rest at time t we must require that

$$v(t) = v_0 + at = 0 \qquad \text{or} \qquad t = -\frac{v_0}{a}$$

The position of the car at time t is equal to

$$d = x\left(-\frac{v_0}{a}\right) = v_0\left(-\frac{v_0}{a}\right) + \frac{1}{2}a\left(-\frac{v_0}{a}\right)^2 = -\frac{1}{2}\frac{v_0^2}{a}$$

The acceleration of the car must thus be equal to

$$a = -\frac{1}{2}\frac{v_0^2}{d}$$

It is always a good idea to check the units of your answer: $[L/T]^2/[L] = [L]/[T]^2$ which indeed is the unit of acceleration. Using the values provided in the problem we can conclude that the acceleration of the car is -4.33 m/s^2. Using this acceleration we can now determine the time at which the car will come to a stop:

$$t = -\frac{v_0}{a} = 6.58 \text{ s}$$

Example 2-5-B: Trying to beat the red light? A car is traveling at a speed of v_0 = 18.6 m/s and is entering an intersection of width d = 12.3 m. The light is yellow and will turn red in Δt = 0.582 s. Assume the driver accelerates with constant acceleration. What does the acceleration of the car need to be so that its front is out of the intersection when the light turns red? What will the velocity of the car be as it leaves the intersection if it has this acceleration?

Approach: In this problem we know the distance the car needs to travel (d = 12.3 m) in a time Δt = 0.582 s. Using the known initial velocity and assuming a constant acceleration we can calculate what acceleration is required to cross the intersection before the light turns red.

Solution: Assume the velocity of the car is a. We will choose the origin of our coordinate system such that it coincides with the position where the car enters the intersection. Time t = 0 s corresponds to the time at which the car enters the intersection. With these assumptions we can determine the position of the car as function of time:

$$x(t) = x_0 + v_0 t + \tfrac{1}{2} a t^2 = v_0 t + \tfrac{1}{2} a t^2$$

In order for the car to cross the intersection at time $t = \Delta t$ we must require that

$$d = x(\Delta t) = v_0 \Delta t + \tfrac{1}{2} a (\Delta t)^2$$

This relation can be used to determine the required acceleration a:

$$a = \frac{2(d - v_0 \Delta t)}{(\Delta t)^2} = \frac{2(12.3 - 18.6 \cdot 0.582)}{(0.582)^2} = 8.71 \text{ m/s}^2$$

Using the equations of motion for constant acceleration we can now calculate the velocity of the car when it has crossed the intersection:

$$v(\Delta t) = v_0 + a\Delta t = \frac{2d}{\Delta t} - v_0 = \frac{2 \cdot 12.3}{0.582} - 18.6 = 28.0 \text{ m/s}$$

As part of our calculation we need to make sure we check our units. The final velocity of 28.0 m/s corresponds to 63 mi/h. The driver may not get a ticket for running a red light, but may get a ticket for breaking the speed limit.

Example 2-5-C: Constant acceleration. A car is traveling with velocity of v_0 = 10.2 m/s and is accelerating with an acceleration a = 1.68 m/s^2 in the direction it is traveling. How long does it take to travel a distance d = 389 m?

Approach: Since the acceleration is constant, we can use the equations of motion for constant acceleration. The only unknown is the time t and this time can be determined from the information provided. We will see that there are two possible times, and we need to make sure that we understand how to determine which of these is the time we care about.

Solution: We will choose our coordinate system such that the origin coincides with the position of the car at time t = 0 s at which point it has a velocity of 10.2 m/s. The position of the car as function of time is equal to

$$x(t) = x_0 + v_0 t + \tfrac{1}{2} a t^2 = v_0 t + \tfrac{1}{2} a t^2$$

The time t at which the distance traveled by the car is equal to d can be found by solving the following equation:

$$\tfrac{1}{2} a t^2 + v_0 t - d = 0$$

There are two solutions for time t:

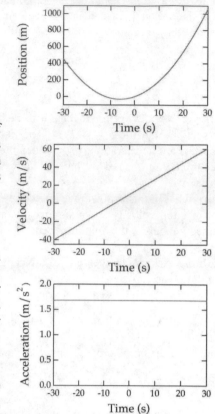

$$t = \frac{-v_0 \pm \sqrt{v_0^2 + 2ad}}{a} = \frac{-10.2 \pm \sqrt{(10.2)^2 + 2(1.68)(389)}}{1.68} =$$

$$= -6.07 \pm 22.4 = 16.3 \text{ s or } -28.4 \text{ s}$$

We can understand the reason for the two solutions if we examine the graph of position versus time, shown on the right. There are indeed two different times at which the position of the car is 389 m away from the origin. One solution is in the past ($t = -28.4$ s); the other solution is in the future ($t = 16.3$ s). It is the latter we care about.

Section 2-6. Solving Problems

It is very tempting to try to solve physics problems by searching for an equation that seems to apply and blindly plug in numbers to get an answer. This method of solving problems might produce some correct answers for relatively simple problems, but is sure to fail for even moderately complicated problems.   Just getting the right answers is not a measurement of your understanding physics; you must understand the logical process of getting to the answer. To assist you in learning this logical thought process, an outline of a general approach to problems is given:

1. Read the problem carefully, forming an image in your mind of the problem. Reread the problem to see if you missed any details and gain further understanding of what the situation is and what is being asked for.
2. Draw a diagram of the situation. Include a set of coordinate axes whenever needed.
3. Write down the quantities that are given in the problem and the quantities that are to be determined.
4. Think about what physical principles apply to the problem.
5. Determine which relationships relate the quantities that are known to those that are unknown. A single equation may not directly relate the knowns to the unknowns. You may need to use several different equations or multiple applications of the same relationship to solve the problem.
6. Carry out calculations if a numerical result is needed. Results of intermediate calculations should be kept to a precision of one or two significant figures more than the final solution to prevent cumulative round-off errors. Round off the final result to the appropriate number of significant figures.
7. Try to decide if the result is reasonable. Does it make sense to your intuition and experience? Make a rough estimate of your calculations using powers of ten (order-of-magnitude calculation). This may not catch all errors, but often catches silly mistakes that have been made.
8. Check to see that the units are correct. If the same units do not occur on both sides of the equation, you have made an error.

Section 2-7. Freely Falling Objects

Near the surface of the earth, objects in free fall have a constant acceleration in the vertical direction. This acceleration is the same for all objects. The **acceleration due to gravity** is 9.80 m/s^2, downward. The constant g is the magnitude of this acceleration: $g = 9.80$ m/s^2. In general, we choose the y axis of our coordinate system to coincide with the vertical direction; a positive y direction corresponds to an upward direction. With this choice of coordinate system we can write down the following equations of motion for the motion of a freely falling object:

$$v = v_0 - gt$$

$$y = y_0 + v_0 t - \frac{1}{2} g t^2$$

Example 2-7-A: Vertical motion near the surface of the earth. An object is thrown upward. It reaches a height of $h_1 = 10.7$ m above its release point and then reaches a height of $h_2 = 35.3$ m a time $\Delta t = 3.24$ s after it was at height of h_1. What was the velocity of the object as it passed height h_1? What was the initial upward velocity of the object at the time it was released?

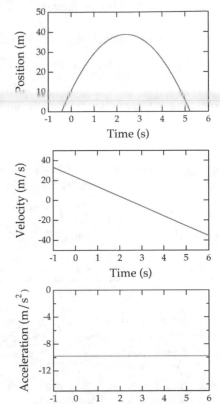

Approach: The problem provides us with information about the time difference between the times at which the object reaches two different heights. Although we are not provided with information about the initial velocity and the absolute times at which the object reaches heights h_1 and h_2, we will find that we have sufficient information to determine the motion of the object, which requires specifying its initial conditions. The starting point of our solution will be the equations of motion for constant acceleration. However, our choice of reference frame and reference time will make a big difference in the complexity of the solution.

Solution: To solve this problem we will assume that our time reference is such that time $t = 0$ s corresponds to the time at which the object is located a distance h_1 above the point of release, which is assumed to be located at the origin of our coordinate system. Note: we do **not** assume that the object is located at the origin of our coordinate system at time $t = 0$ s. Using this convention we can describe the vertical position of the object as follows

$$y(t) = h_1 + v_1 t - \frac{1}{2} g t^2$$

where v_1 is the velocity of the object at height h_1. The problem tells us that at time $t = \Delta t$ the object is located at $y = h_2$. This requires that

$$h_2 = y(\Delta t) = h_1 + v_1 \Delta t - \frac{1}{2} g \Delta t^2$$

The velocity v_1 can be calculated easily:

$$v_1 = \frac{(h_2 - h_1) + \frac{1}{2} g \Delta t^2}{\Delta t} = \frac{(35.3 - 10.7) + \frac{1}{2}(9.80)(3.24)^2}{3.24} = 23.5 \text{ m/s}$$

Using the equation of motion in the vertical direction we can determine at which time the object was thrown up by solving the following equation:

$$y(t_0) = h_1 + v_1 t_0 - \frac{1}{2} g t_0^2 = 0 \text{ m}$$

There are two solutions for time t_0:

$$t_0 = \frac{v_1 \pm \sqrt{v_1^2 + 2 g h_1}}{g} = -0.419 \text{ s or } 5.21 \text{ s}$$

The negative solution corresponds to the time the ball was thrown up while the positive solution corresponds to the time at which the ball returns to the surface (see the position and velocity graphs shown in the Figure to the right). The velocity with which the ball was thrown up can be now be determined:

$$v_0 = v(t_0) = v_1 - g t_0 = \sqrt{v_1^2 + 2 g h_1} = 27.6 \text{ m/s}$$

Section 2-8. Variable Acceleration; Integral Calculus

From the definition of velocity, we can write $dv = a\,dt$. Both sides of this equation can be integrated:

$$\int_{v'=v_0}^{v} dv' = \int_{t'=0}^{t} a\,dt'$$

Completing the integration on the left side of the equation we obtain

$$v - v_0 = \int_{t'=0}^{t} a\,dt' \qquad \text{or} \qquad v = v_0 + \int_{t'=0}^{t} a\,dt'$$

Similarly, from the definition of velocity, we can write $dx = v\,dt$. Again, both sides of this equation can be integrated:

$$\int_{x'=x_0}^{x} dx' = \int_{t'=0}^{t} v\,dt'$$

Completing the integration on the left side of the equation and substituting the above expression for v we obtain

$$x - x_0 = \int_{t'=0}^{t} \left\{ v_0 + \int_{t''=0}^{t'} a\,dt'' \right\} dt' \qquad \text{or} \qquad x = x_0 + v_0 t + \int_{t'=0}^{t} \int_{t''=0}^{t'} a\,dt''\,dt'$$

Example 2-8-A: Motion with a time-dependent acceleration. An object starts from rest at time $t = 0$ s and has an acceleration given by $a = \alpha t^2 + \beta t$, where $\alpha = 3.07$ m/s^4 and $\beta = -4.21$ m/s^3. a) Determine the velocity of the object at time $t = 4.55$ s. b) Determine how far the object is from its starting point at time $t = 4.55$ s.

Approach: Since the acceleration changes as function of time, we cannot use the equations of motion for constant acceleration to solve this problem. Instead we will have to use the integral relations between position, velocity, and acceleration to obtain the equations of motion.

Solution: The velocity of the object at time t can be found by integrating the acceleration between time $t = 0$ s and time t, taking into consideration that the velocity at time $t = 0$ s is 0 m/s:

$$v(t) = v(0) + \int_0^t a\,dt' = \int_0^t \left\{ \alpha t'^2 + \beta t' \right\} dt' = \frac{1}{3}\alpha t^3 + \frac{1}{2}\beta t^2$$

Using the dimensions of α and β we immediately see that the units of the right-hand side is indeed the unit of velocity. Using the values of α and β provided in the problem we find that the velocity at $t = 4.55$ s is 52.8 m/s.

The expression for the velocity as function of time can now be used to determine the position as function time. If we make the assumption that the origin of our coordinate system is chosen such that it coincides with the position of the object at time $t = 0$ s, we find for $x(t)$

$$x(t) = x(0) + v(0)t + \int_0^t \left\{ \frac{1}{3}\alpha t'^3 + \frac{1}{2}\beta t'^2 \right\} dt' = \frac{1}{12}\alpha t^4 + \frac{1}{6}\beta t^3$$

Using the values of α and β provided in the problem we find that the position at $t = 4.55$ s is 43.5 m.

Section 2-9. Graphical Analysis and Numerical Integration

If we have an analytical expression for the position of an object we can obtain the velocity and acceleration by differentiating the expression for the position with respect to time. If we have an analytical expression for the acceleration of an object we can obtain the velocity by integrating the expression of the acceleration with respect to time. The process of integrating the acceleration between two time intervals is equivalent to finding the area under the curve that shows the acceleration versus time. Even if no analytical expression exists for the acceleration or its integral, we can still find the velocity by using numerical integration techniques, assuming we can determine the acceleration at discrete time intervals.

Example 2-9-A: Finding the velocity using numerical integration.
Consider that the velocity of an object depends on time in the following manner (see also the graph shown on the right):

$$v(t) = 5t + 10t^2$$

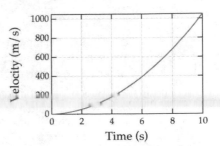

Assuming that the object starts at the origin of our coordinate system at time t = 0 s, calculate its position at t = 10 s using numerical and analytical techniques.

Approach: In this particular case, we have an analytical expression for the velocity that can be integrated easily. The analytical method will be faster in this particular case, but the problem wants you to use both methods to compare the results of these different techniques.

Solution: We will first solve the problem using the analytical approach. The position at time t = 10 s is equal to

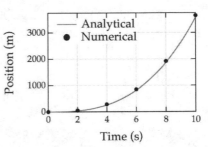

$$x(t = 10\text{ s}) = x(t = 0\text{ s}) + \int_0^{10}\left(5t + 10t^2\right)dt = \left(\frac{5}{2}t^2 + \frac{10}{3}t^3\right)\Bigg|_0^{10} = 3600\text{ m}$$

The dependence of position on time is shown by the solid curve in the Figure to the right.
In order to apply numerical integration of the velocity, we need to divide our time interval in small time steps. In this example we will use 2-second wide intervals. For each interval we will calculate the average velocity based on the initial and the final velocity. The results for each of the five intervals are summarized in the following table.

Interval	1	2	3	4	5
$[t_{\text{left}}, t_{\text{right}}]$	[0, 2] s	[2, 4] s	[4, 6] s	[6, 8] s	[8, 10] s
$[v_{\text{left}}, v_{\text{right}}]$	[0, 50] m/s	[50, 180] m/s	[180, 390] m/s	[390, 680] m/s	[680, 1050] m/s
v_{aver}	25 m/s	115 m/s	285 m/s	535 m/s	865 m/s
$v_{\text{aver}} \Delta t$	50 m	230 m	570 m	1070 m	1730 m
$\Sigma\ v_{\text{aver}} \Delta t$	50 m	280 m	850 m	1920 m	3650 m

In order to determine the position as function of time we use the following approximation:

$$x(t) = \sum_0^t v_{aver}\Delta t$$

The position after 10 s is found to be 3,650 m, which is close to the result of the analytical calculation (3,600 m). The results of the numerical calculation at various times is shown by the solid data points in the position versus time graph shown in the Figure above. The results of the numerical calculation are in good agreement with the results of numerical calculations over the entire time interval.

Practice Quiz

1. Over a particular time interval t, an object has an average speed of 10.67 m/s. What can you conclude about the magnitude of the average velocity of the object during that same time interval?
    a) The magnitude of the average velocity is equal to 10.67 m/s.
    b) The magnitude of the average velocity could be any value.
    c) The magnitude of the average velocity is less than or equal to 10.67 m/s.
    d) The magnitude of the average velocity is zero.

2. At a given instant, the instantaneous speed of a car is 23.45 m/s. What can you conclude about the magnitude of the instantaneous velocity?
    a) The magnitude of the instantaneous velocity is equal to 23.45 m/s.
    b) The magnitude of the instantaneous velocity could be any value.
    c) The magnitude of the instantaneous velocity is less than or equal to 23.45 m/s
    d) The magnitude of the instantaneous velocity is zero.

3. The velocity and the acceleration of an object are pointing in opposite directions. What is the motion of the object at the time those conditions are true?
    a) The object is traveling with constant speed and direction.
    b) The object is traveling with constant speed but is reversing direction.
    c) The object is speeding up.
    d) The object is slowing down.

Questions 4 through 7 refer to the graph of position versus time, shown on the right.

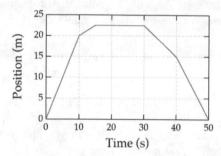

4. During which time interval on the graph is the average speed the greatest?
    a) 0 s to 10 s
    b) 10 s to 15 s
    c) 15 s to 30 s
    d) 30 s to 40 s
    e) 40 s to 50 s

5. During which time interval on the graph is the average speed the least?
    a) 0 s to 10 s
    b) 10 s to 15 s
    c) 15 s to 30 s
    d) 30 s to 40 s
    e) 40 s to 50 s

6. During which time interval is the average velocity the largest?
    a) 0 s to 10 s
    b) 10 s to 15 s
    c) 15 s to 30 s
    d) 30 s to 40 s
    e) 40 s to 50 s

7. During which time interval is the average velocity the smallest?
    a) 0 s to 10 s
    b) 10 s to 15 s
    c) 15 s to 30 s
    d) 30 s to 40 s
    e) 40 s to 50 s

8. A car changes its velocity from 12.6 m/s to 26.7 m/s in a time such that its average acceleration is 4.8 m/s^2. If at a later time the time required for the same velocity change to occur doubled, what would the average acceleration be at that time?
    a) 4.8 m/s^2
    b) 2.4 m/s^2
    c) 1.2 m/s^2
    d) 9.6 m/s^2
    e) 19.2 m/s^2

9. Car A moves from point x_1 to point x_2 with constant velocity in a time t. Car B moves from point x_1 to point x_2 with constant acceleration after starting from rest in the same time t. Which car has the greater average velocity in going from x_1 to x_2?
   a) A
   b) B
   c) Neither; the average velocity is the same for both cars.
   d) Not enough information is given to determine which had the greater average velocity.

10. Which car in problem 9 reaches the greater instantaneous speed in going from point x_1 to x_2?
    a) A
    b) B
    c) Neither; the average velocity is the same for both.
    d) Not enough information is given to determine which had the greater average velocity.

11. A ball is thrown upward vertically in the air with a speed of 24.2 m/s and is released from a height of 2.18 m above the ground. What is the maximum height that the ball reaches above the ground?

12. How long does it take the ball in question number 11 to hit the ground from the time it is released?

13. A race car accelerates from rest with a constant acceleration of 5.31 m/s² for 8.05 s and then travels with constant velocity. A second race car accelerates from rest at the same time as the first car with a constant acceleration of 4.78 m/s² for 8.95 s and then travels with constant velocity. Which car wins the race if the track is 400 m long?

14. A car travels with a constant speed of 30 mph for a distance of 45 miles. The car then travels with a constant speed of 60 mph for another 45 miles. What is the average speed of the car for the 90-mile trip?

15. A ball is thrown downward with an initial speed of 30 m/s from a height of 135 m from the ground. How long does it take to hit the ground?

Responses to Select End-of-Chapter Questions

1. A car speedometer measures only speed, since it gives no indication of the direction in which the car is traveling.

7. Yes. If the velocity and the acceleration have different signs (opposite directions), then the object is slowing down. For example, a ball thrown upward has a positive velocity and a negative acceleration while it is going up. A car traveling in the negative x-direction and braking has a negative velocity and a positive acceleration.

13. Average speed is the displacement divided by the time. If the distances from A to B and from B to C are equal, then you spend more time traveling at 70 km/h than at 90 km/h, so your average speed should be less than 80 km/h. If the distance from A to B (or B to C) is x, then the total distance traveled is $2x$. The total time required to travel this distance is $x/70$ plus $x/90$. Then

$$\bar{v} = \frac{d}{t} = \frac{2x}{x/70 + x/90} = \frac{2(90)(70)}{90 + 70} = 79 \text{ km/h}$$

19. The object begins with a speed of 14 m/s and increases in speed with constant positive acceleration from $t = 0$ until $t = 45$ s. The acceleration then begins to decrease, goes to zero at $t = 50$ s, and then goes negative. The object slows down from $t = 50$ s to $t = 90$ s, and is at rest from $t = 90$ s to $t = 108$ s. At that point the acceleration becomes positive again and the velocity increases from $t = 108$ s to $t = 130$ s.

Solutions to Select End-of-Chapter Problems

1. The distance of travel (displacement) can be found by rearranging Eq. 2-2 for the average velocity. Also note that the units of the velocity and the time are not the same, so the speed units will be converted.

    $$\bar{v} = \frac{\Delta x}{\Delta t} \rightarrow \Delta x = \bar{v}\Delta t = \left(110\,\text{km/h}\right)\left(\frac{1\,\text{h}}{3600\,\text{s}}\right)\left(2.0\,\text{s}\right) = 0.061\,\text{km} = \boxed{61\,\text{m}}$$

7. The distance traveled is 116 km + 0.5(116 km) = 174 km and the displacement is 116 km - 0.5(116 km) = 58 km. The total time is 14.0 s + 4.8 s = 18.8 s.

    (*a*) Average speed $= \dfrac{\text{distance}}{\text{time elapsed}} = \dfrac{174\,\text{m}}{18.8\,\text{s}} = \boxed{9.26\,\text{m/s}}$

    (*b*) Average velocity $= v_{\text{avg}} = \dfrac{\text{displacement}}{\text{time elapsed}} = \dfrac{58\,\text{m}}{18.8\,\text{s}} = \boxed{3.1\,\text{m/s}}$

13. (*a*) The area between the concentric circles is equal to the length times the width of the spiral path.

    $$\pi R_2^2 - \pi R_1^2 = wl \rightarrow$$

    $$l = \frac{\pi\left(R_2^2 - R_1^2\right)}{w} = \frac{\pi\left[\left(0.058\,\text{m}\right)^2 - \left(0.025\,\text{m}\right)^2\right]}{1.6\times10^{-6}\,\text{m}} = 5.378\times10^3\,\text{m} \approx \boxed{5400\,\text{m}}$$

    (*b*) $5.378\times10^3\,\text{m}\left(\dfrac{1\,\text{s}}{1.25\,\text{m}}\right)\left(\dfrac{1\,\text{min}}{60\,\text{s}}\right) = \boxed{72\,\text{min}}$

19. The average speed of sound is given by $v_{\text{sound}} = \Delta x/\Delta t$, and so the time for the sound to travel from the end of the lane back to the bowler is $\Delta t_{\text{sound}} = \dfrac{\Delta x}{v_{\text{sound}}} = \dfrac{16.5\,\text{m}}{340\,\text{m/s}} = 4.85\times10^{-2}\,\text{s}$. Thus, the time for the ball to travel from the bowler to the end of the lane is given by $\Delta t_{\text{ball}} = \Delta t_{\text{total}} - \Delta t_{\text{sound}} = 2.50\,\text{s} - 4.85\times10^{-2}\,\text{s} = 2.4515\,\text{s}$ and the speed of the ball is equal to $v_{\text{ball}} = \dfrac{\Delta x}{\Delta t_{\text{ball}}} = \dfrac{16.5\,\text{m}}{2.4515\,\text{s}} = \boxed{6.73\,\text{m/s}}$.

25. (*a*) $\bar{v} = \dfrac{\Delta x}{\Delta t} = \dfrac{385\,\text{m} - 25\,\text{m}}{20.0\,\text{s} - 3.0\,\text{s}} = \boxed{21.2\,\text{m/s}}$

    (*b*) $\bar{a} = \dfrac{\Delta v}{\Delta t} = \dfrac{45.0\,\text{m/s} - 11.0\,\text{m/s}}{20.0\,\text{s} - 3.0\,\text{s}} = \boxed{2.00\,\text{m/s}^2}$

31. By definition, the acceleration is $a = \dfrac{v - v_0}{t} = \dfrac{21\,\text{m/s} - 12\,\text{m/s}}{6.0\,\text{s}} = \boxed{1.5\,\text{m/s}^2}$. The distance of travel can be found from Eq. 2-12b.

    $$x - x_0 = v_0 t + \tfrac{1}{2}at^2 = \left(12\,\text{m/s}\right)\left(6.0\,\text{s}\right) + \tfrac{1}{2}\left(1.5\,\text{m/s}^2\right)\left(6.0\,\text{s}\right)^2 = \boxed{99\,\text{m}}$$

37. The words "slows down uniformly" implies that the car has a constant acceleration. The distance of travel is found from combining equations 2-2 and 2-9.

    $$x - x_0 = \frac{v_0 + v}{2}t = \left(\frac{18.0\,\text{m/s} + 0\,\text{m/s}}{2}\right)\left(5.00\,\text{sec}\right) = \boxed{45.0\,\text{m}}$$

43. Use the information for the first 180 m to find the acceleration, and the information for the full motion to find the final velocity. For the first segment, the train has $v_0 = 0$ m/s, $v_1 = 23$ m/s, and a displacement of $x_1 - x_0 = 180$ m. Find the acceleration from Eq. 2-12c.

$$v_1^2 = v_0^2 + 2a(x_1 - x_0) \rightarrow a = \frac{v_1^2 - v_0^2}{2(x_1 - x_0)} = \frac{(23\,\text{m/s})^2 - 0}{2(180\ \text{m})} = 1.469\,\text{m/s}^2$$

Find the speed of the train after it has traveled the total distance (total displacement of $x_2 - x_0 = 255$ m) using Eq. 2-12c.

$$v_2^2 = v_0^2 + 2a(x_2 - x_0) \rightarrow v_2 = \sqrt{v_0^2 + 2a(x_2 - x_0)} = \sqrt{2(1.469\,\text{m/s}^2)(255\ \text{m})} = \boxed{27\,\text{m/s}}$$

49. Choose downward to be the positive direction. The initial velocity is $v_0 = 0$ m/s, the final velocity is $v = 55$ km/h = (55 km/h)(1 m/s)/(3.6 km/h) = 15.28 m/s, and the acceleration is $a = 9.80$ m/s^2. The time can be found by solving Eq. 2-12a for the time.

$$v = v_0 + at \rightarrow t = \frac{v - v_0}{a} = \frac{15.28\,\text{m/s} - 0}{9.80\,\text{m/s}^2} = \boxed{1.6\,\text{s}}$$

55. Choose downward to be the positive direction, and take $y_0 = 0$ m to be the height where the object was released. The initial velocity is $v_0 = -5.10$ m/s, the acceleration is $a = 9.80$ m/s^2, and the displacement of the package will be $y = 105$ m. The time to reach the ground can be found from Eq. 2-12b, with x replaced by y.

$$y = y_0 + v_0 t + \tfrac{1}{2}at^2 \rightarrow t^2 + \frac{2v_0}{a}t - \frac{2y}{a} = 0 \rightarrow t^2 + \frac{2(-5.10\,\text{m/s})}{9.80\,\text{m/s}^2}t - \frac{2(105\ \text{m})}{9.80\,\text{m/s}^2} = 0 \rightarrow$$

$$t = 5.18\,\text{s} , \ -4.14\,\text{s}$$

The correct time is the positive answer, $\boxed{t = 5.18\ \text{s}}$.

61. Choose downward to be the positive direction, and $y_0 = 0$ m to be the height from which the stone is dropped. Call the location of the top of the window y_w, and the time for the stone to fall from release to the top of the window is t_w. Since the stone is dropped from rest, using Eq. 2-12b with y substituting for x, we have

$$y_w = y_0 + v_0 t + \tfrac{1}{2}at^2 = 0 + 0 + \tfrac{1}{2}gt_w^2$$

The location of the bottom of the window is $y_w + 22$ m, and the time for the stone to fall from release to the bottom of the window is $t_w + 0.33$ s. Since the stone is dropped from rest, using Eq. 2-12b, we have the following:

$$y_w + 2.2\ \text{m} = y_0 + v_0 + \tfrac{1}{2}at^2 = 0 + 0 + \tfrac{1}{2}g(t_w + 0.33\text{s})^2 .$$

Substitute the first expression for y_w into the second expression.

$$\tfrac{1}{2}gt_w^2 + 2.2\ \text{m} = \tfrac{1}{2}g(t_w + 0.33\ \text{s})^2 \rightarrow t_w = 0.515\ \text{s}$$

Use this time in the first equation to get the height above the top of the window from which the stone fell.

$$y_w = \tfrac{1}{2}gt_w^2 = \tfrac{1}{2}(9.80\,\text{m/s}^2)(0.515\ \text{s})^2 = \boxed{1.3\,\text{m}}$$

67. The displacement is found from the integral of the velocity, over the given time interval.

$$\Delta x = \int_{t_1}^{t_2} v\,dt = \int_{t=1.5\text{s}}^{t=3.1\text{s}} (25 + 18t)\,dt = (25t + 9t^2)\Big|_{t=1.5\text{s}}^{t=3.1\text{s}} = \left[25(3.1) + 9(3.1)^2\right] - \left[25(1.5) + 9(1.5)^2\right]$$

$$= \boxed{106\,\text{m}}$$

73. The initial velocity of the car is $v_0 = 100$ km/h $= (100$ km/h$)(1$ m/s$)/(3.6$ km/h$) = 27.8$ m/s. Choose $x_0 = 0$ m to be the location at which the deceleration begins. We have $v = 0$ m/s and $a = -30g = -294$ m/s^2. Find the displacement from Eq. 2-12c.

$$v^2 = v_0^2 + 2a(x - x_0) \rightarrow x = x_0 + \frac{v^2 - v_0^2}{2a} = 0 + \frac{0 - (27.8 \text{m/s})^2}{2(-2.94 \times 10^2 \text{ m/s}^2)} = 1.31 \text{m} \approx \boxed{1.3 \text{m}}$$

79. First consider the "uphill lie" in which the ball is being putted down the hill. Choose $x_0 = 0$ m to be the ball's original location, and the direction of the ball's travel as the positive direction. The final velocity of the ball is $v = 0$ m/s, the acceleration of the ball is $a = -1.8$ m/s^2, and the displacement of the ball will be $x - x_0 = 6.0$ m for the first case and $x - x_0 = 8.0$ m for the second case. Find the initial velocity of the ball from Eq. 2-12c.

$$v^2 = v_0^2 + 2a(x - x_0) \rightarrow v_0 = \sqrt{v^2 - 2a(x - x_0)} = \begin{cases} \sqrt{0 - 2(-1.8 \text{m/s}^2)(6.0 \text{ m})} = 4.6 \text{m/s} \\ \sqrt{0 - 2(-1.8 \text{m/s}^2)(8.0 \text{ m})} = 5.4 \text{m/s} \end{cases}$$

The range of acceptable velocities for the uphill lie is $\boxed{4.6 \text{ m/s to } 5.4 \text{ m/s}}$, a spread of 0.8 m/s.

Now consider the "downhill lie" in which the ball is being putted up the hill. Use a very similar setup for the problem, with the basic difference being that the acceleration of the ball is now $a = -2.8$ m/s^2. Find the initial velocity of the ball from Eq. 2-12c.

$$v^2 = v_0^2 + 2a(x - x_0) \rightarrow v_0 = \sqrt{v^2 - 2a(x - x_0)} = \begin{cases} \sqrt{0 - 2(-2.8 \text{m/s}^2)(6.0 \text{ m})} = 5.8 \text{m/s} \\ \sqrt{0 - 2(-2.8 \text{m/s}^2)(8.0 \text{ m})} = 6.7 \text{m/s} \end{cases}$$

The range of acceptable velocities for the downhill lie is $\boxed{5.8 \text{ m/s to } 6.7 \text{ m/s}}$, a spread of 0.9 m/s.

Because the range of acceptable velocities is smaller for putting down the hill, more control in putting is necessary, and so putting the ball downhill (the "uphill lie") is more difficult.

85. Choose upward to be the positive direction, and the origin to be at the level where the ball was thrown. The velocity at the top of the ball's path will be $v = 0$ m, and the ball will have an acceleration of $a = -g$. If the maximum height that the ball reaches is $y = H$, then the relationship between the initial velocity and the maximum height can be found from Eq. 2-12c, with x replaced by y.

$$v^2 = v_0^2 + 2a(y - y_0) \rightarrow 0 = v_0^2 + 2(-g)H \rightarrow H = v_0^2/2g$$

It is given that $v_{0 \text{ Bill}} = 1.5 \, v_{0 \text{ Joe}}$, so $\dfrac{H_{\text{Bill}}}{H_{\text{Joe}}} = \dfrac{(v_{0 \text{ Bill}})^2/2g}{(v_{0 \text{ Joe}})^2/2g} = \dfrac{(v_{0 \text{ Bill}})^2}{(v_{0 \text{ Joe}})^2} = 1.5^2 = 2.25 \approx \boxed{2.3}$.

91. The speed of the conveyor belt is given by

$$d = \bar{v}\Delta t \rightarrow \bar{v} = \frac{d}{\Delta t} = \frac{1.1 \text{ m}}{2.5 \text{ min}} = \boxed{0.44 \text{ m/min}}$$

The rate of burger production, assuming the spacing given is center to center, can be found as follows.

$$\left(\frac{1 \text{ burger}}{0.15 \text{ m}} \right) \left(\frac{0.44 \text{ m}}{1 \text{ min}} \right) = \boxed{2.9 \frac{\text{burgers}}{\text{min}}}$$

97. (*a*) For each segment of the path, the time is given by the distance divided by the speed.

$$t = t_{land} + t_{pool} = \frac{d_{land}}{v_{land}} + \frac{d_{pool}}{v_{pool}}$$

$$= \frac{x}{v_R} + \frac{\sqrt{D^2 + (d - x)^2}}{v_S}$$

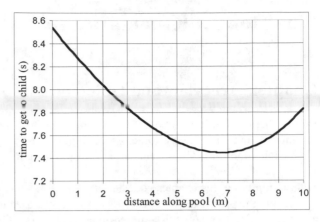

(*b*) The graph is shown here. The minimum time occurs at a distance along the pool of about $x = 6.8\ \text{m}$. An analytic differentiation to solve for the minimum point gives $x = 6.76$ m.

Chapter 3: Kinematics in Two or Three Dimensions; Vectors

Chapter Overview and Objectives

In this chapter, the kinematics of objects moving in two or three dimensions is described. The concepts of vectors are introduced and all the one-dimensional kinematic concepts discussed in Chapter 2 are applied to two- and three-dimensional motion. The special case of projectile motion is discussed in detail.

After completing this chapter you should:
- Know the properties of a vector quantity as opposed to a scalar quantity.
- Know how to resolve two-dimensional vectors into Cartesian components.
- Know how to add and subtract vectors, both graphically and by components.
- Know the vector definitions of velocity and acceleration.
- Know how to solve two- and three-dimensional constant acceleration problems.
- Know how to relate the velocity measurements made by two observers in relative motion with respect to each other.

Summary of Equations

Expressing $\vec{\mathbf{V}}$ into Cartesian components:
$$\vec{\mathbf{V}} = \vec{\mathbf{V}}_x + \vec{\mathbf{V}}_y + \vec{\mathbf{V}}_z \qquad \text{(Section 3-4)}$$

Relating components and vector properties:
$$V_x = V\cos\theta \qquad V_y = V\sin\theta \qquad \text{(Section 3-4)}$$

$$V = \sqrt{V_x^2 + V_y^2} \qquad \theta = \arctan\frac{V_y}{V_x} \qquad \text{(Section 3-4)}$$

Expressing vectors in terms of Cartesian unit vectors:
$$\vec{\mathbf{V}} = V_x\hat{\mathbf{i}} + V_y\hat{\mathbf{j}} + V_z\hat{\mathbf{k}} \qquad \text{(Section 3-5)}$$

Vector definition of instantaneous velocity:
$$\vec{\mathbf{v}} = \frac{d\vec{\mathbf{r}}}{dt} \qquad \text{(Section 3-6)}$$

Vector definition of instantaneous acceleration:
$$\vec{\mathbf{a}} = \frac{d\vec{\mathbf{v}}}{dt} \qquad \text{(Section 3-6)}$$

Motion with constant acceleration:
$$\vec{\mathbf{v}} = \vec{\mathbf{v}}_0 + \vec{\mathbf{a}}t \qquad \text{(Section 3-6)}$$

$$\vec{\mathbf{r}} = \vec{\mathbf{r}}_0 + \vec{\mathbf{v}}_0 t + \tfrac{1}{2}\vec{\mathbf{a}}t^2 \qquad \text{(Section 3-6)}$$

Velocity according to different observers:
$$\vec{\mathbf{v}}_{AC} = \vec{\mathbf{v}}_{AB} + \vec{\mathbf{v}}_{BC} \qquad \text{(Section 3-9)}$$

Chapter Summary

Section 3-1. Vectors and Scalars

A quantity that has a magnitude and a direction is a **vector** quantity. Examples of vector quantities are velocity and displacement. A quantity that is completely specified by only a number is called a **scalar**. Examples of scalar quantities are mass and temperature. The notation that will be used for vectors in the textbook is a boldface letter with an arrow; the velocity of an object may be written as $\vec{v}$.

Section 3-2. Addition of Vectors - Graphical Methods

Vector operations, such as addition, can be carried out using graphical techniques. When we draw a vector, we draw an arrow that points in the direction of the vector quantity and has a length that is proportional to the magnitude of the vector quantity. To add two vectors graphically we use the following procedure:

1. Decide on the scale for the length of the vectors.
2. Draw a set of Cartesian axes to set up a coordinate system.
3. Draw the first of the two vectors to be added. Start the vector at the origin of the coordinate system and draw it to the correct length and in the correct direction relative to the Cartesian axes.
4. Draw the second vector to be added beginning from the end of the first vector. Draw the vector to the correct length and in the appropriate direction.
5. Draw an arrow from the origin of the coordinate system to the end of the second vector. This arrow's length is representative of the vector sum magnitude and its direction is the direction of the vector sum.

Example 3-2-A: Vector addition. An airplane flies 300 km in a direction northeast. It then flies 200 km north. What is the final displacement vector of the airplane?

Approach: In order to find the displacement vector we need to follow the procedure outlined above. It is usually a good idea to make a rough estimate of the maximum distance we need to display. Based on the information provided in the problem, we can conclude that the maximum distance will be less than 500 km.

Solution: We follow the steps outlined above.
1. *Decide on the scale for the length of the vectors.* In order to be able to draw distances of up to 500 km, we will use the following scale:
2. *Draw a set of Cartesian axes to set up a coordinate system.* The Cartesian coordinate system we will be using in this problem is shown in the following Figure on the right. The horizontal axis is directed along the East-West direction and the vertical axis is directed along the North-South direction.
3. *Draw the first of the two vectors to be added. Start the vector at the origin of the coordinate system and draw it to the correct length and in the correct direction relative to the Cartesian axes* (see the left Figure below).
4. *Draw the second vector to be added beginning from the end of the first vector. Draw the vector to the correct length and in the appropriate direction* (see the middle Figure below).
5. *Draw an arrow from the origin of the coordinate system to the end of the second vector. This arrow's length is representative of the vector sum magnitude and its direction is the direction of the vector sum* (see the right Figure below). By using the scale defined in Step 1 we can determine that magnitude of the displacement is about 460 km. A protractor can be used to determine the direction of the displacement.

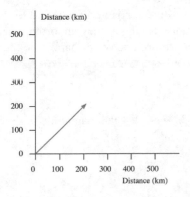
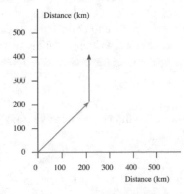
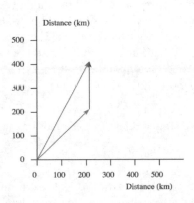

Section 3-3. Subtraction of Vectors, and Multiplication of a Vector by a Scalar

A vector can be multiplied by a scalar. Multiplying a scalar c times vector $\vec{V}$ results in a vector that has a magnitude that is the absolute value of c times as big as the magnitude of $\vec{V}$. The resulting vector is pointing in the same direction as $\vec{V}$ if c is positive and is pointing in the opposite direction of $\vec{V}$ if c is negative.

The difference between two vectors, $\vec{A} - \vec{B}$, is defined as the addition of $\vec{A}$ and the vector $(-1)\vec{B}$:

$$\vec{A} - \vec{B} = \vec{A} + (-1)\vec{B}$$

Example 3-3-A: Vector Subtraction. Graphically determine the difference of the first displacement and the second displacement of the airplane discussed in Example 3-2-A.

Approach: We use the same approach we used in the solution of Example 3-2-A, but instead of adding the second vector in step 4 we add the opposite of this vector. The rest of the solution is the same.

Solution: The relevant steps to be used to determine the difference between the displacement to the north-east and the displacement to the north are shown in the Figure below. By using the scale defined in Step 1 of problem 3-2-A we can determine that magnitude of the displacement is about 213 km. A protractor can be used to determine the direction of the displacement.

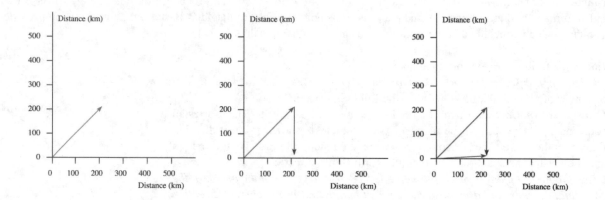

Section 3-4. Adding Vectors by Components

Graphical addition or subtraction of vectors has limited precision because of finite thickness of lines and limited precision of scales and protractors. Sometimes greater precision than graphical methods can provide is necessary. This additional precision can be obtained by using an algebraic method of vector addition.

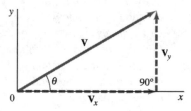

To use the algebraic method of vector addition, we first need to choose a set of Cartesian coordinate axes to work with. This means choosing a set of three orthogonal (perpendicular) directions. We will restrict ourselves to only two dimensions here to help simplify the discussion. We will usually choose the positive x axis horizontal and to the right and the positive y axis vertical and upward, but we are free to choose the directions as long as they are orthogonal. Any vector, $\vec{V}$, that lies in the plane of these two axes can be written uniquely as the sum of one vector parallel to the x axis, $\vec{V}_x$, and one vector parallel to the y axis, $\vec{V}_y$:

$$\vec{V} = \vec{V}_x + \vec{V}_y$$

There are simple relations between the magnitudes of the **component vectors**, $\vec{V}_x$ and $\vec{V}_y$, the magnitude of the vector, $\vec{V}$, and the angle θ:

$$V_x = V \cos\theta \qquad V_y = V \sin\theta \qquad V = \sqrt{V_x^2 + V_y^2} \qquad \theta = \arctan\frac{V_y}{V_x}$$

The x and y components of the vector sum of two vectors are equal to the sum of the x and y components of the two vectors (see the Figure to the right):

$$\vec{V} = \vec{V}_x + \vec{V}_y = \vec{V}_1 + \vec{V}_2 = \left(\vec{V}_{1x} + \vec{V}_{2x}\right) + \left(\vec{V}_{1y} + \vec{V}_{2y}\right)$$

The magnitude and the direction of the vector sum are equal to

$$|\vec{V}| = \sqrt{\left(V_{1x} + V_{2x}\right)^2 + \left(V_{1y} + V_{2y}\right)^2}$$

$$\theta = \arctan\left(\frac{V_{1y} + V_{2y}}{V_{1x} + V_{2x}}\right)$$

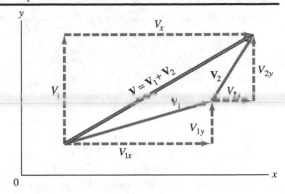

Example 3-4-A. Vector addition. Add the vectors in Example 3-2-A by using the components of their vectors.

Approach: To solve this problem we first need to find the components of the vectors and then add the corresponding components.

Solution: We use a coordinate system with the positive x axis pointed towards the east and the positive y axis pointed towards the north. The x and y components of the displacement to the north-east are

$$V_{1x} = 300\cos 45° = 212\,\text{km}$$

$$V_{1y} = 300\sin 45° = 212\,\text{km}$$

The x and y components of the displacement to the north are

$$V_{2x} = 0\,\text{km}$$

$$V_{2y} = 200\,\text{km}$$

The components of the vector sum of these two vectors will be

$$V_x = V_{1x} + V_{2x} = 212 + 0 = 212\,\text{km}$$

$$V_y = V_{1y} + V_{2y} = 212 + 200 = 412\,\text{km}$$

We can now determine the magnitude of the vector sum

$$V = \sqrt{V_x^2 + V_y^2} = \sqrt{\left(212\right)^2 + \left(412\right)^2} = 463\,\text{km}$$

and its direction

$$\theta = \arctan\left(\frac{V_y}{V_x}\right) = \arctan\left(\frac{412}{212}\right) = 62.8°$$

where the angle θ is the direction towards the north, measured away from east. Note the agreement between these results and the graphical results obtained in Example 3-2-A.

Section 3-5. Unit Vectors

A unit vector is a vector of magnitude one. Any vector can be written as a scalar multiple of a unit vector. Any vector can also be written uniquely as the sum of scalar multiples of a set of three orthogonal unit vectors (in three dimensions). We commonly use the three **unit vectors** that lie along the direction of the positive x, y, and z axes of our three-dimensional Cartesian coordinate system (see Figure). These three unit vectors are called $\hat{\mathbf{i}}$, $\hat{\mathbf{j}}$, and $\hat{\mathbf{k}}$, respectively. Any vector can be written in terms of its components and the three Cartesian unit vectors:

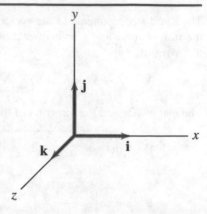

$$\vec{\mathbf{V}} = V_x\hat{\mathbf{i}} + V_y\hat{\mathbf{j}} + V_z\hat{\mathbf{k}}$$

Example 3-5-A: Expressing a vector in terms of the unit vectors. A vector $\vec{\mathbf{A}}$ has a magnitude of 10.5 m and a direction pointing 37.6° counter-clockwise from the positive x axis. Express this vector in terms of the Cartesian unit vectors.

Approach: Since the direction and the magnitude of the vector are provided, we can use the relations between these parameters and the components of the vector discussed in Section 3.4 to determine the components of $\vec{\mathbf{A}}$.

Solution: The x and y components of the vector $\vec{\mathbf{A}}$ can be found by using equations 3-2a and 3-2b in the textbook and the magnitude and the direction of $\vec{\mathbf{A}}$ provided in the problem:

$$A_x = A\cos\theta = (10.5)\cos(37.6°) = 8.32 \text{ m}$$

$$A_y = A\sin\theta = (10.5)\sin(37.6°) = 6.41 \text{ m}$$

We can now write $\vec{\mathbf{A}}$ in terms of the unit vectors:

$$\vec{\mathbf{A}} = A_x\hat{\mathbf{i}} + A_y\hat{\mathbf{j}} = 8.32\hat{\mathbf{i}} + 6.41\hat{\mathbf{j}}$$

Section 3-6. Vector Kinematics

The position of an object in two or three dimensions is a vector quantity. We will denote the position of an object by the vector $\vec{\mathbf{r}}$. The definitions of velocity $\vec{\mathbf{v}}$ and acceleration $\vec{\mathbf{a}}$ in two or three dimensions are similar to the corresponding definitions in one-dimension:

> **Average velocity:** $\qquad \vec{\mathbf{v}}_{avg} = \dfrac{\Delta\vec{\mathbf{r}}}{\Delta t}$

> **Instantaneous velocity:** $\qquad \vec{\mathbf{v}} = \lim\limits_{\Delta t \to 0}\dfrac{\Delta\vec{\mathbf{r}}}{\Delta t} = \dfrac{d\vec{\mathbf{r}}}{dt}$

> **Average acceleration:** $\qquad \vec{\mathbf{a}}_{avg} = \dfrac{\Delta\vec{\mathbf{v}}}{\Delta t}$

> **Instantaneous acceleration:** $\qquad \vec{\mathbf{a}} = \lim\limits_{\Delta t \to 0}\dfrac{\Delta\vec{\mathbf{v}}}{\Delta t} = \dfrac{d\vec{\mathbf{v}}}{dt}$

The velocity and acceleration can be written in terms of the Cartesian unit vectors

$$\vec{\mathbf{v}} = \frac{dx}{dt}\hat{\mathbf{i}} + \frac{dy}{dt}\hat{\mathbf{j}} + \frac{dz}{dt}\hat{\mathbf{k}} = v_x\hat{\mathbf{i}} + v_y\hat{\mathbf{j}} + v_z\hat{\mathbf{k}}$$

$$\vec{\mathbf{a}} = \frac{dv_x}{dt}\hat{\mathbf{i}} + \frac{dv_y}{dt}\hat{\mathbf{j}} + \frac{dv_z}{dt}\hat{\mathbf{k}} = a_x\hat{\mathbf{i}} + a_y\hat{\mathbf{j}} + a_z\hat{\mathbf{k}}$$

The equations of motion with constant acceleration in two or three dimensions are very similar to the equations of motion with constant acceleration in one dimension:

$$\vec{v}(t) = \vec{v}_0 + \vec{a}t$$

$$\vec{r}(t) = \vec{r}_0 + \vec{v}_0 t + \frac{1}{2}\vec{a}t^2$$

where $\vec{r}(t)$ is the position at time t, $\vec{r}_0$ is the position at time $t = 0$ s, $\vec{v}(t)$ is the velocity at time t, $\vec{v}_0$ is the velocity at time $t = 0$ s, and $\vec{a}$ is the constant acceleration.

Example 3-6-A: Three-dimensional motion with constant acceleration.
An object has an initial velocity $\vec{v}_0 = 2.0\hat{i} + 3.0\hat{j} - 1.5\hat{k}$ and a constant acceleration $\vec{a} = -1.0\hat{i} + 1.0\hat{j} + 2.5\hat{k}$. What is the displacement of the object after 3.0 s?

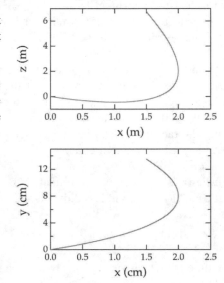

Approach: The equations of motion for constant acceleration apply to the motion of the object. We will define our coordinate system such that the object is located at its origin at time $t = 0$ s. The position of the object at time $t = 3.0$ s is then equal to its displacement after 3.0 s.

Solution: The position of the object at time t is given by

$$\vec{r}(t) = \vec{v}_0 t + \frac{1}{2}\vec{a}t^2$$

$$= \left(2.0t - 0.5t^2\right)\hat{i} + \left(3.0t + 0.5t^2\right)\hat{j} + \left(-1.5t + 1.25t^2\right)\hat{k}$$

At the $t = 3.0$ s the displacement $\Delta\vec{r}$ is equal to

$$\Delta\vec{r} = \vec{r}(t = 3\text{ s}) = (1.5)\hat{i} + (13.5)\hat{j} + (6.75)\hat{k}$$

The figures on the right show the trajectory of the object in the z-x and the y-x plane during the first 3 seconds.

Section 3-7. Projectile Motion

If we consider motion near the surface of the Earth for objects free to fall in the Earth's gravity, the motion undergoes constant acceleration in the downward vertical direction with a magnitude 9.80 m/s². This magnitude is called the local acceleration due to gravity and is given the symbol g. Because the direction of the acceleration of gravity is vertical, horizontal motion occurs with constant velocity under these conditions. This type of motion is called **projectile motion**. If we use a coordinate system with the x axis parallel to the initial horizontal velocity, if any, and the y axis positive in the upward vertical direction, we can write the kinematic equations for this motion with constant acceleration as

Horizontal motion	Vertical motion
$v_x(t) = v_{x0}$	$v_y(t) = v_{y0} - gt$
$r(t) = r_0 + v_{x0}t$	$y(t) = y_0 + v_{y0}t - \frac{1}{2}gt^2$

In some cases, we are given the initial speed, v_0, and angle above horizontal, θ, of the initial motion. The components of the initial horizontal velocity, v_{x0}, and the initial vertical velocity, v_{y0}, are related to the initial speed and angle by

$$v_{x0} = v_0 \cos\theta$$

$$v_{y0} = v_0 \sin\theta$$

Note that θ is negative when the initial velocity is directed below horizontal.

Section 3-8. Solving Problems Involving Projectile Motion

Apply the following general rules to solve problems involving projectile motion:
1. Read the problem carefully, trying to form a picture of the problem in your head. Reread the problem to be sure you have read it correctly.
2. Choose a set of coordinate axes to use and draw a diagram.
3. Make a list of known and unknown quantities, remembering that for projectile motion problems the horizontal acceleration is in general zero, and the vertical acceleration is downward with a magnitude $g = 9.80$ m/s^2.
4. Think before selecting which of the equations of projectile motion to use before beginning to solve equations. Use the equations that allow you to solve for the unknown quantities in terms of the known quantities.

Example 3-8-A: Projectile motion. An airplane is climbing at a $10°$ angle above the horizontal plane at an altitude of 366 m and a speed of 256 km/hr when a package is released from the airplane. How far does the package travel horizontally from the release point before hitting the ground? Assume air resistance is negligible.

Approach: The problem provides us with information about the release position and direction of the package. Given these initial conditions, the motion of the package is completely defined. When we calculate the displacement of the package, we need to make sure that we convert all quantities to consistent units (distance in meters, time in seconds, etc.). We will use the motion in the vertical direction to determine the time at which the package will reach the ground. Using the fact that the horizontal velocity is constant, we can then use this time to determine how far the package will travel in the horizontal direction.

Solution: We choose our coordinate system such that at time $t = 0$ s, the package is located at $(0, h_0)$. The initial speed of the package is v_0 and it is directed at an angle θ. The position in the vertical direction is equal to

$$y(t) = h_0 + (v_0 \sin\theta)t - \frac{1}{2}gt^2$$

The time t at which the package will reach the ground can be found by solving the following equation:

$$0 = h_0 + (v_0 \sin\theta)t - \frac{1}{2}gt^2$$

or

$$t = \frac{2(v_0 \sin\theta) \pm \sqrt{4(v_0 \sin\theta)^2 + 8gh_0}}{2g}$$

We only care about the solution for which time t is positive:

$$t = \frac{2(v_0 \sin\theta) + \sqrt{4(v_0 \sin\theta)^2 + 8gh_0}}{2g} = \left(\frac{v_0 \sin\theta}{g}\right)\left\{1 + \sqrt{1 + \frac{2gh_0}{(v_0 \sin\theta)^2}}\right\} = 9.99 \text{ s}$$

Note: make sure that you convert the velocity of the package from km/hr to m/s. The horizontal displacement of the package during this time is equal to

$$\Delta x = (v_0 \cos\theta)t = 699 \text{ m}$$

Example 3-8-B. Projectile motion. A projectile is launched from a height of 28 m above the ground with a speed of 68 m/s. At what angle must it be launched to reach a target on the ground, a distance of 184 m away horizontally?

Approach: The solution to this problem is similar to the solution of Example 3-8-A, except that in this problem we know the horizontal range and not the launch angle. However, we can work backwards from the solution of Example 3-8-A to find the launch angle.

Solution: The range of the projectile is d. Using the solution of Example 3-8-A we must require that

$$d = \left(v_0 \cos\theta\right)t$$

The time of flight is thus equal to

$$t = \frac{d}{v_0 \cos\theta}$$

Time t must also satisfy the following condition:

$$y\left(t = \frac{d}{v_0 \cos\theta}\right) = 0 = h_0 + v_0 \sin\theta \frac{d}{v_0 \cos\theta} - \frac{1}{2}g\left(\frac{d}{v_0 \cos\theta}\right)^2 = h_0 + d\tan\theta - \frac{1}{2}\frac{gd^2}{v_0^2}\left(1 + \tan^2\theta\right)$$

This equation can be rewritten as

$$\left(h_0 - \frac{1}{2}\frac{gd^2}{v_0^2}\right) + d\tan\theta - \frac{1}{2}\frac{gd^2}{v_0^2}\tan^2\theta = 0$$

We can solve this equation for $\tan\theta$:

$$\tan\theta = \frac{d \pm \sqrt{d^2 + 2\left(h_0 - \frac{1}{2}\frac{gd^2}{v_0^2}\right)\frac{gd^2}{v_0^2}}}{\frac{gd^2}{v_0^2}} = \frac{v_0^2}{gd}\left\{1 \pm \sqrt{1 + 2\left(h_0 - \frac{1}{2}\frac{gd^2}{v_0^2}\right)\frac{g}{v_0^2}}\right\} = 0.0432 \text{ or } 5.09$$

or

$$\theta = 2.5° \text{ or } \theta = 78.9°$$

Section 3-9. Relative Velocity

Velocity measurements made in different inertial reference frames are related by

$$\vec{v}_{AC} = \vec{v}_{AB} + \vec{v}_{BC}$$

where $\vec{v}_{AB}$ is the velocity of the object A measured in reference frame B, $\vec{v}_{AC}$ is the velocity of the object A measured in reference frame C, and $\vec{v}_{BC}$ is the velocity of reference frame B measured in reference frame C.
We will often use the fact that the velocity of reference frame A as measured by an observer in reference frame B is the negative of the velocity of reference frame B as measured by an observer in reference frame A:

$$\vec{v}_{BA} = -\vec{v}_{AB}$$

Example 3-9-A. Calculating the relative velocity of an airplane relative to the ground. An airplane flies at a speed of 300 km/hr directly northward relative to the air. The wind is blowing at 20 m/s toward the southwest. What is the velocity of the plane relative to the ground?

Approach: In order to solve this problem we will look at the airplane from two reference frames: one reference frame is fixed to the ground, the other reference frame is fixed with respect to the air. When we use the relation between the velocity of the airplane in the two different reference frames we must make sure that all velocities are used with the same units (either m/s or km/h).

Solution: We will use the symbols $\vec{v}_{PA}$ for the velocity of the plane relative to the air, $\vec{v}_{AG}$ for the velocity of the air relative to the ground, and $\vec{v}_{PG}$ for the velocity of the plane relative to the ground. These velocities are related as follows

$$\vec{v}_{PG} = \vec{v}_{PA} + \vec{v}_{AG}$$

The velocity of the airplane with respect to the air is 300 km/hr. The direction of the airplane is north. Since this is a two-dimensional problem, we can represent the velocities as two-dimensional vectors; the unit vector along the positive x axis is directed towards the East while the unit vector along the positive y axis is directed towards the North. The velocity of the airplane with respect of the air is given by the following vector:

$$\vec{\mathbf{v}}_{PA} = \left(300 \times \frac{1000}{3600} \right) \hat{\mathbf{j}} = 83.3 \hat{\mathbf{j}}$$

The conversion factor in this equation is used to convert the velocity of the airplane from km/h to m/s. The velocity of the air with respect to the ground is given by the following vector:

$$\vec{\mathbf{v}}_{AG} = 20 \times \frac{1}{2} \sqrt{2} \left(-\hat{\mathbf{i}} - \hat{\mathbf{j}} \right) = 14.1 \left(-\hat{\mathbf{i}} - \hat{\mathbf{j}} \right)$$

The velocity of the airplane with respect to the ground can now be determined:

$$\vec{\mathbf{v}}_{PG} = \vec{\mathbf{v}}_{PA} + \vec{\mathbf{v}}_{AG} = 83.3 \hat{\mathbf{j}} + 14.1 \left(-\hat{\mathbf{i}} - \hat{\mathbf{j}} \right) = -14.1 \hat{\mathbf{i}} + 69.2 \hat{\mathbf{j}}$$

Practice Quiz

1. Which of the following is not a vector quantity?
    a) Velocity
    b) Acceleration
    c) Displacement
    d) Time

2. Which inequality does the vector sum of the two vectors $\vec{\mathbf{A}}$ and $\vec{\mathbf{B}}$ necessarily satisfy?
    a) The magnitude of $\vec{\mathbf{A}} + \vec{\mathbf{B}}$ is greater than or equal to the magnitude of $\vec{\mathbf{A}}$ and greater than or equal to the magnitude of $\vec{\mathbf{B}}$.
    b) The magnitude of $\vec{\mathbf{A}} + \vec{\mathbf{B}}$ is greater than or equal to the magnitude of $\vec{\mathbf{A}}$ plus the magnitude of $\vec{\mathbf{B}}$.
    c) The magnitude of $\vec{\mathbf{A}} + \vec{\mathbf{B}}$ is less than or equal to the magnitude of $\vec{\mathbf{A}}$ and less than or equal to the magnitude of $\vec{\mathbf{B}}$.
    d) The magnitude of $\vec{\mathbf{A}} + \vec{\mathbf{B}}$ is less than or equal to the magnitude of $\vec{\mathbf{A}}$ plus the magnitude of $\vec{\mathbf{B}}$.

3. Given vector $\vec{\mathbf{A}}$, what vector added to it results in the zero vector?
    a) $\vec{\mathbf{A}}$
    b) $-\vec{\mathbf{A}}$
    c) There is no such vector.
    d) Zero

4. A projectile is launched with an initial speed v. Which of the following statements is necessarily true at any point along its trajectory above the initial launch position?
    a) The speed of the projectile is greater than the launch speed.
    b) The speed of the projectile is equal to the launch speed.
    c) The speed of the projectile is less than the launch speed.
    d) The angle of the velocity is steeper than the angle of the launch velocity.

5. What is/are the condition(s) of the motion of a projectile at its highest point along its trajectory?
    a) The speed and the acceleration are zero.
    b) The horizontal velocity and the acceleration are zero.
    c) The vertical velocity and the acceleration are zero.
    d) The vertical velocity is zero.

6. A projectile is launched from the top of a hill with an initial speed v. Which of the following statements is necessarily true at any point along its trajectory below the initial launch position?
   a) The speed of the projectile is greater than the launch speed.
   b) The speed of the projectile is equal to the launch speed.
   c) The speed of the projectile is less than the launch speed.
   d) The angle of the velocity is shallower than the angle of the launch velocity.

7. To observer A, an object is moving with a velocity $\vec{v}$. What would the velocity of observer B be relative to observer A, if observer B sees the object at rest?
   a) $\vec{v}$
   b) $-\vec{v}$
   c) Zero.
   d) That is impossible.

8. Observer B is moving with a constant velocity relative to observer A. Observer A sees an object accelerate with an acceleration $\vec{a}$. What does observer B see as the acceleration of the object?
   a) $\vec{a}$
   b) $-\vec{a}$
   c) Zero.
   d) Depends on the direction of the relative velocity of B to A.

9. In order to take off, a plane must have an airspeed of 60 mi/h. What is the required groundspeed if the plane is taking off with a headwind of 15 mi/h?
   a) 75 mi/h
   b) 60 mi/h
   c) 45 mi/h

10. Three projectiles A, B, and C are launched with the same initial velocities at angles of 30°, 45°, and 60°, respectively. Their launch position and their points of impact are located at sea level. What is the relation between their flight times?
   a) $t_A = t_C > t_B$
   b) $t_B > t_A = t_C$
   c) $t_A > t_B > t_C$
   d) $t_C > t_B > t_A$

11. Graphically find the difference between a vector of magnitude 3.65 pointing 32° north of east and a vector of magnitude 5.88 pointing 12° north of west.

12. Find the vector sum of a vector of magnitude 3.45 pointing 43° south of east and a vector of magnitude 7.63 pointing 21° west of south using vector components.

13. A diver leaps off the end of a diving board 3.00 m above the water with an initial velocity of 5.8 m/s at an angle 80° above horizontal. How long is the diver in the air before striking the water? How far horizontally from the end of the diving board does the diver hit the water? Ignore any changes in the orientation of the diver's body.

14. The earth moves in a nearly circular orbit around the sun once a year. Assuming the earth's orbit is circular, what is the centripetal acceleration of the earth in its orbit around the sun?

15. An airplane flies at a speed of 224 km/hr in a direction 12° east of north. An observer on the ground observes the plane moving at a speed of 241 km/hr in a direction 2° west of north. What is the wind velocity?

Responses to Select End-of-Chapter Questions

1. No. Velocity is a vector quantity, with a magnitude and direction. If two vectors have different directions, they cannot be equal.

7. The maximum magnitude of the sum is 7.5 km, in the case where the vectors are parallel. The minimum magnitude of the sum is 0.5 km, in the case where the vectors are antiparallel.

13. No. The arrow will fall toward the ground as it travels toward the target, so it should be aimed above the target. Generally, the farther you are from the target, the higher above the target the arrow should be aimed, up to a maximum launch angle of 45°. (The maximum range of a projectile that starts and stops at the same height occurs when the launch angle is 45°.)

19. This is a question of relative velocity. From the point of view of an observer on the ground, both trains are moving in the same direction (forward), but at different speeds. From your point of view on the faster train, the slower train (and the ground) will appear to be moving backward. (The ground will be moving backward faster than the slower train!)

Solutions to Select End-of-Chapter Problems

1. The resultant vector displacement of the car is given by

$$\vec{\mathbf{D}}_{R} = \vec{\mathbf{D}}_{west} + \vec{\mathbf{D}}_{south-\atop west}$$

The westward displacement is $225 + 78\cos(45°) = 280.2$ km and the south displacement is $78\sin(45°) = 55.2$ km. The resultant displacement has a magnitude of

$$\sqrt{280.2^2 + 55.2^2} = \boxed{286\,\text{km}}$$

The direction is

$$\theta = \tan^{-1} 55.2/280.2 = \boxed{11°\,\text{south of west}}$$

7. (a) $v_{north} = (835\,\text{km/h})(\cos 41.5°) = \boxed{625\,\text{km/h}}$ $v_{west} = (835\,\text{km/h})(\sin 41.5°) = \boxed{553\,\text{km/h}}$

   (b) $\Delta d_{north} = v_{north}t = (625\ \text{km/h})(2.50\ \text{h}) = \boxed{1560\ \text{km}}$ $\Delta d_{west} = v_{west}t = (553\ \text{km/h})(2.50\ \text{h}) = \boxed{1380\ \text{km}}$

19. From the original position vector, we have $x = 9.60t$, $y = 8.85$ m, and $z = -1.00t^2$. Thus $z = -(x/9.60)^2 = -ax^2$ and $y = 8.85$ m. This is the equation for a $\boxed{\text{parabola}}$ in the x-z plane that has its vertex at coordinate $(0,8.85,0)$ and opens downward.

25. (a) Differentiate the position vector, $\vec{\mathbf{r}} = \left(3.0t^2\,\hat{\mathbf{i}} - 6.0t^3\hat{\mathbf{j}}\right)\text{m}$, with respect to time in order to find the velocity and the acceleration.

   $$\vec{\mathbf{v}} = \frac{d\vec{\mathbf{r}}}{dt} = \boxed{\left(6.0t\,\hat{\mathbf{i}} - 18.0t^2\hat{\mathbf{j}}\right)\text{m/s}} \qquad \vec{\mathbf{a}} = \frac{d\vec{\mathbf{v}}}{dt} = \boxed{\left(6.0\hat{\mathbf{i}} - 36.0t\,\hat{\mathbf{j}}\right)\text{m/s}^2}$$

   (b) $\vec{\mathbf{r}}(2.5\,\text{s}) = \left[3.0(2.5)^2\,\hat{\mathbf{i}} - 6.0(2.5)^3\,\hat{\mathbf{j}}\right]\text{m} = \boxed{\left(19\hat{\mathbf{i}} - 94\hat{\mathbf{j}}\right)\text{m}}$

   $\vec{\mathbf{v}}(2.5\,\text{s}) = \left[6.0(2.5)\hat{\mathbf{i}} - 18.0(2.5)^2\,\hat{\mathbf{j}}\right]\text{m/s} = \boxed{\left(15\,\hat{\mathbf{i}} - 110\hat{\mathbf{j}}\right)\text{m/s}}$

31. Apply the range formula from Example 3-10.

$$R = \frac{v_0^2 \sin 2\theta_0}{g} \quad \rightarrow$$

$$\sin 2\theta_0 = \frac{Rg}{v_0^2} = \frac{(2.5\,\text{m})(9.80\,\text{m/s}^2)}{(6.5\,\text{m/s})^2} = 0.5799$$

$$2\theta_0 = \sin^{-1} 0.5799 \quad \rightarrow \quad \theta_0 = \boxed{18°, 72°}$$

There are two angles because each angle gives the same range. If one angle is $\theta = 45° + \delta$, then $\theta = 45° - \delta$ is also a solution. The two paths are shown in the graph.

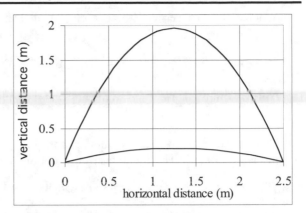

37. When shooting the gun vertically, half the time of flight is spent moving upwards. Thus the upwards flight takes 2.0 s. Choose upward as the positive y direction. Since at the top of the flight, the vertical velocity is zero, find the launching velocity from Eq. 2-12a.

$$v_y = v_{y0} + at \quad \rightarrow \quad v_{y0} = v_y - at = 0 - \left(-9.80\,\text{m/s}^2\right)(2.0\,\text{s}) = 19.6\,\text{m/s}$$

Using this initial velocity and an angle of 45° in the range formula (from Example 3-10) will give the maximum range for the gun.

$$R = \frac{v_0^2 \sin 2\theta_0}{g} = \frac{(19.6\,\text{m/s})^2 \sin(90°)}{9.80\,\text{m/s}^2} = \boxed{39\,\text{m}}$$

43. Choose downward to be the positive y direction. The origin is the point where the supplies are dropped. In the vertical direction, $v_{y0} = 0$ m/s, $a_y = 9.80$ m/s^2, $y_0 = 0$ m, and the final position is $y = 150$ m. The time of flight is found from applying Eq. 2-12b to the vertical motion.

$$y = y_0 + v_{y0}t + \tfrac{1}{2}a_y t^2 \quad \rightarrow \quad 160\,\text{m} = 0 + 0 + \tfrac{1}{2}\left(9.80\,\text{m/s}^2\right)t^2 \quad \rightarrow$$

$$t = \sqrt{\frac{2(150\,\text{m})}{9.80\,\text{m/s}^2}} = \boxed{5.5\,\text{s}}$$

Note that the horizontal speed of the airplane does not enter into this calculation.

49. Choose the origin to be the location from which the balloon is fired, and choose upward as the positive y direction. Assume the boy in the tree is a distance H up from the point at which the balloon is fired, and that the tree is a distance d horizontally from the point at which the balloon is fired. The equations of motion for the balloon and boy are as follows, using constant acceleration relationships.

$$x_{\text{Balloon}} = v_0 \cos\theta_0 t \quad y_{\text{Balloon}} = 0 + v_0 \sin\theta_0 t - \tfrac{1}{2}gt^2 \quad y_{\text{Boy}} = H - \tfrac{1}{2}gt^2$$

Use the horizontal motion at constant velocity to find the elapsed time after the balloon has traveled d to the right.

$$d = v_0 \cos\theta_0 t_D \quad \rightarrow \quad t_D = \frac{d}{v_0 \cos\theta_0}$$

Where is the balloon vertically at that time?

$$y_{\text{Balloon}} = v_0 \sin\theta_0 t_D - \tfrac{1}{2}gt_D^2 = v_0 \sin\theta_0 \frac{d}{v_0 \cos\theta_0} - \tfrac{1}{2}g\left(\frac{d}{v_0 \cos\theta_0}\right)^2 = d\tan\theta_0 - \tfrac{1}{2}g\left(\frac{d}{v_0 \cos\theta_0}\right)^2$$

Where is the boy vertically at that time? Note that $H = d \tan \theta_0$.

$$y_{\text{Boy}} = H - \tfrac{1}{2} g t_D^2 = H - \tfrac{1}{2} g \left(\frac{d}{v_0 \cos \theta_0} \right)^2 = d \tan \theta_0 - \tfrac{1}{2} g \left(\frac{d}{v_0 \cos \theta_0} \right)^2$$

Note that $y_{\text{Balloon}} = y_{\text{Boy}}$, and so the boy and the balloon are at the same height and the same horizontal location at the same time. Thus they collide!

55. Choose the origin to be at the bottom of the hill, just where the incline starts. The equation of the line describing the hill is $y_2 = x \tan \phi$. The equations of the motion of the object are

$$y_1 = v_{0y} t + \tfrac{1}{2} a_y t^2 \text{ and } x = v_{0x} t, \text{ with } v_{0x} = v_0 \cos \theta \text{ and } v_{0y} = v_0 \sin \theta$$

Solve the horizontal equation for the time of flight, and insert that into the vertical projectile motion equation.

$$t = \frac{x}{v_{0x}} = \frac{x}{v_0 \cos \theta} \;\rightarrow\; y_1 = v_0 \sin \theta \frac{x}{v_0 \cos \theta} - \tfrac{1}{2} g \left(\frac{x}{v_0 \cos \theta} \right)^2 = x \tan \theta - \frac{g x^2}{2 v_0^2 \cos^2 \theta}$$

Equate the y-expressions for the line and the parabola to find the location where the two x-coordinates intersect.

$$x \tan \phi = x \tan \theta - \frac{g x^2}{2 v_0^2 \cos^2 \theta} \;\rightarrow\; \tan \theta - \tan \phi = \frac{g x}{2 v_0^2 \cos^2 \theta} \;\rightarrow\;$$

$$x = \frac{\left(\tan \theta - \tan \phi \right)}{g} 2 v_0^2 \cos^2 \theta$$

This intersection x-coordinate is related to the desired quantity d by $x = a \cos \phi$.

$$d \cos \phi = \left(\tan \theta - \tan \phi \right) \frac{2 v_0^2 \cos^2 \theta}{g} \;\rightarrow\; d = \frac{2 v_0^2}{g \cos \phi} \left(\sin \theta \cos \theta - \tan \phi \cos^2 \theta \right)$$

To maximize the distance, set the derivative of d with respect to θ equal to 0, and solve for θ.

$$\frac{d(d)}{d\theta} = \frac{2 v_0^2}{g \cos \phi} \frac{d}{d\theta} \left(\sin \theta \cos \theta - \tan \phi \cos^2 \theta \right)$$

$$= \frac{2 v_0^2}{g \cos \phi} \left[\sin \theta \left(-\sin \theta \right) + \cos \theta \left(\cos \theta \right) - \tan \phi \left(2 \right) \cos \theta \left(-\sin \theta \right) \right]$$

$$= \frac{2 v_0^2}{g \cos \phi} \left[-\sin^2 \theta + \cos^2 \theta + 2 \tan \phi \cos \theta \sin \theta \right] = \frac{2 v_0^2}{g \cos \phi} \left[\cos 2\theta + \sin 2\theta \tan \phi \right] = 0$$

$$\cos 2\theta + \sin 2\theta \tan \phi = 0 \;\rightarrow\; \theta = \tfrac{1}{2} \tan^{-1} \left(-\frac{1}{\tan \phi} \right)$$

This expression can be confusing, because it would seem that a negative sign enters the solution. In order to get appropriate values, $180°$ or π radians must be added to the angle resulting from the inverse tangent operation, to have a positive angle. Thus a more appropriate expression would be the following:

$$\theta = \tfrac{1}{2} \left[\pi + \tan^{-1} \left(-\frac{1}{\tan \phi} \right) \right]$$

This can be shown to be equivalent to $\boxed{\theta = \frac{\phi}{2} + \frac{\pi}{4}}$, because

$$\tan^{-1} \left(-\frac{1}{\tan \phi} \right) = \tan^{-1} \left(-\cot \phi \right) = \cot^{-1} \cot \phi - \frac{\pi}{2} = \phi - \frac{\pi}{2}.$$

61. The lifeguard will be carried downstream at the same rate as the child. Thus only the horizontal motion need be considered. To cover 45 meters horizontally at a rate of 2 m/s takes $(45 \text{ m})/(2 \text{ m/s}) = \boxed{22.5 \text{ s}}$ for the lifeguard to reach the child. During this time they would both be moving downstream at 1.0 m/s, and so would travel $(1.0 \text{ m/s})(22.5 \text{ s}) = \boxed{22.5 \text{ m}}$ downstream.

67. Call the direction of the flow of the river the x direction, and the direction straight across the river the y direction. Call the location of the swimmer's starting point the origin.

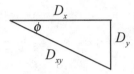

$$\vec{\mathbf{v}}_{\substack{\text{swimmer} \\ \text{rel. shore}}} = \vec{\mathbf{v}}_{\substack{\text{swimmer} \\ \text{rel. water}}} + \vec{\mathbf{v}}_{\substack{\text{water rel.} \\ \text{shore}}} = 0.60 \,\text{m/s}\,\hat{\mathbf{j}} + 0.50 \,\text{m/s}\,\hat{\mathbf{i}}$$

(a) Since the swimmer starts from the origin, the distances covered in the x and y directions will be exactly proportional to the speeds in those directions.

$$\frac{\Delta x}{\Delta y} = \frac{v_x t}{v_y t} = \frac{v_x}{v_y} \quad \rightarrow \quad \frac{\Delta x}{55 \text{ m}} = \frac{0.50 \,\text{m/s}}{0.60 \,\text{m/s}} \quad \rightarrow \quad \Delta x = \boxed{46 \text{ m}}$$

(b) The time is found from the constant velocity relationship for either the x or y directions.

$$\Delta y = v_y t \quad \rightarrow \quad t = \frac{\Delta y}{v_y} = \frac{55 \text{ m}}{0.60 \,\text{m/s}} = \boxed{92 \text{ s}}$$

73. Let east be the positive x-direction, north be the positive y-direction, and up be the positive z-direction. Then the plumber's resultant displacement in component notation is $\boxed{\vec{\mathbf{D}} = 66 \,\text{m}\,\hat{\mathbf{i}} - 35 \,\text{m}\,\hat{\mathbf{j}} - 12 \,\text{m}\,\hat{\mathbf{k}}}$. Since this is a 3-dimensional problem, it requires 2 angles to determine his location (similar to latitude and longitude on the surface of the Earth). For the x-y (horizontal) plane, see the first figure.

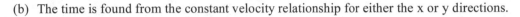

$$\phi = \tan^{-1}\frac{D_y}{D_x} = \tan^{-1}\frac{-35}{66} = -28° = 28° \text{ south of east}$$

$$D_{xy} = \sqrt{D_x^2 + D_y^2} = \sqrt{(66)^2 + (-35)^2} = 74.7 \text{ m} \approx 75 \text{ m}$$

For the vertical motion, consider another right triangle, made up of D_{xy} as one leg, and the vertical displacement D_z as the other leg. See the second figure, and the following calculations.

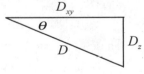

$$\theta_2 = \tan^{-1}\frac{D_z}{D_{xy}} = \tan^{-1}\frac{-12 \text{ m}}{74.7 \text{ m}} = -9° = 9° \text{ below the horizontal}$$

$$D = \sqrt{D_{xy}^2 + D_z^2} = \sqrt{D_x^2 + D_y^2 + D_z^2} = \sqrt{(66)^2 + (-35)^2 + (-12)^2} = 76 \text{ m}$$

The result is that the displacement is $\boxed{76 \text{ m}}$, at an angle of $\boxed{28° \text{ south of east}}$, and $\boxed{90° \text{ below the horizontal}}$.

79. Assume that the golf ball takes off and lands at the same height, so that the range formula derived in Example 3-10 can be applied. The only variable is to be the acceleration due to gravity.

$$R_{\text{Earth}} = v_0^2 \sin 2\theta_0 / g_{\text{Earth}} \qquad R_{\text{Moon}} = v_0^2 \sin 2\theta_0 / g_{\text{Moon}}$$

$$\frac{R_{\text{Earth}}}{R_{\text{Moon}}} = \frac{v_0^2 \sin 2\theta_0 / g_{\text{Earth}}}{v_0^2 \sin 2\theta_0 / g_{\text{Moon}}} = \frac{1/g_{\text{Earth}}}{1/g_{\text{Moon}}} = \frac{g_{\text{Moon}}}{g_{\text{Earth}}} = \frac{32 \text{ m}}{180 \text{ m}} = 0.18 \quad \rightarrow$$

$$g_{\text{Moon}} = 0.18 g_{\text{Earth}} = 0.18(9.80 \,\text{m/s}^2) \approx \boxed{1.8 \,\text{m/s}^2}$$

85. Work in the frame of reference in which the car is at rest at ground level. In this reference frame, the helicopter is moving horizontally with a speed of 208 km/h - 156 km/h = 52 km/h = 14.4 m/s. For the vertical motion, choose the level of the helicopter to be the origin, and downward to be positive. Then the package's y displacement is $y = 78.0$ m, $v_{y0} = 0$ m/s, and $a_y = g$. The time for the package to fall is calculated from Eq. 2-12b.

$$y = y_0 + v_{y0}t + \tfrac{1}{2}a_y t^2 \quad \rightarrow \quad 78.0\,\text{m} = \tfrac{1}{2}\left(9.80\,\text{m/s}^2\right)t^2 \quad \rightarrow \quad t = \sqrt{\frac{2\left(78.0\,\text{m}\right)}{9.80\,\text{m/s}^2}} = 3.99\,\text{sec}$$

The horizontal distance that the package must move, relative to the "stationary" car, is found from the horizontal motion at constant velocity.

$$\Delta x = v_x t = \left(14.44\,\text{m/s}\right)\left(3.99\,\text{s}\right) = 57.6\,\text{m}$$

Thus the angle under the horizontal for the package release will be as follows.

$$\theta = \tan^{-1}\left(\frac{\Delta y}{\Delta x}\right) = \tan^{-1}\left(\frac{78.0\,\text{m}}{57.6\,\text{m}}\right) = 53.6° \approx \boxed{54°}$$

91. First, we find the direction of the straight-line path that the boat must take to pass 150 m to the east of the buoy. See the first diagram (not to scale). We find the net displacement of the boat in the horizontal and vertical directions, and then calculate the angle.

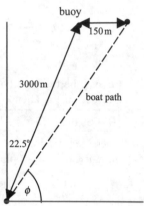

$$\Delta x = \left(3000\,\text{m}\right)\sin 22.5° + 150\,\text{m} \qquad \Delta y = \left(3000\,\text{m}\right)\cos 22.5°$$

$$\phi = \tan^{-1}\frac{\Delta y}{\Delta x} = \frac{\left(3000\,\text{m}\right)\cos 22.5°}{\left(3000\,\text{m}\right)\sin 22.5° + 150\,\text{m}} = 64.905°$$

This angle gives the direction that the boat must travel, so it is the direction of the velocity of the boat with respect to the shore, $\vec{\mathbf{v}}_{\text{boat rel. shore}}$.

So $\vec{\mathbf{v}}_{\text{boat rel. shore}} = v_{\text{boat rel. shore}}\left(\cos\phi\,\hat{\mathbf{i}} + \sin\phi\,\hat{\mathbf{j}}\right)$. Then, using the second diagram (also not to scale), we can write the relative velocity equation relating the boat's travel and the current. The relative velocity equation gives us the following. See the second diagram.

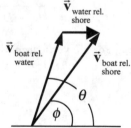

$$\vec{\mathbf{v}}_{\text{boat rel. shore}} = \vec{\mathbf{v}}_{\text{boat rel. water}} + \vec{\mathbf{v}}_{\text{water rel. shore}} \quad \rightarrow$$

$$v_{\text{boat rel. shore}}\left(\cos\phi\,\hat{\mathbf{i}} + \sin\phi\,\hat{\mathbf{j}}\right) = 2.1\left(\cos\theta\,\hat{\mathbf{i}} + \sin\theta\,\hat{\mathbf{j}}\right) + 0.2\hat{\mathbf{i}} \quad \rightarrow$$

$$v_{\text{boat rel. shore}}\cos\phi = 2.1\cos\theta + 0.2 \;;\; v_{\text{boat rel. shore}}\sin\phi = 2.1\sin\theta$$

These two component equations can then be solved for $v_{\text{boat rel. shore}}$ and θ. One technique is to isolate the terms with θ in each equation, and then square those equations and add them. That gives a quadratic equation for $v_{\text{boat rel. shore}}$ which is solved by $v_{\text{boat rel. shore}} = 2.177$ m/s. Then the angle is found to be equal to $\boxed{\theta = 69.9°\text{ N of E}}$.

97. Choose the origin to be the point at the top of the building from which the ball is shot, and call upwards the positive y direction. The initial velocity is $v_0 = 18$ m/s at an angle of $\theta_0 = 42°$. The acceleration due to gravity is $a_y = -g$.

(a) $v_x = v_0 \cos\theta_0 = (18\,\text{m/s}) \cos 42° = 13.38 \approx \boxed{13\,\text{m/s}}$

$v_{y0} = v_0 \sin\theta_0 = (18\,\text{m/s}) \sin 42° = 12.04 \approx \boxed{12\,\text{m/s}}$

(b) Since the horizontal velocity is known and the horizontal distance is known, the time of flight can be found from the constant velocity equation for horizontal motion.

$$\Delta x = v_x t \quad \rightarrow \quad t = \frac{\Delta x}{v_x} = \frac{55\,\text{m}}{13.38\,\text{m/s}} = 4.111\,\text{s}$$

With that time of flight, calculate the vertical position of the ball using Eq. 2-12b.

$$y = y_0 + v_{y0}t + \tfrac{1}{2}a_y t^2 = (12.04\,\text{m/s})(4.111\,\text{s}) + \tfrac{1}{2}(-9.80\,\text{m/s}^2)(4.111\,\text{s})^2$$

$$= -33.3 = \boxed{-33\,\text{m}}$$

So the ball will strike 33 m below the top of the building.

Chapter 4. Dynamics: Newton's Laws of Motion

Chapter Overview and Objectives

In this chapter, Newton's laws of motion are discussed. The force laws relate the forces acting on an object to its acceleration. The use of a free-body diagram is discussed and problem-solving strategies are presented.

After completing this chapter you should:
- Know and understand Newton's three laws of motion.
- Have an understanding of the concepts of mass and force.
- Understand the difference between mass and weight.
- Know the SI units of mass and force.
- Be able to apply Newton's laws of motion to solve for unknown forces, accelerations, or masses.

Summary of Equations

Newton's Second Law of Motion: $$\sum \vec{\mathbf{F}} = m\vec{\mathbf{a}}$$ (Section 4-4)

Newton's Third Law of Motion: $$\vec{\mathbf{F}}_{12} = -\vec{\mathbf{F}}_{21}$$ (Section 4-5)

Weight: $$\vec{\mathbf{F}}_G = m\vec{\mathbf{g}}$$ (Section 4-6)

Chapter Summary

Section 4-1. Force

Force is an interaction that in everyday life we would describe as a push or pull. For an object to accelerate, it must have a force applied to it. Force is a vector quantity because force has a magnitude and a direction. Spring scales can be used to measure forces. The net force acting on a body is the vector sum of each of the forces acting on the body.

Section 4-2. Newton's First Law of Motion

Every body continues in its state of rest or of uniform speed in a straight line unless acted on by a non-zero net force. This law of motion, also called **Newton's first law of motion**, is only valid in inertial reference frames and can be viewed as a definition of inertial reference frames.

Section 4-3. Mass

Mass is a measurement of a body's inertia against a change in motion. The standard of mass in the SI unit system is the kilogram (kg). Mass is a property of an object. **Weight** is the force of gravity acting on the object; it depends on both the object's mass and the location of the object near a gravitating body.

Section 4-4. Newton's Second Law of Motion

The acceleration of a body is proportional to the net force acting on it and inversely proportional to the body's mass. The direction of the acceleration is the direction of the net force acting on the body. Written as a mathematical equation, **Newton's second law of motion** is

$$\sum \vec{\mathbf{F}} = m\vec{\mathbf{a}}$$

Example 4-4-A: Constant-force motion. A runner with a mass of $m = 80.6$ kg runs a $d = 50$ m dash, starting from rest, with constant acceleration. If the runner finishes in a time of $t = 6.2$ s, what was the force of the ground acting on the runner?

Approach: In this problem we assume that a constant force F acts on the runner. Using the information provided, we can determine the acceleration of the runner. Since we know the mass of the runner, we can use this information to determine the force acting on the runner.

Solution: We will choose our coordinate system in such a away that the runner is located at the origin of our coordinate system: $x = 0$ m ($x_0 = 0$ m). Since the runner starts from rest, the velocity at time $t = 0$ s is 0 m/s ($v_0 = 0$ m/s). The position of the runner as function of time is given by the following equation:

$$x(t) = \frac{1}{2}at^2$$

We know that at time t the runner is at position $x = d$. We must thus require that

$$x(t) = d = \frac{1}{2}at^2$$

The problem provides us with the distance d and the time t, and we can thus determine the acceleration a:

$$a = \frac{2d}{t^2}$$

Knowing the acceleration a and the mass m, Newton's second law can be used to determine the force:

$$F = ma = \frac{2md}{t^2} = \frac{2(80.6)(2.60)}{(6.2)^2} = 210\,\text{N}$$

Example 4-4-B: Towing a car. A truck pulls a car on level ground. Starting from rest, the truck and car accelerate with a constant acceleration $a = 2.34$ m/s^2. The horizontal force with which the ground pushes on the truck is constant and equal to $F = 1.41 \times 10^4$ N; the mass of the truck is $M = 3400$ kg. a) What is the mass of the car? b) How much tension is in the cable that tows the car?

Approach: The problem provides information about the acceleration of the car and truck. The problem also provides information about the external force that acts on the car-truck system. Using Newton's second law, we can determine the total mass of the system. Since we know the mass of the truck, we can use the total mass to determine the mass of the car. Since the only force acting on the car is the tension in the cable, we can use Newton's second law to determine the tension using the known acceleration and the calculated mass of the car.

Solution: The Figure below shows a schematic of the forces acting on the car and the truck in the horizontal direction. Additional forces exist in the vertical direction.

The total mass of the system is $m + M$ and it can be determined using the information about the force F and acceleration a provided:

$$m + M = \frac{F}{a}$$

Since we know the mass of the truck, M, we can determine the mass of the car, m:

$$m = \frac{F}{a} - M = \frac{1.41 \times 10^4}{2.34} - 3400 = 2600\,\text{kg}$$

The car has an acceleration a and since we now know its mass we can calculate the force acting on this car. This is the tension in the cable:

$$T = ma = F - Ma = 1.41 \times 10^4 - 3400 \times 2.34 = 6.14 \times 10^3 \text{ N}$$

Section 4-5. Newton's Third Law of Motion

Force is the result of an interaction between two bodies. During the interaction, a force is acting on each body. The magnitude of the force on each body is identical and the forces on the two bodies are directed in opposite directions. This is called **Newton's third law of motion**. If we use the convention that $\vec{F}_{AB}$ means the force on object A from object B, then Newton's third law can be written:

$$\vec{F}_{BA} = -\vec{F}_{AB}$$

Example 4-5-A: Applications of Newton's third law. A jet airplane's engines accelerates a mass $m = 50$ kg of air from rest to a speed of $v_{air} = 200$ m/s as it passes through the engine a distance of $d = 3.67$ m. What is the force of the air on the airplane?

Approach: We know what force the engine must exert on the air in order to accelerate it as indicated. Since forces come in pairs, the air will exert a force on the engine that is equal in magnitude, but directed in an opposite direction. In order to solve this problem, we will assume that the air accelerates with constant acceleration.

Solution: We will describe the motion of the air in a coordinate system that has its origin located at the position of the air when it is at rest (time $t = 0$ s). The x axis of the coordinate system is chosen along the direction in which the air will be moving. The position of the air as function of time is given by

$$x(t) = \frac{1}{2} a t^2$$

The time at which the air has moved a distance d can now be determined:

$$t = \sqrt{\frac{2d}{a}}$$

The velocity at this time is given by

$$v(t) = at = \sqrt{2ad} = v_{air}$$

The acceleration of the air is thus equal to

$$a = \frac{v_{air}^2}{2d}$$

The force that must act on the air in order to produce this acceleration is thus equal to

$$F_{air} = ma = \frac{m v_{air}^2}{2d} = \frac{50 \times (200)^2}{2 \times 3.67} = 2.7 \times 10^5 \text{ N}$$

The force that the air exerts on the engine has the same magnitude, but will be directed in the opposite direction.

Section 4-6. Weight - the Force of Gravity; and the Normal Force

The force of gravity on an object is called the object's **weight**. The unit of weight is the unit of force (the Newton). Since the weight of an object depends on the gravitational acceleration, the weight of an object will change when the gravitational acceleration changes. The weight of an object is thus not an intrinsic property of an object. A 1-kg object will have a weight of 9.8 N on the surface of the earth and 1.8 N on the surface of the moon.
Consider an object that is at rest on the surface of a table. If the object remains at rest, the net force on it must be zero. Thus, we conclude that the net force on the object in the vertical direction must be zero. What are the forces that act in

the vertical direction? The gravitational force acts in the negative vertical direction. In order to ensure that the net force in the vertical direction will be zero, there must be an additional force acting on the block in the vertical direction, which cancels the gravitational force. The additional force is called the **normal force**. It is directed in a direction perpendicular to the surface of the table.

Example 4-6-A: The normal force. A box with a mass $m = 8.7$ kg sits on a table. A person pushes down on the box with a force $F = 80$ N at an angle $\theta = 30°$ below the horizon. Determine the normal force of the table on the box.

Approach: We make the assumption that the box does not move in the vertical direction when the force F is applied. The vertical component of the net force on the box must thus be zero. There are three forces that have vertical components: 1) the gravitational force, 2) the applied force F, and 3) the normal force N. The only unknown is the normal force N.

Solution: The diagram on the right shows the three forces that act on the box. To solve this problem we are only concerned with the components of the force in the vertical direction. The net force in the vertical direction is equal to

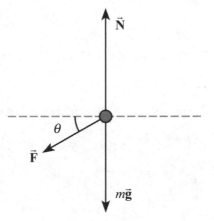

$$\sum_i \vec{F}_{i,y} = N - mg - F\sin\theta$$

Since the net force in the vertical direction must be equal to zero, we can now calculate the normal force N:

$$N = mg + F\sin\theta = 8.7 \times 9.8 + 80 \times \sin(30°) = 125 \text{ N}$$

The normal force is directed perpendicular to the surface of the table. In vector form:

$$\vec{N} = 125\hat{j}$$

Section 4-7. Solving Problems with Newton's Laws: Free-Body Diagrams

The **net force** on a body is the vector sum of all the forces acting on the body. One of the most common mistakes made by students solving physics problems is to forget to include all forces acting on the object. **Free-body diagrams** are used to help you visualize all the forces acting on a given body. In a free-body diagram, the forces acting on the body are represented by arrows. The arrows are drawn in the direction of the forces, and are labeled clearly to make it obvious which forces they represent. Although graphical techniques can be used to determine the net force acting on the object, we will use the free-body diagram as a tool to assemble analytical expressions for the net force acting on the object.

The diagram we used to solve Example 4-6-A is an example of a free-body diagram. The problem focused only on the vertical component of all the forces, but you can immediately conclude that the net force in the horizontal direction is not equal to 0 N. The object will thus move to the left. If the problem had stated that the box remained at rest, we would immediately conclude that we did not include all forces acting on the box. This is a good example of how free-body diagrams can be used to determine if there are additional forces that must be acting on the body.

The following steps are recommended when solving problems using Newton's Laws of Motion:

1. Draw a sketch of the situation.
2. For each object in the problem, draw a separate free-body diagram. Make sure that you include only the forces that act on that particular object the diagram represents.
3. Choose a set of Cartesian coordinate axes and resolve each force into its components along the axes.
4. For each object, write down Newton's second law for the components along the coordinate axes for each of the objects in the problem.
5. Solve the set of equations resulting from this procedure for the unknown parameters.

Example 4-7-A: Rocket Motion. A satellite of mass $m = 156$ kg is traveling horizontally, directly toward the east. The controllers need to divert the satellite so that it has an acceleration a at an angle $\theta = 14.7°$ north of east. The orientation of the satellite is such that one rocket on the satellite exhausts gas to the west with a force of $F_W = 12.8$ N. A second rocket exhausts gas toward the south with an adjustable force F_S. To achieve the desired acceleration direction, what should be the force on the exhaust gas for the south-pointing rocket? What will the magnitude of the satellite's acceleration be?

Approach: The problem indicates the direction of the acceleration of the satellite. According to Newton's second law, the direction of the acceleration of the satellite is the direction of the net force acting on the satellite. There are two forces acting on the satellite: the force exerted on the satellite by the west-pointing rocket, which is directed towards the east, and the force exerted on the satellite by the south-pointing rocket, which is directed towards the north. Knowing the direction of the net force we can determine the magnitude of the force exerted by the south-pointing rocket.

Solution: The two forces acting on rocket are shown in the Figure on the right. The direction of the net force, specified by the angle θ, is related to the magnitude of the forces exerted by the rockets:

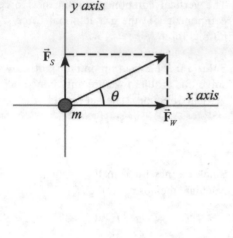

$$\tan\theta = \frac{F_S}{F_w}$$

The force that must be exerted by the south-pointing rocket is thus equal to

$$F_S = F_w \tan\theta$$

The magnitude of the net force acting on the satellite is equal to

$$F_{tot} = \sqrt{F_S^{\,2} + F_W^{\,2}} = F_W \sqrt{1 + \tan^2\theta} = 13.2 \text{ N}$$

The acceleration of the satellite can now be determined

$$a = \frac{F_{tot}}{m} = 8.48 \times 10^{-2} \text{ m/s}^2$$

Example 4-7-B: Calculating the net force and acceleration. Three forces act on an object. The first force has a magnitude $F_1 = 24.6$ N and acts in a direction $\theta_1 = 23.6°$ east of north. The second force has a magnitude $F_2 = 38.7$ N and acts in a direction $\theta_2 = 87.2°$ north of west. The third force has a magnitude $F_3 = 18.6$ N and acts in a direction $\theta_3 = 45.7°$ south of west. What is the net force acting on the object? If the mass of the object is $m = 16.8$ kg, what is the acceleration of the object?

Approach: In order to determine the net force on an object, we first must identify all the forces acting on it. The net force is the vector sum of all these forces; the easiest way to determine the vector sum is to determine the components of the forces along the coordinate axis and add these components. The solution will not be affected by the choice of the coordinate system, but the choice of the coordinate system may have a big impact on the complexity of the solution. If several forces are pointing in the same direction, choosing one coordinate axis to point in that direction may simplify the solution. However, in this case the easiest choice of coordinate system is a system with the coordinate axes parallel to the east-west line and the north-south line since the angles are specified with respect to these directions.

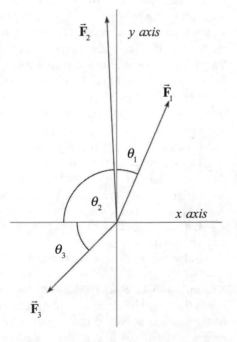

Solution: We start the solution by drawing a free-body diagram of the forces that act on the object and defining the coordinate axis to be used. The proper free-body diagram is shown in the Figure to the right. Using the angles provided we can express the forces into their components:

$$\vec{F}_1 = F_1 \sin\theta_1 \hat{\mathbf{i}} + F_1 \cos\theta_1 \hat{\mathbf{j}}$$
$$\vec{F}_2 = -F_2 \cos\theta_2 \hat{\mathbf{i}} + F_2 \sin\theta_2 \hat{\mathbf{j}}$$
$$\vec{F}_3 = -F_3 \cos\theta_3 \hat{\mathbf{i}} - F_3 \sin\theta_3 \hat{\mathbf{j}}$$

The net force on the object is equal to

$$\vec{F}_{net} = \left(F_1 \sin\theta_1 - F_2 \cos\theta_2 - F_3 \cos\theta_3\right)\hat{i} +$$
$$+\left(F_1 \cos\theta_1 + F_2 \sin\theta_2 - F_3 \sin\theta_3\right)\hat{j}$$

Using the forces and angles provided, we can calculate the net force:

$$\vec{F}_{net} = -5.03\hat{i} + 47.9\hat{j}$$

The magnitude of the net force is equal to

$$\left|\vec{F}_{net}\right| = \sqrt{\left(-5.03\right)^2 + \left(47.9\right)^2} = 48.2 \text{ N}$$

The acceleration of the object can now be determined:

$$\left|\vec{a}\right| = \frac{\left|\vec{F}_{net}\right|}{m} = \frac{48.2}{16.8} = 2.87 \text{ m/s}^2$$

Example 4-7-C. Tension. A sign of mass $m = 6.34$ kg is supported by two ropes, arranged as shown in the diagram on the right. Determine the tension in each rope. The angles of the ropes are $\theta_1 = 60°$ and $\theta_2 = 45°$.

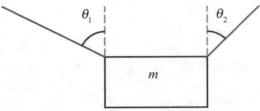

Approach: The ropes exert forces on the sign in the direction of the ropes. In addition to these forces, there is a gravitational force acting on the sign. We need to start this problem by creating a free-body diagram. Since the sign is at rest, the net force acting on the sign must be zero.

Solution: A free-body diagram of the forces acting on the sign is shown in the Figure on the right. We choose our coordinate system such that the x axis is directed along the horizontal direction and the y axis is directed along the vertical direction. Using the angles provided in the problem we can determine the x and y components of the net force

$$\vec{F}_{net} = \vec{T}_1 + \vec{T}_2 + m\vec{g} = \left(-T_1 \sin\theta_1 + T_2 \sin\theta_2\right)\hat{i} + \left(T_1 \cos\theta_1 + T_2 \cos\theta_2 - mg\right)\hat{j}$$

Since the net force on the sign is zero we must require that

$$-T_1 \sin\theta_1 + T_2 \sin\theta_2 = 0$$

$$T_1 \cos\theta_1 + T_2 \cos\theta_2 - mg = 0$$

We have two equations with two unknowns and we can determine the tensions:

$$T_1 = \frac{\sin\theta_2}{\sin\theta_1 \cos\theta_2 + \cos\theta_1 \sin\theta_2} mg = 45.5 \text{ N}$$

$$T_2 = \frac{\sin\theta_1}{\sin\theta_1 \cos\theta_2 + \cos\theta_1 \sin\theta_2} mg = 55.7 \text{ N}$$

Section 4-8. Problem Solving - A General Approach

The approach we have used to solve problems related to Newton's laws can be applied to most problems you will encounter in your physics courses. The following steps should be followed when you work on your physics problems:
1. **Read and reread the problem carefully**.

2. **Draw an accurate picture or diagram** of the situation described in the problem. Use arrows to represent all of the vector quantities involved in the problem. Make a separate free-body diagram for each of the objects in the problem, being careful to only include those forces acting on that particular body.

3. **Choose a convenient coordinate system** for resolving vector quantities into components so that vector additions can be carried out. The choice of coordinate axes doesn't change the solution to the problem, but a good choice of coordinate system will minimize the algebra required to reach the final solution. Pick the coordinate axes so that as many of the vector quantities as possible are directed along the coordinate axes.

4. **Determine which quantities are known and which are unknown**. Let the conceptual principles that relate the known and the unknown quantities guide you to writing down the equations that relate the known and unknown quantities. Make sure the relationships you use are valid under the circumstances described in the problem.

5. **Make a rough calculation** to see if the approach you intend to take is sufficient for solving the problem. You may discover that an additional equation or more information is necessary. Also, a rough calculation can be used to check the values of the actual solution.

6. **Solve the equation or system of equations** obtained in step 4. Solve these equations in terms of the variables provided (do not substitute numbers at this stage).

7. **Keep track of the units**.

8. **Consider whether your answer is reasonable or not**. Use dimensional analysis to determine if your solution has the correct dimensions.

Practice Quiz

1. A box with a weight of 300 N remains at rest while on a level floor, even though you push on it with a horizontal force of 200 N. Which statement is a physically correct explanation of why the box remains at rest, even though you are applying a force to it?
   a) The object is pushing back with an equal and opposite force according to Newton's third law and so the total force is zero.
   b) The weight of the box exceeds the force you are pushing on it with.
   c) The floor also pushes horizontally on the object with a force of 200 N in a direction opposite to your push.
   d) The sum of your push, the normal force of the floor, and the weight of the box is zero.

2. Which of the following objects has a mass of about 1.0 kg?
   a) A marble
   b) A book
   c) A car
   d) The Earth

3. An airplane flies straight in level flight at constant speed. Newton's second law implies that the net force on the airplane is zero. How can this be if the gravitational force of the Earth is pulling downward on the plane?
   a) There is no gravitational force at the altitude planes fly at.
   b) The airplane is flying fast enough that it is actually falling around the curvature of the earth.
   c) The downward gravitational pull of the earth is balanced by the upward gravitational pull of the moon.
   d) There is an upward force on the plane from the air beneath its wings.

4. Knowing that an object accelerates at 3.000 m/s^2 when the force acting on it is 396 N, how much force must be applied to the object so that the magnitude of its acceleration is 6.000 m/s^2?
   a) 198 N
   b) 99 N
   c) 792 N
   d) 1584 N

5. What are the SI units of force divided by acceleration?
   a) kg
   b) m/s
   c) kg m/s^2
   d) m/s^2

6. If only two forces act on an object but the object has no acceleration, what must be true about these two forces?
    a) Both forces are zero.
    b) The forces are equal in magnitude, but point in opposite directions.
    c) The forces are equal in magnitude, but point in perpendicular directions.
    d) The forces can be of any magnitude and pointing in any direction.

7. An object starts from rest and travels a distance d in time t when a constant force F acts on the object. What constant force must act on the object in order for it to start from rest and travel the same distance in a time $t/2$?
    a) $4F$
    b) $F/4$
    c) $2F$
    d) $F/2$

8. When a person with a mass of 98.4 kg stands on a spring scale in an elevator, the scale reads 0 N. Which statement about the elevator is true?
    a) The elevator is stationary.
    b) The elevator is moving downward with a velocity of 9.8 m/s.
    c) The elevator is accelerating upward with an acceleration of 9.8 m/s^2.
    d) The elevator is accelerating downward with an acceleration of 9.8 m/s^2.

9. When you jump upward from the ground, the Earth is applying an upward force on you. Newton's third law implies that you are applying a force, equal in magnitude but opposite in direction, on the Earth. Why doesn't everyone around you feel the Earth accelerate downward when you jump?
    a) It happens in too short a time for other people to notice.
    b) The Earth's mass is so great that its acceleration is unnoticeable.
    c) Everyone does notice and your jump is recorded on seismographs around the world.
    d) This is an incorrect application of Newton's third law.

10. Two forces act on an object of mass 3.249 kg and cause it to accelerate with an acceleration of 1.000 m/s^2 toward the east. What is the magnitude and direction of a third force applied to the object so that its acceleration is zero?
    a) Impossible. Two forces are needed: one to cancel each of the other two forces.
    b) Need to know the other two forces to determine the third force.
    c) 1.000 m/s^2 toward the west.
    d) 3.249 N toward the west.

11. An object with a mass of 3.411 kg has three forces acting on it: 240 N in a direction east, 322 N in a direction 47° north of west, and 568 N in a direction 12° west of south. What is the acceleration of the object?

12. An object with a mass of 3.42 kg has a time-varying force acting on it given by

    $$F(t) = 1.0t - 2.5t^2$$

    The direction of the force is constant. If the object starts from rest at $t = 0$ s, how far from its initial position is it after 10 s?

13. A 28.4-kg block rests on a frictionless inclined plane that is tilted 26.5° from the horizontal. A rope is attached to the block that pulls parallel to and up the incline. The acceleration of the block is 0.893 m/s^2 down the incline. What is the tension in the rope?

14. What is the tension in the rope in question 13 if the acceleration is 0.893 m/s^2 up the incline instead of down the incline?

15. An elevator is at rest and begins descending with constant acceleration. After it descends 27.2 m, it has a downward velocity of 19.2 m/s. If the weight of the elevator is 9340 N, what is the tension in the cable that is lowering the elevator?

Responses to Select End-of-Chapter Questions

1. When you give the wagon a sharp pull forward, the force of friction between the wagon and the child acts on the child to move her forward. But the force of friction acts at the contact point between the child and the wagon – either the feet, if the child is standing, or her bottom, if sitting. In either case, the lower part of the child begins to move forward, while the upper part, following Newton's first law (the law of inertia), remains almost stationary, making it seem as if the child falls backward.

7. As you take a step on the log, your foot exerts a force on the log in the direction opposite to the direction in which *you* want to move, which pushes the log "backwards." (The log exerts an equal and opposite force forward on you, by Newton's third law.) If the log had been on the ground, friction between the ground and the log would have kept the log from moving. However, the log is floating in water, which offers little resistance to the movement of the log as you push it backwards.

13. If a person gives a sharp pull on the dangling thread, the thread is likely to break below the stone. In the short time interval of a sharp pull, the stone barely begins to accelerate because of its great mass (inertia), and so does not transmit the force to the upper string quickly. The stone will not move much before the lower thread breaks. If a person gives a slow and steady pull on the thread, the thread is most likely to break *above* the stone because the tension in the upper thread is the applied force *plus* the weight of the stone. Since the tension in the upper thread is greater, it is likely to break first.

19. A weight of 1 N corresponds to 0.225 lb. That's about the weight of (*a*) an apple.

Solutions to Select End-of-Chapter Problems

1. Use Newton's second law to calculate the force.

$$\sum F = ma = (55 \text{ kg})(1.4 \text{ m/s}^2) = \boxed{77 \text{ N}}$$

7. Find the average acceleration from Eq. 2-12c, and then find the force needed from Newton's second law.

$$a_{avg} = \frac{v^2 - v_0^2}{2(x - x_0)} \rightarrow$$

$$F_{avg} = ma_{avg} = m\frac{v^2 - v_0^2}{2(x - x_0)} = (7.0 \text{ kg})\left[\frac{(13 \text{ m/s})^2 - 0}{2(2.8 \text{ m})}\right] = 211.25 \text{ N} \approx \boxed{210 \text{ N}}$$

13. Choose up to be the positive direction. Write Newton's second law for the vertical direction, and solve for the acceleration.

$$\sum F = F_T - mg = ma$$

$$a = \frac{F_T - mg}{m} = \frac{163 \text{ N} - (14.0 \text{ kg})(9.80 \text{ m/s}^2)}{14.0 \text{ kg}} = \boxed{1.8 \text{ m/s}^2}$$

Since the acceleration is positive, the bucket has an $\boxed{\text{upward}}$ acceleration.

19. (*a*) To calculate the time to accelerate from rest, use Eq. 2-12a.

$$v = v_0 + at \quad \rightarrow \quad t = \frac{v - v_0}{a} = \frac{9.0 \,\text{m/s} - 0}{1.2 \,\text{m/s}^2} = 7.5 \,\text{s}$$

The distance traveled during this acceleration is found from Eq. 2-12b.

$$x - x_0 = v_0 t + \tfrac{1}{2} at^2 = \tfrac{1}{2}\left(1.2 \,\text{m/s}^2\right)\left(7.5 \,\text{s}\right)^2 = 33.75 \,\text{m}$$

To calculate the time to decelerate to rest, use Eq. 2-12a.

$$v = v_0 + at \quad \rightarrow \quad t = \frac{v - v_0}{a} = \frac{0 - 9.0 \,\text{m/s}}{-1.2 \,\text{m/s}^2} = 7.5 \,\text{s}$$

The distance traveled during this deceleration is found from Eq. 2-12b.

$$x - x_0 = v_0 t + \tfrac{1}{2} at^2 = \left(9.0 \,\text{m/s}\right)\left(7.5 \,\text{s}\right) + \tfrac{1}{2}\left(-1.2 \,\text{m/s}^2\right)\left(7.5 \,\text{s}\right)^2 = 33.75 \,\text{m}$$

To distance traveled at constant velocity is 180 m - 2(33.75 m) = 112.5 m. To calculate the time spent at constant velocity, use Eq. 2-8.

$$x = x_0 + \bar{v}t \quad \rightarrow \quad t = \frac{x - x_0}{\bar{v}} = \frac{112.5 \,\text{m/s}}{9.0 \,\text{m/s}} = 12.5 \,\text{s} \approx 13 \,\text{s}$$

Thus the times for each stage are:

$$\boxed{\text{Accelerating: } 7.5 \,\text{s} \quad \text{Constant Velocity: } 13 \,\text{s} \quad \text{Decelerating: } 7.5 \,\text{s}}$$

(*b*) The normal force when at rest is *mg*. From the free-body diagram, if up is the positive direction, we have that $F_N - mg = ma$. Thus the change in normal force is the difference in the normal force and the weight of the person, or *ma*.

Accelerating:
$$\frac{\Delta F_N}{F_N} = \frac{ma}{mg} = \frac{a}{g} = \frac{1.2 \,\text{m/s}^2}{9.80 \,\text{m/s}^2} \times 100 = \boxed{12\%}$$

Constant velocity:
$$\frac{\Delta F_N}{F_N} = \frac{ma}{mg} = \frac{a}{g} = \frac{0}{9.80 \,\text{m/s}^2} \times 100 = \boxed{0\%}$$

Decelerating:
$$\frac{\Delta F_N}{F_N} = \frac{ma}{mg} = \frac{a}{g} = \frac{-1.2 \,\text{m/s}^2}{9.80 \,\text{m/s}^2} \times 100 = \boxed{-12\%}$$

(*c*) The normal force is not equal to the weight during the accelerating and deceleration phases.

$$\frac{7.5 \,\text{s} + 7.5 \,\text{s}}{7.5 \,\text{s} + 12.5 \,\text{s} + 7.5 \,\text{s}} = \boxed{55\%}$$

25. We break the race up into two portions. For the acceleration phase, we call the distance d_1 and the time t_1. For the constant speed phase, we call the distance d_2 and the time t_2. We know that $d_1 = 45$ m, $d_2 = 55$ m, and $t_2 = 10.0$ s $- t_1$. Eq. 2-12b is used for the acceleration phase and Eq. 2-2 is used for the constant speed phase. The speed during the constant speed phase is the final speed of the acceleration phase, found from Eq. 2-12a.

$$x - x_0 = v_0 t + \tfrac{1}{2}at^2 \;\rightarrow\; d_1 = \tfrac{1}{2}at_1^2 \;;\; \Delta x = vt \;\rightarrow\; d_2 = vt_2 = v\left(10.0\,\text{s} - t_1\right) \;;\; v = v_0 + at_1$$

This set of equations can be solved for the acceleration and the velocity.

$$d_1 = \tfrac{1}{2}at_1^2 \;;\; d_2 = v\left(10.0\,\text{s} - t_1\right) \;;\; v = at_1 \;\rightarrow\; 2d_1 = at_1^2 \;;\; d_2 = at_1\left(10.0 - t_1\right) \;\rightarrow$$

$$a = \frac{2d_1}{t_1^2} \;;\; d_2 = \frac{2d_1}{t_1^2}t_1\left(10.0 - t_1\right) = \frac{2d_1}{t_1}\left(10.0 - t_1\right) \;\rightarrow\; d_2 t_1 = 2d_1\left(10.0 - t_1\right) \;\rightarrow$$

$$t_1 = \frac{20.0d_1}{\left(d_2 + 2d_1\right)} \;\rightarrow\; a = \frac{2d_1}{t_1^2} = \frac{2d_1}{\left[\dfrac{20.0d_1}{\left(d_2 + 2d_1\right)}\right]^2} = \frac{\left(d_2 + 2d_1\right)^2}{\left(200\,\text{s}^2\right)d_1}$$

$$v = at_1 = \frac{\left(d_2 + 2d_1\right)^2}{200d_1}\frac{20.0d_1}{\left(d_2 + 2d_1\right)} = \frac{\left(d_2 + 2d_1\right)}{10.0\,\text{s}}$$

(a) The horizontal force is the mass of the sprinter times their acceleration.

$$F = ma = m\frac{\left(d_2 + 2d_1\right)^2}{\left(200\,\text{s}^2\right)d_1} = \left(66\,\text{kg}\right)\frac{\left(145\,\text{m}\right)^2}{\left(200\,\text{s}^2\right)\left(45\,\text{m}\right)} = 154\,\text{N} \approx \boxed{150\,\text{N}}$$

(b) The velocity for the second portion of the race was found above.

$$v = \frac{\left(d_2 + 2d_1\right)}{10.0\,\text{s}} = \frac{145\,\text{m}}{10.0\,\text{s}} = \boxed{14.5\,\text{m/s}}$$

31. (a) We draw a free-body diagram for the piece of the rope that is directly above the person. That piece of rope should be in equilibrium. The person's weight will be pulling down on that spot, and the rope tension will be pulling away from that spot towards the points of attachment. Write Newton's second law for that small piece of the rope.

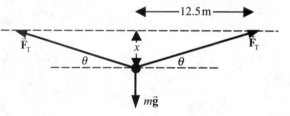

$$\sum F_y = 2F_\text{T}\sin\theta - mg = 0 \;\rightarrow\; \theta = \sin^{-1}\frac{mg}{2F_\text{T}} = \sin^{-1}\frac{\left(72.0\,\text{kg}\right)\left(9.80\,\text{m/s}^2\right)}{2\left(2900\,\text{N}\right)} = 6.988°$$

$$\tan\theta = \frac{x}{12.5\,\text{m}} \;\rightarrow\; x = \left(12.5\,\text{m}\right)\tan 6.988° = 1.532\,\text{m} \approx \boxed{1.5\,\text{m}}$$

(b) Use the same equation to solve for the tension force with a sag of only ¼ that found above.

$$x = \tfrac{1}{4}\left(1.532\,\text{m}\right) = 0.383\,\text{m} \;;\; \theta = \tan^{-1}\frac{0.383\,\text{m}}{12.5\,\text{m}} = 1.755°$$

$$F_\text{T} = \frac{mg}{2\sin\theta} = \frac{\left(72.0\,\text{kg}\right)\left(9.80\,\text{m/s}^2\right)}{2\left(\sin 1.755°\right)} = \boxed{11.5\,\text{kN}}$$

The $\boxed{\text{rope will not break}}$, but it exceeds the recommended tension by a factor of about 4.

37. The net force in each case is found by vector addition with components.

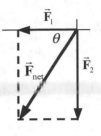

(a) $F_{Net\,x} = -F_1 = -10.2\ N \qquad F_{Net\,y} = -F_2 = -16.0\ N$

$$F_{Net} = \sqrt{(-10.2)^2 + (-16.0)^2} = 19.0\ N \qquad \theta = \tan^{-1}\frac{-16.0}{-10.2} = 57.48°$$

The actual angle from the x axis is then 237.48°. Thus the net force is

$$F_{Net} = \boxed{19.0\ N\ at\ 237.5°}$$

$$a = \frac{F_{Net}}{m} = \frac{19.0\ N}{18.5\ kg} = \boxed{1.03\ m/s^2\ at\ 237.5°}$$

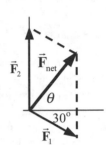

(b) $F_{Net\,x} = F_1\cos 30° = 8.833\ N \qquad F_{Net\,y} = F_2 - F_1\sin 30° = 10.9\ N$

$$F_{Net} = \sqrt{(8.833\ N)^2 + (10.9\ N)^2} = 14.03N \approx \boxed{14.0\ N}$$

$$\theta = \tan^{-1}\frac{10.9}{8.833} = \boxed{51.0°} \qquad a = \frac{F_{Net}}{m} = \frac{14.03\ N}{18.5\ kg} = \boxed{0.758\ m/s^2}\ at\ \boxed{51.0°}$$

43. From the free-body diagram, the net force along the plane on the skater is $mg\sin\theta$ and so the acceleration along the plane is $g\sin\theta$. We use the kinematical data and Eq. 2-12b to write an equation for the acceleration, and then solve for the angle.

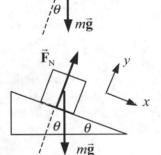

$$x - x_0 = v_0 t + \tfrac{1}{2}at^2 = v_0 t + \tfrac{1}{2}gt^2\sin\theta \rightarrow$$

$$\theta = \sin^{-1}\left(\frac{2\Delta x - v_0 t}{gt^2}\right) = \sin^{-1}\left(\frac{2(18m) - 2(2.0\,m/s)(3.3s)}{(9.80\,m/s^2)(3.3s)^2}\right) = \boxed{12°}$$

49. (a) Consider the free-body diagram for the block on the frictionless surface. There is no acceleration in the y direction. Write Newton's second law for the x direction.

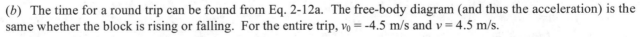

$$\sum F_x = mg\sin\theta = ma \rightarrow a = g\sin\theta$$

Use Eq. 2-12c with $v_0 = -4.5$ m/s and $v = 0$ m/s to find the distance that it slides before stopping.

$$v^2 - v_0^2 = 2a(x - x_0) \rightarrow$$

$$(x - x_0) = \frac{v^2 - v_0^2}{2a} = \frac{0 - (-4.5\,m/s)^2}{2(9.80\,m/s^2)\sin 22.0°} = -2.758\,m \approx \boxed{2.8\,m\ up\ the\ plane}$$

(b) The time for a round trip can be found from Eq. 2-12a. The free-body diagram (and thus the acceleration) is the same whether the block is rising or falling. For the entire trip, $v_0 = -4.5$ m/s and $v = 4.5$ m/s.

$$v = v_0 + at \rightarrow t = \frac{v - v_0}{a} = \frac{(4.5\,m/s) - (-4.5\,m/s)}{(9.80\,m/s^2)\sin 22°} = 2.452s \approx \boxed{2.5s}$$

55. If m doesn't move on the incline, it doesn't move in the vertical direction, and so has no vertical component of acceleration. This suggests that we analyze the forces parallel and perpendicular to the floor. See the force diagram for the small block, and use Newton's second law to find the acceleration of the small block.

$$\sum F_y = F_N \cos\theta - mg = 0 \;\; \rightarrow \;\; F_N = \frac{mg}{\cos\theta}$$

$$\sum F_x = F_N \sin\theta = ma \;\; \rightarrow \;\; a = \frac{F_N \sin\theta}{m} = \frac{mg \sin\theta}{m\cos\theta} = g\tan\theta$$

Since the small block doesn't move on the incline, the combination of both masses has the same horizontal acceleration of $g\tan\theta$. That can be used to find the applied force.

$$F_{applied} = (m+M)a = \boxed{(m+M)g\tan\theta}$$

Note that this gives the correct answer for the case of $\theta = 0°$, where it would take no applied force to keep m stationary. It also gives a reasonable answer for the limiting case of $\theta \rightarrow 90°$, where no force would be large enough to keep the block from falling, since there would be no upward force to counteract the force of gravity.

61. We assume that the pulley is small enough that the part of the cable that is touching the surface of the pulley is negligible, and so we ignore any force on the cable due to the pulley itself. We also assume that the cable is uniform, so that the mass of a portion of the cable is proportional to the length of that portion. We then treat the cable as two masses, one on each side of the pulley. The masses are given by $m_1 = y\, M/l$ and $m_2 = (l-y)\, M/l$. Free-body diagrams for the masses are shown.
(*a*) We take downward motion of m_1 to be the positive direction for m_1 and upward motion of m_2 to be the positive direction for m_2. Newton's second law for the masses gives the following.

$$F_{net\,1} = m_1 g - F_T = m_1 a \;;\; F_{net\,2} = F_T - m_2 g \;\; \rightarrow \;\; a = \frac{m_1 - m_2}{m_1 + m_2}g$$

$$a = \frac{\dfrac{y}{l}M - \dfrac{l-y}{l}M}{\dfrac{y}{l}M + \dfrac{l-y}{l}M}g = \frac{y-(l-y)}{y+(l-y)}g = \frac{2y-l}{l}g = \boxed{\left(\frac{2y}{l}-1\right)g}$$

(*b*) Use the hint supplied with the problem to set up the equation for the velocity. The cable starts with a length y_0 (assuming $y_0 > l/2$) on the right side of the pulley, and finishes with a length l on the right side of the pulley.

$$a = \left(\frac{2y}{l}-1\right)g = \frac{dv}{dt} = \frac{dv}{dy}\frac{dy}{dt} = v\frac{dv}{dy} \;\; \rightarrow \;\; \left(\frac{2y}{l}-1\right)g\,dy = v\,dv \;\; \rightarrow$$

$$\int_{y_0}^{l}\left(\frac{2y}{l}-1\right)g\,dy = \int_{0}^{v_f} v\,dv \;\; \rightarrow \;\; g\left(\frac{y^2}{l}-y\right)\Big|_{y_0}^{l} = \left(\tfrac{1}{2}v^2\right)\Big|_{0}^{v_f} \;\; \rightarrow \;\; gy_0\left(1-\frac{y_0}{l}\right) = \tfrac{1}{2}v_f^2 \;\; \rightarrow$$

$$\boxed{v_f = \sqrt{2gy_0\left(1-\frac{y_0}{l}\right)}}$$

(*c*) For $y_0 = (2/3)l$, we have

$$v_f = \sqrt{2gy_0\left(1-\frac{y_0}{l}\right)} = \sqrt{2g\left(\tfrac{2}{3}\right)l\left(1-\frac{\tfrac{2}{3}l}{l}\right)} = \boxed{\tfrac{2}{3}\sqrt{gl}}$$

67. (*a*) Draw a free-body diagram for each block. Write Newton's second law for each block. Notice that the acceleration of block A in the y_A direction will be zero, since it has no motion in the y_A direction.

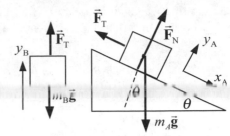

$$\sum F_{yA} = F_N - m_A g \cos\theta = 0 \quad \rightarrow \quad F_N = m_A g \cos\theta$$

$$\sum F_{xA} = m_A g \sin\theta - F_T = m_A a_{xA}$$

$$\sum F_{yB} = F_T - m_B g = m_B a_{yB} \quad \rightarrow \quad F_T = m_B \left(g + a_{yB}\right)$$

Since the blocks are connected by the cord, $a_{yB} = a_{xA} = a$. Substitute the expression for the tension force from the last equation into the x direction equation for block 1, and solve for the acceleration.

$$m_A g \sin\theta - m_B\left(g + a\right) = m_A a \quad \rightarrow \quad m_A g \sin\theta - m_B g = m_A a + m_B a$$

$$\boxed{a = g \frac{\left(m_A \sin\theta - m_B\right)}{\left(m_A + m_B\right)}}$$

(*b*) If the acceleration is to be down the plane, it must be positive. That will happen if $\boxed{m_A \sin\theta > m_B}$. The acceleration will be up the plane (negative) if $\boxed{m_A \sin\theta < m_B}$. If $\boxed{m_A \sin\theta = m_B}$ then the system will not accelerate. It will move with a constant speed if set in motion by a push.

73. (*a*) The value of the constant c can be found from the free-body diagram, knowing that the net force is 0 when coasting downhill at the specified speed.

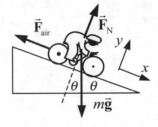

$$\sum F_x = mg \sin\theta - F_{air} = 0 \quad \rightarrow \quad F_{air} = mg \sin\theta = cv \quad \rightarrow$$

$$c = \frac{mg \sin\theta}{v} = \frac{\left(80.0\,\text{kg}\right)\left(9.80\,\text{m/s}^2\right)\sin 5.0^\circ}{\left(6.0\,\text{km/h}\right)\left(\dfrac{1\,\text{m/s}}{3.6\,\text{km/h}}\right)} = 40.998\,\frac{\text{N}}{\text{m/s}}$$

$$\approx \boxed{41\,\frac{\text{N}}{\text{m/s}}}$$

(*b*) Now consider the cyclist with an added pushing force $\vec{F}_P$ directed along the plane. The free-body diagram changes to reflect the additional force the cyclist must exert. The same axes definitions are used as in part (*a*).

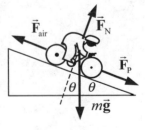

$$\sum F_x = F_P + mg \sin\theta - F_{air} = 0 \quad \rightarrow$$

$$F_P = F_{air} - mg \sin\theta = cv - mg \sin\theta$$

$$= \left(40.998\,\frac{\text{N}}{\text{m/s}}\right)\left(\left(18.0\,\text{km/h}\right)\left(\frac{1\,\text{m/s}}{3.6\,\text{km/h}}\right)\right)$$

$$- \left(80.0\,\text{kg}\right)\left(9.80\,\text{m/s}^2\right)\sin 5.0^\circ = 136.7\,\text{N} \approx \boxed{140\,\text{N}}$$

79 (*a*) We assume that the maximum horizontal force occurs when the train is moving very slowly, and so the air resistance is negligible. Thus the maximum acceleration is given by the following.

$$a_{max} = \frac{F_{max}}{m} = \frac{4 \times 10^5\,\text{N}}{6.4 \times 10^5\,\text{kg}} = 0.625\,\text{m/s}^2 \approx \boxed{0.6\,\text{m/s}^2}$$

(*b*) At top speed, we assume that the train is moving at constant velocity. Therefore the net force on the train is 0 N, and so the air resistance and friction forces together must be of the same magnitude as the horizontal pushing force, which is $\boxed{1.5 \times 10^5\,\text{N}}$.

85. Use the free-body diagram to find the net force in the x direction, and then find the acceleration. Then Eq. 2-12c can be used to find the final speed at the bottom of the ramp.

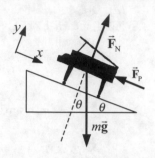

$$\sum F_x = mg \sin\theta - F_P = ma \ \rightarrow$$

$$a = \frac{mg \sin\theta - F_P}{m} = \frac{(450\,\text{kg})(9.80\,\text{m/s}^2)\sin 22° - 1420\,\text{N}}{450\,\text{kg}}$$

$$= 0.516\,\text{m/s}^2$$

$$v^2 = v_0^2 + 2a(x - x_0) \ \rightarrow \ v = \sqrt{2a(x - x_0)} = \sqrt{2(0.516\,\text{m/s}^2)(11.5\,\text{m})} = \boxed{3.4\,\text{m/s}}$$

Chapter 5:

Using Newton's Laws: Friction, Circular Motion, Drag Forces

Chapter Overview and Objectives

In this chapter, the application of Newton's second law to some particular situations and types of forces are described. Friction, motion along circular paths, and velocity-dependent forces are introduced.

After completing this chapter you should:
- Be able to determine frictional forces on objects.
- Know the dynamics of uniform and non-uniform circular motion.
- Be able to solve problems involving circular motion.
- Know that drag forces in fluids depend on the relative speed of objects to the fluid.

Summary of Equations

Relationship between kinetic friction and the normal force:	$F_{\text{fr}} = \mu_k F_N$	(Section 5-1)
Relationship between static friction and the normal force:	$F_{\text{fr}} \leq \mu_s F_N$	(Section 5-1)
Centripetal acceleration of an object in circular motion:	$a_R = \dfrac{v^2}{r}$	(Section 5-2)
Magnitude of tangential acceleration:	$a_{\tan} = \dfrac{dv}{dt}$	(Section 5-5)
Total acceleration of an object in circular motion:	$\vec{a} = \vec{a}_R + \vec{a}_{\tan}$	(Section 5-5)
Viscous drag force on an object moving in a fluid:	$F_D = -bv$	(Section 5-6)
Inertial drag force on an object moving in a fluid:	$F_D = -Cv^2$	(Section 5-6)

Chapter Summary

Section 5-1. Applications of Newton's Laws Involving Friction

Whenever there is relative motion between two objects, whose surfaces are in contact, there are friction forces acting on each object, parallel to their contact surfaces. These forces act on the objects in a direction opposite to their direction of the relative motion. These friction forces are called **kinetic friction forces**. The magnitude of the kinetic friction force is proportional to the normal force of contact of one surface on the other:

$$F_{\text{fr}} = \mu_k F_N$$

where F_{fr} is the magnitude of the friction force and F_N is the magnitude of the normal force. The **coefficient of kinetic friction**, μ_k, depends on the materials of which the two surfaces are made, the roughness of these surfaces, and conditions such as temperature and humidity.

There can also be friction forces between two objects in contact when they are not in relative motion if the net force on at least one of the objects would cause a relative acceleration of the objects in the absence of friction. These friction forces are called **static friction forces**. Since each object remains at rest, the net force must be equal to zero, and the static friction force must thus be equal in magnitude but opposite in direction to the total of all other forces acting on the object, parallel to the surface of contact. However, the static friction force has an upper limit and satisfies the following relation

$$F_{fr} \leq \mu_s F_N$$

where μ_s is the coefficient of static friction. When the static friction force reaches its maximum value, the object will start to slide (and accelerate). The coefficient of static friction is larger than the coefficient of kinetic friction, and as a result, the kinetic friction force is less than the maximum static friction force. A graph of the friction force as function of the applied force is shown in Figure 5-3 of the textbook. A table of typical values of the coefficients of static and kinetic friction between two different types of materials is shown in Table 5-1 of the textbook.

Example 5-1-A: Maintaining constant velocity with a non-zero applied force. A box of mass $m = 17.2$ kg is slid across a level floor with constant velocity. The coefficient of kinetic friction between the floor and the box is $\mu_k = 0.32$. With what force must the box be pushed to maintain this constant velocity?

Approach: Since the box is moving with constant velocity, its acceleration is equal to 0 m/s^2. The net force acting on the box must thus be equal to 0 N. In order to solve this problem, we must identify all the forces acting on the box and require that the net force (the sum of all these forces) is equal to 0 N.

Solution: A free-body diagram that can be used to analyze this problem is shown to the right. Our choice of coordinate system is indicated in the free-body diagram. Newton's second law in the horizontal and vertical directions can be written as:

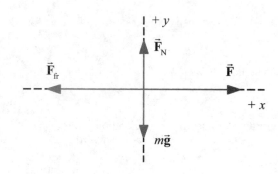

$$\sum F_x = F - F_{fr} = ma_x \qquad \text{and} \qquad \sum F_y = F_N - mg = ma_y$$

Since the box is moving with constant velocity across the floor, its acceleration in both the horizontal and vertical directions must be zero. Newton's second law along the vertical direction immediately tells us that the normal force $F_N = mg$. Since the box is sliding across the floor, the friction force is the kinetic friction force. The magnitude of the kinetic friction force can be calculated immediately based on the known normal force

$$F_{fr} = \mu_k F_N = \mu_k mg$$

Rewriting the equation for Newton's second law in the horizontal direction we conclude that

$$\sum F_x = F - \mu_k F_N = F - \mu_k mg = 0$$

or

$$F = \mu_k mg = (0.32)(17.2\,\text{kg})(9.80\,\text{m/s}^2) = 54\,\text{N}$$

Example 5-1-B: How to start sliding motion? A box of mass $m = 3.23$ kg rests on a ramp with a slope of $\theta = 34.2°$. The coefficient of static friction between the box and the ramp is $\mu_s = 0.774$. What additional force, directed directly down the ramp, will start the box sliding down the ramp?

Approach: Just before the box starts sliding down the incline it is at rest ($a = 0$ m/s^2), and the net force acting on it must be 0 N. As the additional force increases, the static friction force will also increase to try to keep the box at rest. However, at one point, the static friction force will reach its maximum value and at that point the box will start to slide. In order to solve this problem, we will determine at what value of the applied force the static friction force will reach its maximum value.

Solution: A free-body diagram that can be used to solve this problem is shown to the next page. In this diagram, $\vec{F}$ is the additional force required to start the box moving, $m\vec{g}$ is the gravitational force on the box, $\vec{F}_N$ is the normal force of the ramp on the box, and $\vec{F}_{fr}$ is the static friction force on the box. The coordinate system used has the x-direction parallel and down the ramp, and the y-direction perpendicular to the ramp and angled upward.

The problem asks us to determine the minimum force that will start the box moving. The force necessary will be that force that just overcomes the maximum static friction force. The maximum value of the static friction force is $\mu_s F_N$. In order to calculate the friction force we must determine the magnitude of the normal force exerted by the ramp on the box. Since the box is not moving along the y axis, its acceleration along this axis is 0 m/s². Applying Newton's second law along the y axis we obtain

$$\sum F_y = -mg\cos\theta + F_N = ma_y = 0$$

The normal force is thus equal to

$$F_N = mg\cos\theta$$

The box will be on the verge of moving when the sum of the maximum static friction force and the x components of the other forces acting on the box is equal to 0 N:

$$\sum F_x = mg\sin\theta + F - \mu_s mg\cos\theta = 0$$

The applied force $\vec{F}$ thus has a magnitude equal to

$$F = mg\left(\mu_s\cos\theta - \sin\theta\right) = \left(3.23\,\text{kg}\right)\left(9.80\,\text{m/s}^2\right)\left(\left(0.774\right)\cos34.2° - \sin34.2°\right) = 2.47\,\text{N}$$

When the box starts to move, the friction force changes from static friction to kinetic friction, and the net force along the x axis is not equal to zero anymore (remember that the kinetic friction force is smaller than the maximum static friction force). The net force along the x axis is equal to

$$\sum F_x = mg\sin\theta + F - \mu_k mg\cos\theta = ma_x$$

The acceleration along the x axis, a_x, is equal to

$$a_x = g\sin\theta + \frac{F}{m} - \mu_k g\cos\theta$$

Example 5-1-C: The net acceleration of a complex system. Two boxes are in contact with each other on a horizontal surface (see diagram to the right). The larger box has a mass $m_1 = 4.55$ kg and a coefficient of kinetic friction of $\mu_1 = 0.365$ with the surface. The smaller box has a mass $m_2 = 3.21$ kg and a coefficient of kinetic friction of $\mu_2 = 0.822$ with the surface. A force of $F = 90.6$ N pushes on the larger box at an angle $\theta = 10.2°$ below the horizontal, as shown in the diagram. What is the acceleration of the boxes?

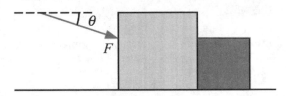

Approach: The solution of this problem requires us to determine all the forces acting on the two boxes. Since the blocks are moving, we need to use kinetic friction to solve this problem. The friction forces between the surface and the two boxes will depend on the normal force acting on each box. In this problem we need to take into consideration the fact that the applied force is acting at an angle with respect to the horizontal direction, and its vertical component will affect the normal force acting on the larger box, and thus the kinetic friction force between this box and the surface. Although there are contact forces acting between the blocks, these contact forces will cancel when we consider the net force acting on both boxes. It is this net force we need to determine in order to calculate the acceleration of the boxes.

Solution: The free-body diagrams for each box are shown in the diagram to the right. We have used the symbols with subscript 1 for the larger box and subscript 2 for the smaller box. The forces $\vec{F}_{2\,on\,1}$ and $\vec{F}_{1\,on\,2}$ are the force of the second box on the first box and the force of the first box on the second box, respectively.

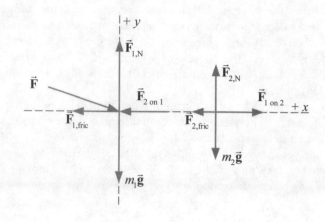

In order to determine the acceleration of the boxes along the x axis we need to determine the friction forces first. Since the boxes are moving as a result of the applied force, the friction will be kinetic friction. These friction forces can be calculated once we have determined the normal forces exerted on the boxes by the surface. The net acceleration of each box in the vertical direction is equal to zero, and we can determine the normal forces by applying Newton's second law for each box along the vertical direction

$$m_1 a_{1,y} = F_{1,N} - m_1 g - F \sin\theta = 0$$

$$m_2 a_{2,y} = F_{2,N} - m_2 g = 0$$

The only unknown in these two equations are the normal forces, and they can now be determined. Once we have determined the normal forces, we can easily determine the friction forces acting on the two boxes.

$$F_{1,N} = m_1 g + F \sin\theta \quad \Rightarrow \quad F_{1,fric} = \mu_{1,k} F_{1,N} = \mu_{1,k}\left(m_1 g + F \sin\theta\right)$$

$$F_{2,N} = m_2 g \quad \Rightarrow \quad F_{2,fric} = \mu_{2,k} F_{2,N} = \mu_{2,k} m_2 g$$

Since each box will have the same acceleration in the x direction, we know that the net force in this direction must be equal to $(m_1 + m_2)a$. Using the force diagrams, we see that the net force on the boxes is also equal to

$$\sum F_x = F \cos\theta - F_{1,fric} - F_{2,fric} = F \cos\theta - \mu_{1,k}\left(m_1 g + F \sin\theta\right) - \mu_{2,k} m_2 g = \left(m_1 + m_2\right)a$$

The acceleration of the boxes is thus equal to

$$a = \frac{F \cos\theta - \mu_{1,k}\left(m_1 g + F \sin\theta\right) - \mu_{2,k} m_2 g}{\left(m_1 + m_2\right)} = 5.31\,\text{m/s}^2$$

Section 5-2. Uniform Circular Motion - Kinematics

An object that moves along a circular path of radius r with constant speed v is said to carry out **uniform circular motion**. Although the speed of the object is constant, the direction of its velocity is changing continuously. As a result, the acceleration of the object is non-zero. The acceleration is directed toward the center of the circular path and has a magnitude of

$$a_R = \frac{v^2}{r}$$

This acceleration is called the **centripetal acceleration**, centripetal meaning toward the center.

Example 5-2-A: Gravitational acceleration on board of the space shuttle. The space shuttle flies at an average altitude of $h = 300$ km around the earth. The orbital period at this altitude is $T = 90$ minutes. What is the acceleration experienced by an astronaut on board of the space shuttle?

Approach: In this example we determine the gravitational acceleration experienced by astronauts on board of the space shuttle when it is in orbit around the earth. Although the astronauts are in a weightless environment, they experience a net acceleration since they carry out uniform circular motion. In this example we will show that this acceleration is not so different from the gravitational acceleration we experience on the surface of the earth.

Solution: The radius of the orbit of the space shuttle is equal to $r = R_E + h$. The velocity of the shuttle in this orbit is equal to $2\pi r/T$. Assuming the motion of the shuttle is uniform circular motion, we can use the centripetal acceleration to determine the acceleration experienced by an astronaut on board of the shuttle:

$$a_R = \frac{v^2}{r} = \frac{\left(\dfrac{2\pi r}{T}\right)^2}{r} = \frac{4\pi^2 r}{T^2} = \frac{4\pi^2\left(h + R_E\right)}{T^2} = \frac{4\pi^2\left(3\times10^5 + 64\times10^6\right)}{\left(90\times60\right)^2} = 9.1 \text{ m/s}^2$$

Section 5-3. Dynamics of Uniform Circular Motion

Consider an object of mass m carrying out uniform circular motion along a circle of radius r and with speed v. Applying Newton's second law to this motion requires that the net force on this object satisfies the following relation:

$$\sum\vec{F} = -m\frac{v^2}{r}\hat{r} \qquad \text{or} \qquad \sum F_R = m\frac{v^2}{r}$$

where $\hat{r}$ is the unit vector pointing outward from the center of the circle to the position of the object and the F_R's are the radial components of each of the forces acting on the object. An object that moves with constant speed along a circular path must thus have a net inward force acting on it in order to continue its circular motion. This force is called the **centripetal force**. A common mistake is to treat the centripetal force as a real force, and for example add it to the force diagrams. The centripetal force however only provides us with information about the magnitude and direction of the net force required to act on an object carrying out uniform circular motion.

Example 5-3-A: Spinning a rock at an angle. A rock is attached to a string and spun in a circle of radius r. The plane of the circle is inclined at an angle θ to the horizontal. At the highest point of the circle the speed of the rock is v. What is the tension in the string when the rock is at the highest point on the circle?

Approach: In this example we will study circular motion in a plane that makes an angle with the horizontal direction. In this case, neither the tension in the string nor the gravitational force acting on the rock will be directed towards the center of the circle. Since the relative angle between the tension and the gravitational force will be different at different positions along the path of the rock, the tension force will vary continuously. In this example, we focus on the calculation of the tension at one particular position along this path.

Solution: Since the rock carries out circular motion in a plane tilted at an angle θ to the horizontal, the direction of the acceleration will be at an angle θ to the horizontal also. When the object is at its highest point, the net force will be directed at an angle θ below the horizontal. The free-body diagram of the rock at its highest point is shown in the Figure to the right. Note that the force exerted by the string on the rock is directed at an angle α below horizontal, which will be different from the angle θ of the net force. Newton's second law in the horizontal (x) and vertical (y) directions can be written as

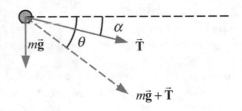

$$\sum F_x = ma_x \quad \Rightarrow \quad T\cos\alpha = m\frac{v^2}{r}\cos\theta$$

$$\sum F_y = ma_y \quad \Rightarrow \quad -T\sin\alpha - mg = -m\frac{v^2}{r}\sin\theta$$

This set of two equations has two unknowns, T and α. We solve this set of equations by moving the mg term in the second equation to the right-hand side of that equation, squaring both sides of each equation, and adding the two resulting equations together to get

$$T^2\left(\cos^2\alpha + \sin^2\alpha\right) = \left(m\frac{v^2}{r}\cos\theta\right)^2 + \left(-m\frac{v^2}{r}\sin\theta + mg\right)^2$$

Using the trigonometry identity $\cos^2\alpha + \sin^2\alpha = 1$, this can be simplified to

$$T^2 = \left(\frac{mv^2}{r}\right)^2 - \frac{2m^2v^2g}{r}\sin\theta + m^2g^2$$

Taking the square root of each side of this equation we can determine the tension T

$$T = \sqrt{\left(\frac{mv^2}{r}\right)^2 - \frac{2m^2v^2g}{r}\sin\theta + m^2g^2}$$

Section 5-4. Highway Curves: Banked and Unbanked

A car traveling around a curve on a horizontal road must have a centripetal force provided by the friction force between the tires and the road. The forces acting on the car are shown in the free-body diagram on the right. As the point of contact of the tires with the road is at rest with respect to the road, the friction force is the static friction force. There is therefore a maximum centripetal force and a limit on the centripetal acceleration. If the speed exceeds a certain critical value, the maximum static friction force of the tires on the road is exceeded and the car can no longer travel without its tires skidding. Therefore, there is always a maximum speed at which a given curve can be rounded by a car without skidding. This maximum speed changes with a change in the coefficient of static friction between the tires and the road. Under wet or icy road conditions, the coefficient of static friction is reduced and so is the maximum speed with which the car can round the curve without skidding.

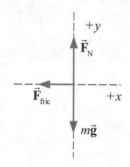

Similar arguments apply if the roadway in the turn is banked rather than horizontal. However, for a banked curve there will not only be a component of the frictional force in the centripetal direction but also an additional component due to the normal force of the roadway (see the free-body diagram on the right; the dashed vector indicates the required centripetal force). Certainly, in these cases there is the possibility that no frictional force is required. That happens when the component of the normal force in the centripetal direction is equal to the required centripetal force.

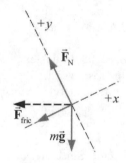

Example 5-4-A: Rounding a badly banked curve. A car goes around a curve that is banked outward at $\theta = 13.4°$ instead of inward as it should be. The radius of the curve is $r = 65.4$ m and the speed of the car is $v = 17.3$ m/s. What is the minimum coefficient of static friction that will enable the car to stay on the road?

Approach: In this problem we study a banked roadway where the horizontal component of the normal force points in a direction opposite to the direction of the required centripetal force. As a result, the friction force must be large enough to compensate for the horizontal component of the normal force and provide the required centripetal force. Please remember to be very careful when you drive on a roadway that is banked in the wrong direction; even at low velocities you may not be able to make a turn!

Solution: The free-body diagram for this problem is shown to the right. The curve is banked outward which implies that the normal force is pointing away from the center of the circle. The required centripetal force is indicated by the dashed vector shown in the free-body diagram, and has a magnitude equal to mv^2/r. The coordinate system we will be using to solve this problem is indicated in the diagram. Applying Newton's second law along the coordinate axes we obtain the following equations:

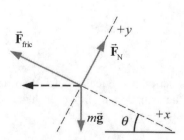

$$ma_x = \sum F_x \quad \Rightarrow \quad -m\frac{v^2}{r}\cos\theta = -\mu_s F_N + mg\sin\theta$$

$$ma_y = \sum F_y \quad \Rightarrow \quad -m\frac{v^2}{r}\sin\theta = F_N - mg\cos\theta$$

In the first equation we have replaced the friction force with the maximum static friction force. The second equation can be used to determine the normal force:

$$F_N = mg\cos\theta - m\frac{v^2}{r}\sin\theta$$

The first equation can now be used to determine μ_s.

$$\mu_s = \frac{mg\sin\theta + m\frac{v^2}{r}\cos\theta}{F_N} = \frac{g\sin\theta + \frac{v^2}{r}\cos\theta}{g\cos\theta - \frac{v^2}{r}\sin\theta} = \frac{\left(9.80\ \text{m/s}^2\right)\sin13.4^\circ + \frac{\left(17.3\ \text{m/s}\right)^2}{65.4\ \text{m}}\cos13.4^\circ}{\left(9.80\ \text{m/s}^2\right)\cos13.4^\circ - \frac{\left(17.3\text{m/s}\right)^2}{65.4\ \text{m}}\sin13.4^\circ} = 0.793$$

Section 5-5. Non-uniform Circular Motion

Non-uniform circular motion means motion in a circle of fixed radius, but with varying speed. At any instant, the acceleration in the centripetal direction is still equal to v^2/r. However, there can now be a component of the acceleration perpendicular to the centripetal direction or tangent to the circle. We call this acceleration the **tangential acceleration**. The magnitude of the tangential acceleration, a_{tan}, is equal to

$$a_{tan} = \frac{dv}{dt}$$

where v is the speed of the object. The direction of the tangential acceleration is in the direction of the velocity of the object. The acceleration vector of the object, $\vec{a}$, is the sum of the centripetal acceleration and the tangential acceleration:

$$\vec{a} = \vec{a}_R + \vec{a}_{tan}$$

This concept can be applied to an object traveling along an arbitrary curved path. At each point along any curved path there is a unique circle that is tangent to the path with the same radius of curvature as the path and in the plane of the path.

Example 5-5-A: Examining non-uniform circular motion. A rock of mass $m = 1.27$ kg is tied to a string and spun in a circle as it slides on a frictionless horizontal surface. The radius of the circle the rock follows is $r = 1.04$ m. At a given moment, the string lies along the direction of the arrow in the diagram when the rock is in the position shown. The magnitude of the tension in the string is $T = 18.1$ N. What are the speed and rate of change of the speed of the rock at that moment?

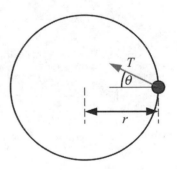

Approach: Since the only force acting on the rock is the tension in the string, there will be both a non-zero radial and a non-zero tangential acceleration. The radial component of the acceleration is responsible for the change in the direction of the rock, and the tangential component of the acceleration is responsible for a change in the speed of the rock.

Solution: We can resolve the force vector on mass m into a component in the radial direction and a component in the tangential direction:

$$F_R = T\cos\theta$$
$$F_{tan} = T\sin\theta$$

Newton's second law in the radial direction provides us with the following requirement

$$F_R = ma_R = \frac{mv^2}{r} = T\cos\theta \quad \Rightarrow \quad v = \sqrt{\frac{rT\cos\theta}{m}} = 3.82\ \text{m/s}$$

Newton's second law in the tangential direction tells us

$$F_{tan} = ma_{tan} = m\frac{dv}{dt} \quad \Rightarrow \quad \frac{dv}{dt} = \frac{F_{tan}}{m} = \frac{T\sin\theta}{m} = 2.47 \text{ m/s}^2$$

Section 5-6. Velocity-Dependent Forces; Drag and Terminal Velocity

When objects move through fluids, such as water or air, they feel a frictional force called the **drag force**. The drag force depends on the velocity of the object relative to the fluid and is directed opposite to the direction of motion. In the limit of low speeds (speed approaching zero), the drag force F_D is proportional to the speed of the object:

$$F_D = -bv$$

The constant b depends on the shape and the orientation of the object, and on the viscosity of the fluid. This type of drag force is called **viscous drag**.

At higher speeds, the drag force becomes proportional to the square of the speed of the object:

$$F_D = -Cv^2$$

The constant C depends on the shape and the orientation of the object, and on the density of the fluid. This type of drag force is called **inertial drag**.

Example 5-6-A: Motion with inertial drag. An object enters a fluid at a speed v. The only force acting on it is the inertial drag force $F_D = -Cv^2$. As long as the inertial drag dependence is valid, how does the speed of the object depend on time?

Approach: The inertial drag force is proportional to v^2. In order to determine the time dependence of the velocity v we need to use Newton's second law to obtain a differential equation for v. Solving this equation with the appropriate boundary conditions will provide us with the information about the time dependence of v. Note: this is an important example of how to deal with a non-constant force.

Solution: We write Newton's second law in terms of drag force and write the acceleration as the time derivative of the velocity

$$\sum F = ma \quad \Rightarrow \quad -Cv^2 = m\frac{dv}{dt} \quad \Rightarrow \quad -\frac{C}{m}dt = \frac{1}{v^2}dv$$

Both sides of the last expression can be integrated

$$\int_0^T \frac{-C}{m}dt = \int_{v(0)}^{v(T)} \frac{1}{v^2}dv \quad \Rightarrow \quad \frac{-C}{m}t\Big|_0^T = -\frac{1}{v}\Big|_{v(0)}^{v(T)} \quad \Rightarrow \quad \frac{-CT}{m} = -\frac{1}{v(T)} + \frac{1}{v(0)}$$

Solving the last equation for $v(T)$ we obtain

$$v(T) = \frac{1}{\dfrac{1}{v(0)} + \dfrac{CT}{m}}$$

The time dependence of the velocity is shown by the dashed curve in the Figure on the right. The dashed curve was calculated under the assumption that the initial velocity is 10 m/s and $C/m = 1$ m^{-1}.

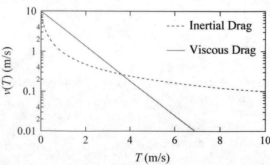

Example 5-6-B: Motion with viscous drag. An object enters a fluid at a speed v. The only force acting on it is the viscous drag force $F_D = -bv$. As long as the viscous drag dependence is valid, how does the speed of the object depend on time?

Approach: The viscous drag force is proportional to v. In order to determine the time dependence of the velocity v we need to use Newton's second law to obtain a differential equation for v. Solving this equation with the appropriate boundary conditions will provide us with the information about the time dependence of v. Note: this is an important example of how to deal with a non-constant force.

Solution: We write Newton's second law in terms of drag force and write the acceleration as the time derivative of the velocity

$$\sum F = ma \quad \Rightarrow \quad -bv = m\frac{dv}{dt} \quad \Rightarrow \quad -\frac{b}{m}dt = \frac{1}{v}dv$$

Both sides of the last expression can be integrated

$$\int_0^T \frac{-b}{m}dt = \int_{v(0)}^{v(T)} \frac{1}{v}dv \quad \Rightarrow \quad \frac{-b}{m}t\Big|_0^T = \ln v\Big|_{v(0)}^{v(T)} \quad \Rightarrow \quad \frac{-bT}{m} = \ln v(T) - \ln v(0) = \ln\left(\frac{v(T)}{v(0)}\right)$$

Solving the last equation for $v(T)$ we obtain

$$v(T) = v(0)e^{-\frac{bT}{m}}$$

The time dependence of the velocity is shown by the solid curve in the graph shown as part of the solution of Example 5-6-A. The solid curve was calculated under the assumption that the initial velocity is 10 m/s and $b/m = 1$ s^{-1}.

Practice Quiz

1. Three identical cars are connected to each other, one to the next by a towrope. The car in front starts to accelerate, pulling all the cars behind it. How does the minimum coefficient of static friction between the car's tires and the road compare to that for a single car with the same acceleration?
   a) The same minimum coefficient of static friction is required.
   b) A coefficient of static friction three times greater is required.
   c) A minimum coefficient of static friction one third as great is required.
   d) Not enough information is given to answer the question.

2. Many of the problems in this chapter involved a car going around a curve at constant speed. What would happen to the required minimum coefficient of static friction, μ_s, if the car were changing speed?
   a) The minimum μ_s is bigger if speeding up and is less if slowing down.
   b) The minimum μ_s is less if speeding up and is bigger if slowing down.
   c) The minimum μ_s is the same whether the car is changing speed or not.
   d) The minimum μ_s is bigger than if the car were not changing speed.

3. Your uncle tries to impress you by pulling the tablecloth out from under the place settings on the table without everything crashing to the floor. Which of the following statements about the coefficients of friction between the tablecloth and the dinnerware is required to be true for the trick to work successfully?
   a) The coefficient of static friction must be equal to the coefficient of kinetic friction.
   b) The coefficient of static friction must be less than the coefficient of kinetic friction.
   c) The coefficient of static friction must be greater than the coefficient of kinetic friction.
   d) Both coefficients of friction must be zero.

4. An object is speeding up as it goes around a circle. Which statement is necessarily true about the net force acting on the object?
   a) The force on the object acts directly toward the center of the circle.
   b) The force on the object acts directly away from the center of the circle.
   c) The net force on the object acts in a direction tangent to the circle at the position of the object.
   d) The net force cannot be in any of the three directions above.

5. Anti-lock braking systems on cars are designed to keep the tires rolling on the road rather than sliding. Why is this advantageous to stopping and controlling a vehicle?
    a) It isn't advantageous. The car will take longer to stop.
    b) The car manufacturer can charge more money for the car.
    c) The coefficient of static friction of the tires on the road is higher than the coefficient of kinetic friction.
    d) The tires won't squeal when you stop.

6. A car travels around a curve of radius r at constant speed v. Its acceleration has a magnitude a. If the car rounds the same curve so that its acceleration is $2a$, what will its speed be?
    a) $v/2$
    b) $2v$
    c) $4v$
    d) $\sqrt{2}\,v$

7. A car travels around a curve of radius r at constant speed v. Its acceleration has a magnitude a. If the car rounds a different curve at the same speed such that its acceleration is $2a$, what will the radius of the curve be?
    a) $2r$
    b) $r/2$
    c) $4r$
    d) $\sqrt{2}\,r$

8. If you place a toy boat in a stream, it begins to move. Why doesn't the drag force stop the toy boat?
    a) The drag force is what makes the boat move.
    b) The force of gravity on the boat overcomes the drag force.
    c) The drag force is small and eventually the drag force will stop the boat.
    d) The boat's hull is designed to make the drag force negligible.

9. An object of mass 5.61 kg is dropped from an airplane. At the highest speeds attained by the object in free fall, the magnitude of the drag force on the object is given by $F_D = (0.634 \text{ Ns/m})\, v$. What is the terminal speed of the object in free-fall?
    a) 86.7 m/s
    b) 9.8 m/s
    c) 34.9 m/s
    d) 8.85 m/s

10. Two objects are moving through a fluid at the same speed initially and they experience the same drag force at a given moment. If object A is experiencing viscous drag ($F_D \propto v$) and object B is experiencing inertial drag ($F_D \propto v^2$), which object will come to a rest more quickly?
    a) A
    b) B
    c) Neither; the velocity will be the same for both.
    d) Not enough information is given to determine which will slow down more quickly.

11. A box of mass $m = 3.44$ kg rests on a sloped ramp angled at $\theta = 29.1°$ above the horizontal plane (see diagram). The coefficient of friction between the ramp and the box is $\mu_k = 0.697$. At what horizontal acceleration of the ramp will the box begin sliding down the ramp?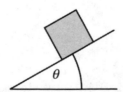

12. A car of mass $m = 680$ kg travels with a speed of $v = 12.6$ m/s around a 3.05° banked curve (angle of inclination is $\theta = 3.05°$). The radius of the curve is $r = 61.2$ m. What is the force of friction acting on the car?

13. Two boxes rest in contact with each other while resting on the ramp as shown in the diagram. The slope of the ramp is $\theta = 14.0°$ above horizontal. The lower box has a mass of $m_1 = 2.77$ kg and a coefficient of static friction with the ramp of $\mu_s = 0.718$. The upper box has a coefficient of static friction $\mu_s = 0.152$ with the ramp. What is the upper limit on the mass of the upper box so the boxes will not begin sliding down the ramp?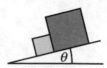

14. A racetrack is circular and has a banking of $\theta = 18.5°$ and a radius of $r = 140$ m. What is the minimum time that a car can complete a lap around the track if the coefficient of static friction between tires and road is $\mu_s = 0.794$?

15. Two blocks are connected by a string, as shown in the diagram on the right. The block on the slope has a mass $m_1 = 10.5$ kg and its coefficient of kinetic friction with the surface of the slope is $\mu_k = 0.559$. The slope is at an angle of $\theta = 22.6°$ above horizontal. The mass hanging on the vertical part of the rope is $m_2 = 15.8$ kg. What is the acceleration of the blocks?

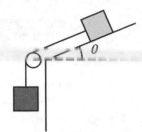

Responses to Select End-of-Chapter Questions

1. Static friction between the crate and the truck bed causes the crate to accelerate.

7. (b). If the car comes to a stop without skidding, the force that stops the car is the force of kinetic friction between the brake mechanism and the wheels. This force is designed to be large. If you slam on the brakes and skid to a stop, the force that stops the car will be the force of kinetic friction between the tires and the road. Even with a dry road, this force is likely to be less than the force of kinetic friction between the brake mechanism and the wheels. The car will come to a stop more quickly if the tires continue to roll, rather than skid. In addition, once the wheels lock, you have no steering control over the car.

13. As the child and sled come over the crest of the hill, they are moving in an arc. There must be a centripetal force, pointing inward toward the center of the arc. The combination of gravity (down) and the normal force (up) provides this centripetal force, which must be greater than or equal to zero. (At the top of the arc, $F_y = mg - N = mv^2/r \geq 0$.) The normal force must therefore be less than the child's weight.

19. At the top of bucket's arc, the gravitational force and normal forces from the bucket provide the centripetal force needed to keep the water moving in a circle. (If we ignore the normal forces, $mg = mv^2/r$, so the bucket must be moving with speed $v \geq \sqrt{(gr)}$ or the water will spill out of the bucket.) At the top of the arc, the water has a horizontal velocity. As the bucket passes the top of the arc, the velocity of the water develops a vertical component. But the bucket is traveling with the water, with the same velocity, and contains the water as it falls through the rest of its path.

Solutions to Select End-of-Chapter Problems

1. A free-body diagram for the crate is shown. The crate does not accelerate vertically, and so $F_N = mg$. The crate does not accelerate horizontally, and so $F_P = F_{fr}$.

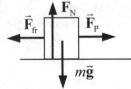

$$F_P = F_{fr} = \mu_k F_N = \mu_k mg = (0.30)(22\,\text{kg})(9.80\,\text{m/s}^2) = \boxed{65\,\text{N}}$$

If the coefficient of kinetic friction is zero, then the horizontal force required is $\boxed{0\,\text{N}}$ since there is no friction to counteract. Of course, it would take a force to START the crate moving, but once it was moving, no further horizontal force would be necessary to maintain the motion.

7. Start with a free-body diagram. Write Newton's second law for each direction.

$$\sum F_x = mg\sin\theta - F_{fr} = ma_x$$
$$\sum F_y = F_N - mg\cos\theta = ma_y = 0$$

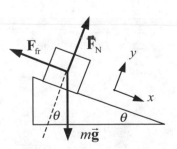

Notice that the sum in the y direction is 0, since there is no motion (and hence no acceleration) in the y direction. Solve for the force of friction.

$$mg\sin\theta - F_{fr} = ma_x \rightarrow F_{fr} = mg\sin\theta - ma_x =$$
$$= (25.0\,\text{kg})\left[(9.80\,\text{m/s}^2)(\sin 27°) - 0.30\,\text{m/s}^2\right] = 103.7\,\text{N} \approx \boxed{1.0\times 10^2\,\text{N}}$$

Now solve for the coefficient of kinetic friction. Note that the expression for the normal force comes from the y direction force equation above.

$$F_{fr} = \mu_k F_N = \mu_k mg \cos\theta \quad \rightarrow \quad \mu_k = \frac{F_{fr}}{mg\cos\theta} = \frac{103.7\,\text{N}}{(25.0\,\text{kg})(9.80\,\text{m/s}^2)(\cos 27^\circ)} = \boxed{0.48}$$

13. We draw three free-body diagrams – one for the car, one for the trailer, and then "add" them for the combination of car and trailer. Note that since the car pushes against the ground, the ground will push against the car with an equal but oppositely directed force. $\vec{F}_{CG}$ is the force on the car due to the ground, $\vec{F}_{TC}$ is the force on the trailer due to the car, and $\vec{F}_{CT}$ is the force on the car due to the trailer. Note that by Newton's third law, $\left|\vec{F}_{CT}\right| = \left|\vec{F}_{TC}\right|$.

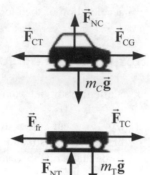

From consideration of the vertical forces in the individual free-body diagrams, it is apparent that the normal force on each object is equal to its weight. This leads to the conclusion that

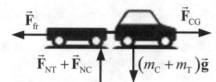

$$F_{fr} = \mu_k F_{NT} = \mu_k m_T g = (0.15)(350\,\text{kg})(9.80\,\text{m/s}^2) = 514.5\,\text{N}$$

Now consider the combined free-body diagram. Write Newton's second law for the horizontal direction. This allows the calculation of the acceleration of the system.

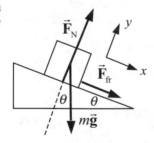

$$\sum F = F_{CG} - F_{fr} = (m_C + m_T)a \quad \rightarrow$$

$$a = \frac{F_{CG} - F_{fr}}{m_C + m_T} = \frac{3600\,\text{N} - 514.5\,\text{N}}{1630\,\text{kg}} = 1.893\,\text{m/s}^2$$

Finally, consider the free-body diagram for the trailer alone. Again write Newton's second law for the horizontal direction, and solve for F_{TC}:

$$\sum F = F_{TC} - F_{fr} = m_T a \quad \rightarrow$$

$$F_{TC} = F_{fr} + m_T a = 514.5\,\text{N} + (350\,\text{kg})(1.893\,\text{m/s}^2) = 1177\,\text{N} \approx \boxed{1200\,\text{N}}$$

19. (*a*) Consider the free-body diagram for the crate on the surface. There is no motion in the y direction and thus no acceleration in the y direction. Write Newton's second law for both directions, and find the acceleration.

$$\sum F_y = F_N - mg\cos\theta = 0 \quad \rightarrow \quad F_N = mg\cos\theta$$
$$\sum F_x = mg\sin\theta + F_{fr} = ma$$
$$ma = mg\sin\theta + \mu_k F_N = mg\sin\theta + \mu_k mg\cos\theta$$
$$a = g(\sin\theta + \mu_k\cos\theta)$$

Now use Eq. 2-12c, with an initial velocity of -3.0 m/s and a final velocity of 0 m/s to find the distance the crate travels up the plane.

$$v^2 - v_0^2 = 2a(x - x_0) \quad \rightarrow$$

$$x - x_0 = \frac{-v_0^2}{2a} = \frac{-(-3.0\,\text{m/s})^2}{2(9.80\,\text{m/s}^2)(\sin 25.0^\circ + 0.17\cos 25.0^\circ)} = -0.796\,\text{m}$$

The crate travels $\boxed{0.80\,\text{m}}$ up the plane.

(*b*) We use the acceleration found above with the initial velocity in Eq. 2-12a to find the time for the crate to travel up the plane.

$$v = v_0 + at \quad \rightarrow \quad t_{up} = -\frac{v_0}{a_{up}} = -\frac{(-3.0\,\text{m/s})}{(9.80\,\text{m/s}^2)(\sin 25.0^\circ + 0.17\cos 25.0^\circ)} = 0.5308\,\text{s}$$

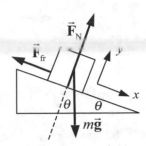

The total time is NOT just twice the time to travel up the plane, because the acceleration of the block is different for the two parts of the motion. The second free-body diagram applies to the block sliding down the plane. A similar analysis will give the acceleration, and then Eq. 2-12b with an initial velocity of 0 is used to find the time to move down the plane.

$$\sum F_y = F_N - mg\cos\theta = 0 \quad \rightarrow \quad F_N = mg\cos\theta$$
$$\sum F_x = mg\sin\theta - F_{fr} = ma$$
$$ma = mg\sin\theta - \mu_k F_N = mg\sin\theta - \mu_k mg\cos\theta$$
$$a = g(\sin\theta - \mu_k \cos\theta)$$

$$x - x_0 = v_0 t + \tfrac{1}{2}at^2 \quad \rightarrow$$

$$t_{down} = \sqrt{\frac{2(x - x_0)}{a_{down}}} = \sqrt{\frac{2(0.796\,\text{m})}{(9.80\,\text{m/s}^2)(\sin 25.0^\circ - 0.17\cos 25.0^\circ)}} = 0.7778\,\text{s}$$

$$t = t_{up} + t_{down} = 0.5308\,\text{s} + 0.7778\,\text{s} = \boxed{1.3\,\text{s}}$$

It is worth noting that the final speed is about 2.0 m/s, significantly less than the 3.0 m/s original speed.

25. (*a*) Consider the free-body diagram for the block on the surface. There is no motion in the *y* direction and thus no acceleration in the *y* direction. Write Newton's second law for both directions, and find the acceleration.

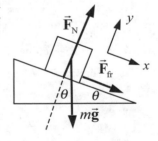

$$\sum F_y = F_N - mg\cos\theta = 0 \quad \rightarrow \quad F_N = mg\cos\theta$$
$$\sum F_x = mg\sin\theta + F_{fr} = ma$$
$$ma = mg\sin\theta + \mu_k F_N = mg\sin\theta + \mu_k mg\cos\theta$$
$$a - g(\sin\theta + \mu_k \cos\theta)$$

Now use Eq. 2-12c, with an initial velocity of v_0, a final velocity of 0 m/s, and a displacement of *-d* to find the coefficient of kinetic friction.

$$v^2 - v_0^2 = 2a(x - x_0) \quad \rightarrow \quad 0 - v_0^2 = 2g(\sin\theta + \mu_k\cos\theta)(-d) \quad \rightarrow$$

$$\mu_k = \boxed{\frac{v_0^2}{2gd\cos\theta} - \tan\theta}$$

(*b*) Now consider the free-body diagram for the block at the top of its motion. We use a similar force analysis, but now the magnitude of the friction force is given by $F_{fr} \le \mu_s F_N$, and the acceleration is 0 m/s^2.

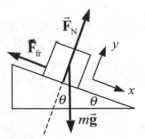

$$\sum F_y = F_N - mg\cos\theta = 0 \quad \rightarrow \quad F_N = mg\cos\theta$$
$$\sum F_x = mg\sin\theta - F_{fr} = ma = 0 \quad \rightarrow \quad F_{fr} = mg\sin\theta$$
$$F_{fr} \le \mu_s F_N \quad \rightarrow \quad mg\sin\theta \le \mu_s mg\cos\theta \quad \rightarrow \quad \boxed{\mu_s \ge \tan\theta}$$

31. Draw a free-body diagram for each block.

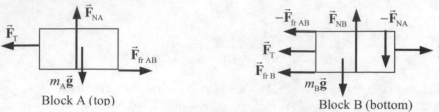

Block A (top)   Block B (bottom)

$\vec{F}_{fr\,AB}$ is the force of friction between the two blocks, $\vec{F}_{NA}$ is the normal force of contact between the two blocks, $\vec{F}_{fr\,B}$ is the force of friction between the bottom block and the floor, and $\vec{F}_{NB}$ is the normal force of contact between the bottom block and the floor.

Neither block is accelerating vertically, and so the net vertical force on each block is zero.

$$\text{top:} \quad F_{NA} - m_A g = 0 \quad \rightarrow \quad F_{NA} = m_A g$$

$$\text{bottom:} \quad F_{NB} - F_{NA} - m_B g = 0 \quad \rightarrow \quad F_{NB} = F_{NA} + m_B g = \left(m_A + m_B\right)g$$

Take the positive horizontal direction to be the direction of motion of each block. Thus for the bottom block, positive is to the right, and for the top block, positive is to the left. Then, since the blocks are constrained to move together by the connecting string, both blocks will have the same acceleration. Write Newton's second law for the horizontal direction for each block.

$$\text{top:}\ F_T - F_{fr\,AB} = m_A a \qquad \text{bottom:}\ F - F_T - F_{fr\,AB} - F_{fr\,B} = m_B a$$

(*a*) If the two blocks are just to move, then the force of static friction will be at its maximum, and so the friction forces are as follows.

$$F_{fr\,AB} = \mu_s F_{NA} = \mu_s m_A g \quad ; \quad F_{fr\,B} = \mu_s F_{NB} = \mu_s \left(m_A + m_B\right)g$$

Substitute into Newton's second law for the horizontal direction with $a = 0 \text{ m/s}^2$ and solve for F.

$$\text{top:}\ F_T - \mu_s m_A g = 0 \quad \rightarrow \quad F_T = \mu_s m_A g$$

$$\text{bottom:}\ F - F_T - \mu_s m_A g - \mu_s \left(m_A + m_B\right)g = 0 \quad \rightarrow$$

$$F = F_T + \mu_s m_A g + \mu_s \left(m_A + m_B\right)g = \mu_s m_A g + \mu_s m_A g + \mu_s \left(m_A + m_B\right)g$$

$$= \mu_s \left(3m_A + m_B\right)g = \left(0.60\right)\left(14\,\text{kg}\right)\left(9.80\,\text{m/s}^2\right) = 82.32\,\text{N} \approx \boxed{82\,\text{N}}$$

(*b*) Multiply the force by 1.1 so that $F = 1.1(82.32\ \text{N}) = 90.55\ \text{N}$. Again use Newton's second law for the horizontal direction, but with $a \neq 0 \text{ m/s}^2$ and using the coefficient of kinetic friction.

$$\text{top:} \quad F_T - \mu_k m_A g = m_A a$$

$$\text{bottom:} \quad F - F_T - \mu_k m_A g - \mu_k \left(m_A + m_B\right)g = m_B a$$

$$\text{sum:} \quad F - \mu_k m_A g - \mu_k m_A g - \mu_k \left(m_A + m_B\right)g = \left(m_A + m_B\right)a \quad \rightarrow$$

$$a = \frac{F - \mu_k m_A g - \mu_k m_A g - \mu_k \left(m_A + m_B\right)g}{\left(m_A + m_B\right)} = \frac{F - \mu_k \left(3m_A + m_B\right)g}{\left(m_A + m_B\right)}$$

$$= \frac{90.55\,\text{N} - \left(0.40\right)\left(14.0\,\text{kg}\right)\left(9.80\,\text{m/s}^2\right)}{\left(8.0\,\text{kg}\right)} = 4.459\,\text{m/s}^2 \approx \boxed{4.5\,\text{m/s}^2}$$

37. We assume the water is rotating in a vertical circle of radius r. When the bucket is at the top of its motion, there would be two forces on the water (considering the water as a single mass). The weight of the water would be directed down, and the normal force of the bottom of the bucket pushing on the water would also be down. See the free-body diagram. If the water is moving in a circle, then the net downward force would be a centripetal force.

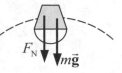

$$\sum F = F_N + mg = ma = mv^2/r \quad \rightarrow \quad F_N = m\left(v^2/r - g\right)$$

The limiting condition of the water falling out of the bucket means that the water loses contact with the bucket, and so the normal force becomes 0.

$$F_N = m\left(v^2/r - g\right) \quad \rightarrow \quad m\left(v^2_{critical}/r - g\right) = 0 \quad \rightarrow \quad v_{critical} = \sqrt{rg}$$

From this, we see that $\boxed{\text{yes}}$, it is possible to whirl the bucket of water fast enough. The minimum speed is $\boxed{\sqrt{rg}}$.

43. The orbit radius will be the sum of the Earth's radius plus the 400 km orbit height. The orbital period is about 90 minutes. Find the centripetal acceleration from these data.

$$r = 6380\,\text{km} + 400\,\text{km} = 6780\,\text{km} = 6.78 \times 10^6\,\text{m} \qquad T = 90\,\text{min}\left(\frac{60\,\text{sec}}{1\,\text{min}}\right) = 5400\,\text{sec}$$

$$a_R = \frac{4\pi^2 r}{T^2} = \frac{4\pi^2\left(6.78 \times 10^6\,\text{m}\right)}{\left(5400\,\text{sec}\right)^2} = \left(9.18\,\text{m/s}^2\right)\left(\frac{1\,g}{9.80\,\text{m/s}^2}\right) = 0.937 \approx \boxed{0.9\,g\text{'s}}$$

Notice how close this is to g, because the shuttle is not very far above the surface of the Earth, relative to the radius of the Earth.

49. The radius of either skater's motion is 0.80 m, and the period is 2.5 sec. Thus their speed is given by

$$v = 2\pi r/T = \frac{2\pi\left(0.80\,\text{m}\right)}{2.5\,\text{s}} = 2.0\,\text{m/s}$$

Since each skater is moving in a circle, the net radial force on each one is given by Eq. 5-3.

$$F_R = mv^2/r = \frac{\left(60.0\,\text{kg}\right)\left(2.0\,\text{m/s}\right)^2}{0.80\,\text{m}} = \boxed{3.0 \times 10^2\,\text{N}}$$

55. A free-body diagram of Tarzan at the bottom of his swing is shown. The upward tension force is created by Tarzan pulling down on the vine. Write Newton's second law in the vertical direction. Since he is moving in a circle, his acceleration will be centripetal, and points upward when he is at the bottom.

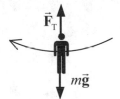

$$\sum F = F_T - mg = ma = mv^2/r \quad \rightarrow \quad v = \sqrt{\frac{\left(F_T - mg\right)r}{m}}$$

The maximum speed will be obtained with the maximum tension.

$$v_{max} = \sqrt{\frac{\left(F_{T\,max} - mg\right)r}{m}} = \sqrt{\frac{\left(1350\,\text{N} - \left(78\,\text{kg}\right)\left(9.80\,\text{m/s}^2\right)\right)5.2\,\text{m}}{78\,\text{kg}}} = \boxed{6.2\,\text{m/s}}$$

61. Apply uniform acceleration relationships to the tangential motion to find the tangential acceleration. Use Eq. 2-12b.

$$\Delta x_{\tan} = v_0{}_{\tan} t + \tfrac{1}{2} a_{\tan} t^2 \quad\rightarrow\quad a_{\tan} = \frac{2\Delta x_{\tan}}{t^2} = \frac{2\left[\tfrac{1}{4}(2\pi r)\right]}{t^2} = \frac{\pi(2.0\,\text{m})}{(2.0\,\text{s})^2} = (\pi/2)\,\text{m/s}^2$$

The tangential acceleration is constant. The radial acceleration is found from $a_{\text{rad}} = \dfrac{v_{\tan}^2}{r} = \dfrac{\left(a_{\tan} t\right)^2}{r}$.

(a) $a_{\tan} = \boxed{(\pi/2)\,\text{m/s}^2}$, $a_{\text{rad}} = \dfrac{\left(a_{\tan} t\right)^2}{r} = \dfrac{\left[(\pi/2)\,\text{m/s}^2(0\,\text{s})\right]^2}{2.0\,\text{m}} = \boxed{0}$

(b) $a_{\tan} = \boxed{(\pi/2)\,\text{m/s}^2}$, $a_{\text{rad}} = \dfrac{\left(a_{\tan} t\right)^2}{r} = \dfrac{\left[(\pi/2)\,\text{m/s}^2(1.0\,\text{s})\right]^2}{2.0\,\text{m}} = \boxed{(\pi^2/8)\,\text{m/s}^2}$

(c) $a_{\tan} = \boxed{(\pi/2)\,\text{m/s}^2}$, $a_{\text{rad}} = \dfrac{\left(a_{\tan} t\right)^2}{r} = \dfrac{\left[(\pi/2)\,\text{m/s}^2(2.0\,\text{s})\right]^2}{2.0\,\text{m}} = \boxed{(\pi^2/2)\,\text{m/s}^2}$

67. (a) We choose downward as the positive direction. Then the force of gravity is in the positive direction, and the resistive force is upwards. We follow the analysis given in Example 5-17.

$$F_{\text{net}} = mg - bv = ma \quad\rightarrow\quad a = \frac{dv}{dt} = g - \frac{b}{m}v = -\frac{b}{m}\left(v - \frac{mg}{b}\right) \quad\rightarrow$$

$$\frac{dv}{v - \dfrac{mg}{b}} = -\frac{b}{m}dt \quad\rightarrow\quad \int_{v_0}^{v} \frac{dv}{v - \dfrac{mg}{b}} = -\frac{b}{m}\int_0^t dt \quad\rightarrow\quad \ln\left[v - \frac{mg}{b}\right]_{v_0}^{v} = -\frac{b}{m}t \quad\rightarrow$$

$$\ln\left[\frac{v - \dfrac{mg}{b}}{v_0 - \dfrac{mg}{b}}\right] = -\frac{b}{m}t \quad\rightarrow\quad \frac{v - \dfrac{mg}{b}}{v_0 - \dfrac{mg}{b}} = e^{-\frac{b}{m}t} \quad\rightarrow\quad \boxed{v = \frac{mg}{b}\left(1 - e^{-\frac{b}{m}t}\right) + v_0 e^{-\frac{b}{m}t}}$$

Note that this motion has a terminal velocity of $v_{\text{terminal}} = mg/b$.

(b) We choose upwards as the positive direction. Then both the force of gravity and the resistive force are in the negative direction.

$$F_{\text{net}} = -mg - bv = ma \quad\rightarrow\quad a = \frac{dv}{dt} = -g - \frac{b}{m}v = -\frac{b}{m}\left(v + \frac{mg}{b}\right) \quad\rightarrow$$

$$\frac{dv}{v + \dfrac{mg}{b}} = -\frac{b}{m}dt \quad\rightarrow\quad \int_{v_0}^{v} \frac{dv}{v + \dfrac{mg}{b}} = -\frac{b}{m}\int_0^t dt \quad\rightarrow\quad \ln\left[v + \frac{mg}{b}\right]_{v_0}^{v} = -\frac{b}{m}t \quad\rightarrow$$

$$\ln\left[\frac{v + \dfrac{mg}{b}}{v_0 + \dfrac{mg}{b}}\right] = -\frac{b}{m}t \quad\rightarrow\quad \frac{v + \dfrac{mg}{b}}{v_0 + \dfrac{mg}{b}} = e^{-\frac{b}{m}t} \quad\rightarrow\quad \boxed{v = \frac{mg}{b}\left(e^{-\frac{b}{m}t} - 1\right) + v_0 e^{-\frac{b}{m}t}}$$

After the object reaches its maximum height $\left[t_{\text{rise}} = \dfrac{m}{b}\ln\left(1 + \dfrac{bv_0}{mg}\right)\right]$, at which point the speed will be 0, it will then start to fall. The equation from part (a) will then describe its falling motion.

73. From problem 72, we have that

$$v = \left(v_0^{\frac{1}{2}} - \frac{bt}{2m} \right)^2 \quad \text{and} \quad x = \left(v_0 t - \frac{v_0^{\frac{1}{2}}b}{2m}t^2 + \frac{b^2}{12m^2}t^3 \right)$$

The maximum distance will occur at the time when the velocity is 0 m/s. From the equation for the velocity, we see that happens at

$$t_{max} = \frac{2mv_0^{\frac{1}{2}}}{b}.$$

Use this time in the expression for distance to find the maximum distance.

$$x\left(t = t_{max} \right) = v_0 \frac{2mv_0^{\frac{1}{2}}}{b} - \frac{v_0^{\frac{1}{2}}b}{2m}\left(\frac{2mv_0^{\frac{1}{2}}}{b} \right)^2 + \frac{b^2}{12m^2}\left(\frac{2mv_0^{\frac{1}{2}}}{b} \right)^3 = \frac{2mv_0^{\frac{3}{2}}}{b} - \frac{2mv_0^{\frac{3}{2}}}{b} + \frac{2mv_0^{\frac{3}{2}}}{3b} = \boxed{\frac{2mv_0^{\frac{3}{2}}}{3b}}$$

79. Consider a free-body diagram of the box. Write Newton's second law for both directions. The net force in the y direction is 0 because there is no acceleration in the y direction.

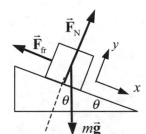

$$\sum F_y = F_N - mg\cos\theta = 0 \quad \rightarrow \quad F_N = mg\cos\theta$$
$$\sum F_x = mg\sin\theta - F_{fr} = ma$$

Now solve for the force of friction and the coefficient of friction.

$$\sum F_y = F_N - mg\cos\theta = 0 \quad \rightarrow \quad F_N = mg\cos\theta$$
$$\sum F_x = mg\sin\theta - F_{fr} = ma$$

$$F_{fr} = mg\sin\theta - ma = m\left(g\sin\theta - a \right) = \left(18.0\,\text{kg} \right)\left[\left(9.80\,\text{m/s}^2 \right)\left(\sin 37.0^\circ \right) - 0.220\,\text{m/s}^2 \right]$$
$$= 102.2\,\text{N} \approx \boxed{102\,\text{N}}$$

$$F_{fr} = \mu_k F_N = \mu_k mg\cos\theta \quad \rightarrow \quad \mu_k = \frac{F_{fr}}{mg\cos\theta} = \frac{102.2\,\text{N}}{\left(18.0\,\text{kg} \right)\left(9.80\,\text{m/s}^2 \right)\cos 37.0^\circ} = \boxed{0.725}$$

85. The radial force is given by Eq. 5-3.

$$F_R = m\frac{v^2}{r} = \left(1150\,\text{kg} \right)\frac{\left(27\,\text{m/s} \right)^2}{450\,\text{m/s}} = 1863\,\text{N} \approx \boxed{1900\,\text{N}}$$

The tangential force is the mass times the tangential acceleration. The tangential acceleration is the change in tangential speed divided by the elapsed time.

$$F_T = ma_T = m\frac{\Delta v_T}{\Delta t} = \left(1150\,\text{kg} \right)\frac{\left(27\,\text{m/s} \right)}{\left(9.0\,\text{s} \right)} = 3450\,\text{N} \approx \boxed{3500\,\text{N}}$$

91. (*a*) The horizontal component of the lift force will produce a centripetal acceleration. Write Newton's second law for both the horizontal and vertical directions, and combine those equations to solve for the time needed to reverse course (a half-period of the circular motion). Note that $T = 2\pi r/v$.

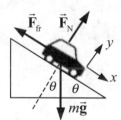

$$\sum F_{\text{vertical}} = F_{\text{lift}} \cos\theta = mg \;\; ; \;\; \sum F_{\text{horizontal}} = F_{\text{lift}} \sin\theta = m\frac{v^2}{r}$$

Divide these two equations.

$$\frac{F_{\text{lift}} \sin\theta}{F_{\text{lift}} \cos\theta} = \frac{mv^2}{rmg} \;\; \rightarrow \;\; \tan\theta = \frac{v^2}{rg} = \frac{v^2}{\dfrac{Tv}{2\pi}g} = \frac{2\pi v}{gT} \;\; \rightarrow$$

$$\frac{T}{2} = \frac{\pi v}{g\tan\theta} = \frac{\pi\left[(480\,\text{km/h})\left(\dfrac{1.0\,\text{m/s}}{3.6\,\text{km/h}} \right) \right]}{(9.80\,\text{m/s}^2)\tan 38°} = \boxed{55\,\text{s}}$$

(*b*) The passengers will feel a change in the normal force that their seat exerts on them. Prior to the banking, the normal force was equal to their weight. During banking, the normal force will increase, so that

$$F_{\substack{\text{normal} \\ \text{banking}}} = \frac{mg}{\cos\theta} = 1.27 mg$$

Thus they will feel "pressed down" into their seats, with about a 25% increase in their apparent weight. If the plane is banking to the left, they will feel pushed to the right by that extra 25% in their apparent weight.

97. We include friction from the start, and then for the no-friction result, set the coefficient of friction equal to 0. Consider a free-body diagram for the car on the hill. Write Newton's second law for both directions. Note that the net force on the *y* direction will be zero, since there is no acceleration in the *y* direction.

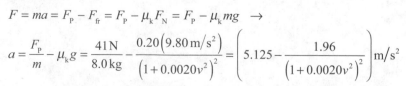

$$\sum F_y = F_{\text{N}} - mg\cos\theta = 0 \;\; \rightarrow \;\; F_{\text{N}} = mg\cos\theta$$
$$\sum F_x = mg\sin\theta - F_{\text{fr}} = ma \;\; \rightarrow$$

$$a = g\sin\theta - \frac{F_{\text{fr}}}{m} = g\sin\theta - \frac{\mu_k mg\cos\theta}{m} = g(\sin\theta - \mu_k \cos\theta)$$

Use Eq. 2-12c to determine the final velocity, assuming that the car starts from rest.

$$v^2 - v_0^2 = 2a(x - x_0) \;\; \rightarrow \;\; v = \sqrt{0 + 2a(x - x_0)} = \sqrt{2g(x - x_0)(\sin\theta - \mu_k \cos\theta)}$$

The angle is given by $\sin\theta = 1/4 \;\; \rightarrow \;\; \theta = \sin^{-1} 0.25 = 14.5°$

(*a*) $\mu_k = 0 \;\; \rightarrow \;\; v = \sqrt{2g(x - x_0)x\sin\theta} = \sqrt{2(9.80\,\text{m/s}^2)(55\,\text{m})\sin 14.5°} = \boxed{16\,\text{m/s}}$

(*b*) $\mu_k = 0.10 \;\; \rightarrow \;\; v = \sqrt{2(9.80\,\text{m/s}^2)(55\,\text{m})(\sin 14.5° - 0.10\cos 14.5°)} = \boxed{13\,\text{m/s}}$

103. Use the free body diagram to write Newton's second law for the block, and solve for the acceleration.

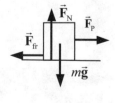

$$F = ma = F_{\text{P}} - F_{\text{fr}} = F_{\text{P}} - \mu_k F_{\text{N}} = F_{\text{P}} - \mu_k mg \;\; \rightarrow$$

$$a = \frac{F_{\text{P}}}{m} - \mu_k g = \frac{41\,\text{N}}{8.0\,\text{kg}} - \frac{0.20(9.80\,\text{m/s}^2)}{(1 + 0.0020v^2)^2} = \left(5.125 - \frac{1.96}{(1 + 0.0020v^2)^2} \right)\text{m/s}^2$$

For $t = 0$, $x(0) = x_0 = 0$, $v(0) = v_0 = 0$, and $a(0) = a_0 = 3.165$ m/s^2. Assume this acceleration is constant over the next time interval, and so

$$x_1 = x_0 + v_0 \Delta t + \tfrac{1}{2} a_0 \left(\Delta t \right)^2$$

$$v_1 = v_0 + a_0 \Delta t$$

$$a_1 = \left(5.125 - \frac{1.96}{\left(1 + 0.0020 v_1^2 \right)^2} \right) \text{m/s}^2$$

This continues for each successive interval. We apply this method first for a time interval of 1 second, and get the speed and position at $t = 5.0$ s. Then we reduce the interval to 0.5 s and again find the speed and position at $t = 5.0$ s. We compare the results from the smaller time interval with those of the larger time interval to see if they agree within 2%. If not, a smaller interval is used, and the process repeated. For this problem, the results for position and velocity for time intervals of 1.0 s and 0.5 s agree to within 2%.

(*a*) The speed at 5.0 s, from the numeric integration, is 18.0 m/s. The velocity–time graph is shown, along with a graph for a constant coefficient of friction, $\mu_k = 0.20$. We see that the varying friction (which decreases with speed) gives a higher speed than the constant friction.

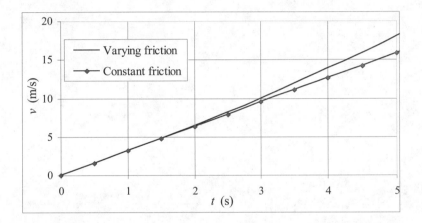

(*b*) The position at 5.0 s, from the numeric integration, is 42.4 m. The position–time graph is shown, along with a graph for a constant coefficient of friction, $\mu_k = 0.20$. We see that the varying friction (which decreases with speed) gives a larger distance than the constant friction.

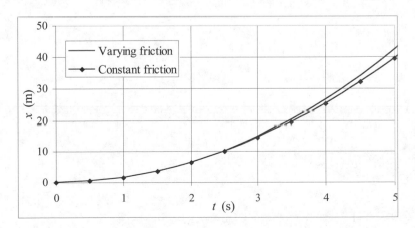

(*c*) If the coefficient of friction is constant, then $a = 3.165$ m/s^2. Constant acceleration relationships can find the speed and position at $t = 5.0$ s.

$$v = v_0 + at = 0 + at \;\; \rightarrow \;\; v_{\text{final}} = \left(3.165 \, \text{m/s}^2\right)\left(5.0 \, \text{s}\right) = 15.8 \, \text{m/s}$$

$$x = x_0 + v_0 t + \tfrac{1}{2} at^2 = 0 + 0 + \tfrac{1}{2} at^2 \;\; \rightarrow \;\; x_{\text{final}} = \tfrac{1}{2}\left(3.165 \, \text{m/s}^2\right)\left(5.0 \, \text{s}\right)^2 = 39.6 \, \text{m}$$

We compare the variable friction results to the constant friction results.

$$v: \quad \% \, \text{diff} = \frac{v_{\mu \, \text{constant}} - v_{\mu \, \text{variable}}}{v_{\mu \, \text{variable}}} = \frac{15.8 \, \text{m/s} - 18.0 \, \text{m/s}}{18.0 \, \text{m/s}} = \boxed{-12\%}$$

$$x: \quad \% \, \text{diff} = \frac{x_{\mu \, \text{constant}} - x_{\mu \, \text{variable}}}{x_{\mu \, \text{variable}}} = \frac{39.6 \, \text{m/s} - 42.4 \, \text{m/s}}{42.4 \, \text{m/s}} = \boxed{-6.6\%}$$

Chapter 6: Gravitation and Newton's Synthesis

Chapter Overview and Objectives

In this chapter, Newton's law of universal gravitation is introduced. Kepler's laws of planetary motion and Einstein's principle of equivalence are discussed.

After completing this chapter you should:
- Know Newton's universal law of gravitation.
- Know how to obtain the surface gravitational acceleration of an object from Newton's universal law of gravitation.
- Know Kepler's laws of planetary motion.

Summary of Equations

Magnitude of gravitational force:
$$F = \frac{Gm_1 m_2}{r^2}$$
(Section 6-1)

Vector form of Newton's law of gravitation:
$$\vec{F}_{12} = -\frac{Gm_1 m_2}{r_{21}^2}\, \hat{r}_{21}$$
(Section 6-2)

Magnitude of acceleration due to gravity:
$$g = \frac{Gm_E}{r_E^2}$$
(Section 6-3)

Kepler's Third Law:
$$\frac{T_1^2}{T_2^2} = \frac{s_1^3}{s_2^3}$$
(Section 6-5)

Gravitational field due to a point mass M:
$$\vec{g} = -\frac{GM}{r^2}\, \hat{r}$$
(Section 6-6)

Chapter Summary

Section 6-1. Newton's Law of Universal Gravitation

Every particle in the universe attracts every other particle with a force that is proportional to the product of their masses and inversely proportional to the square of the distance between them. This force acts along the line joining the two particles. The magnitude of this force, F, is

$$F = \frac{Gm_1 m_2}{r^2}$$

where G is the universal gravitation constant and is equal to 6.67×10^{-11} Nm^2/kg^2. The masses of the two particles are m_1 and m_2, and the distance between the two particles is r. This relation is known as **Newton's law of universal gravitation**.

Example 6-1-A: Orbital motion. Two small asteroids orbit each other in circular orbits. The distance between the centers of the two asteroids is 31.2 m. The mass of the larger asteroid, m_{large}, is 2.86×10^6 kg and the mass of the smaller asteroid, m_{small}, is 1.13×10^5 kg. What is the magnitude of the gravitational force between the two asteroids? What is the acceleration of each of the asteroids? What is the period of the circular motion of each of the two asteroids? Ignore the gravitational force between the Sun and the asteroids.

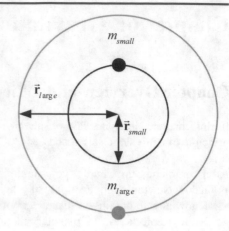

Approach: The system will only be stable if each asteroid has the same orbital period and the distance between the asteroids remains constant. The configuration of the system is shown schematically in the Figure to the right. Since the only force acting on each asteroid is the gravitational force, it must be directed towards the center of the orbit.

Solution: The gravitational force between the asteroids is equal to:

$$F = \frac{G m_{small} m_{large}}{d^2} = 2.214 \times 10^{-2} \text{ N}$$

The acceleration of each asteroid can be determined from Newton's second law:

$$a_{small} = \frac{F}{m_{small}} = \frac{G m_{large}}{d^2} = 1.96 \times 10^{-7} \text{ m/s}^2$$

$$a_{large} = \frac{F}{m_{large}} = \frac{G m_{small}}{d^2} = 7.74 \times 10^{-9} \text{ m/s}^2$$

Since the force is directed in the radial direction, these accelerations are the centripetal accelerations of the asteroids. The orbital period of the smaller asteroid can be expressed in terms of its acceleration and the radius of its orbit:

$$T = \frac{2\pi r_{small}}{v_{small}} = \sqrt{\frac{4\pi^2 r_{small}^2}{v_{small}^2}} = \sqrt{\frac{4\pi^2 r_{small}}{\left(\frac{v_{small}^2}{r_{small}}\right)}} = \sqrt{\frac{4\pi^2 r_{small}}{a_{small}}}$$

where a_{small} is the centripetal acceleration of the smaller asteroid. In a similar fashion we can determine the orbital period of the larger asteroid. Since the periods of the orbital motion of the asteroids must be the same, we must require that

$$T_{small} = \sqrt{\frac{4\pi^2 r_{small}}{a_{small}}} = T_{large} = \sqrt{\frac{4\pi^2 r_{large}}{a_{large}}} \quad \text{or} \quad \frac{r_{small}}{a_{small}} = \frac{r_{large}}{a_{large}} = \frac{d - r_{small}}{a_{large}}$$

In the last step we have used the fact that the sum of the orbital radii must add up to the distance between the centers of the asteroids. Solving the last equation for r_{small} we find

$$r_{small} = \frac{a_{small}}{a_{small} + a_{large}} d$$

The orbital period can now be determined

$$T = \sqrt{\frac{4\pi^2 r_{small}}{a_{small}}} = \sqrt{\frac{4\pi^2 d}{a_{small} + a_{large}}} = \sqrt{\frac{4\pi^2 d}{\frac{G m_{large}}{d^2} + \frac{G m_{small}}{d^2}}} = \sqrt{\frac{4\pi^2 d^3}{G\left(m_{large} + m_{small}\right)}} = 7.78 \times 10^4 \text{ s}$$

It is always important to make sure that the units of your answer are correct. The left-hand side of the last equation obviously has the unit of time. Checking the units of the expression on the right-hand side we find that it also has the unit of time:

$$\left[\sqrt{\frac{4\pi^2 d^3}{G\left(m_{\text{small}} + m_{\text{large}}\right)}} \right] = \sqrt{\frac{[L]^3}{\dfrac{[L]^3}{[T]^2 [M]}\left([M]\right)}} = \sqrt{[t]^2} = [t]$$

Section 6-2. Vector Form of Newton's Law of Universal Gravitation

Newton's law of universal gravitation can be written in vector form as:

$$\vec{\mathbf{F}}_{12} = -\frac{G m_1 m_2}{r_{21}^2}\,\hat{\mathbf{r}}_{21}$$

where the vector $\vec{\mathbf{F}}_{12}$ is the force on particle 1 of mass m_1 caused by particle 2 of mass m_2. The vector $\hat{\mathbf{r}}_{21}$ is the unit vector that points from particle 2 in the direction of particle 1. The minus sign appears because the force is attractive, which makes the direction of the force opposite to the direction of $\hat{\mathbf{r}}_{21}$.

Section 6-3. Gravity Near the Earth's Surface; Geophysical Applications

Comparing the expressions for the magnitude of the gravitational force in Newton's universal law of gravitation and the expression mg, which has been used for the weight of an object at the Earth's surface, we get:

$$mg = \frac{G m_E m}{r_E^2}$$

where m_E is the mass of the Earth and r_E is the radius of the Earth. Solving for g, the acceleration due to gravity at the Earth's surface, we obtain:

$$g = \frac{G m_E}{r_E^2}$$

Example 6-3-A: Gravitational acceleration on the surface of the Moon. Express the gravitational acceleration on the surface of the Moon in terms of the gravitational acceleration on the surface of the Earth, and the masses and radii of the Moon and the Earth.

Approach: Use the expression for g discussed in Section 6.3 to relate the gravitational acceleration on the surface of the Moon to the gravitational acceleration on the surface of the Earth.

Solution: The gravitational acceleration on the surface of the Moon can be found using an expression similar to the one derived in Section 6.3 for the gravitational acceleration on the surface of the Earth:

$$g_m = \frac{G m_m}{r_m^2}$$

The gravitational constant G can be expressed in terms of the gravitational acceleration on the surface of the Earth and the mass and radius of the Earth:

$$G = g_E \frac{r_E^2}{m_E}$$

Combining the last two equations we obtain

$$g_m = \left(g_E \frac{r_E^2}{m_E} \right) \frac{m_m}{r_m^2} = g_E \left(\frac{m_m}{m_E} \right) \left(\frac{r_E^2}{r_m^2} \right) = 9.80 \left(\frac{7.35 \times 10^{22}}{5.97 \times 10^{24}} \right) \left(\frac{6.38 \times 10^3}{1.74 \times 10^3} \right)^2 = 1.62 \text{ m/s}^2$$

Note that in calculating the gravitational acceleration on the Moon, we have left the radii of the Earth and Moon in terms of kilometers. Since only the ratio of radii appears, the units are not important as long as the same units are used for both radii. The same applies to the ratio of the masses.

Section 6-4. Satellites and "Weightlessness"

Satellites that travel in circular orbits have a centripetal acceleration v^2/r, where v is the speed of the satellite in its orbit and r is the orbital radius. The only force acting on the satellite is the gravitational force, and a stable orbit thus requires that

$$m_s \frac{v^2}{r} = \frac{G m_E m_s}{r^2}$$

where m_s is the mass of the satellite and r is the orbital radius.

The only external force acting on satellites in orbit around a planet or on any body in free fall is the force of gravity. Even though a gravitational force is acting on the body in these situations, they experience a condition commonly referred to as **weightlessness**. Objects in free fall appear to be weightless since there are no contact forces acting between them in the vertical direction. If you hold a book in your hands while standing on the surface of the Earth you must apply an upward force on the book to prevent the weight of the book from accelerating it in the downward direction. If you hold the same book in your hands while in free fall, you will not need to apply any force on the book to keep it from accelerating relative to your body because both you and the book are accelerating with the same acceleration; to you the book appears to have no weight. This is often called **apparent weightlessness** because of the apparent lack of downward acceleration observed by the freely falling observer. The word apparent is used because there actually is a gravitational force acting on the body; its weight is not zero.

Section 6-5. Kepler's Laws and Newton's Synthesis

Kepler's laws of planetary motion are empirical laws describing the observed behavior of the motion of planets around the Sun:

- *Kepler's first law*: The path of each planet around the Sun is an ellipse. The Sun is located at one of the two foci of the ellipse.
- *Kepler's second law*: Each planet moves so that the line joining the planet and the Sun sweeps out an area at a constant rate as the planet moves around the Sun.
- *Kepler's third law*: The ratio of the square of the period of a planet is proportional to the cube of the semi-major axis of the elliptical trajectory of the planet's orbit. This relationship can be written as

$$\left(\frac{T_1}{T_2} \right)^2 = \left(\frac{s_1}{s_2} \right)^3$$

Kepler's laws are a direct consequence of Newton's laws of motion and gravitation, but Kepler obtained his laws about half a century before Newton proposed his laws. Newton showed that Kepler's laws are consistent only with force laws that depend on distance as $1/r^2$.

Example 6-5-A: Applying Kepler's third law. A planet is in a circular orbit around a star. The orbital radius is $r_1 = 2.00 \times 10^8$ m and the orbital period is $T_1 = 3.87$ years. A second planet is observed to have an orbital period $T_2 = 29.7$ years. How far is the second planet from the star?

Approach: Kepler's third law relates the ratio of the periods to the ratio of the length of the semi-major axes. For a circle, the length of the semi-major axis is equal to the radius of the circle. The problem provides the ratio of the periods, and using the length of the semi-major axis of planet 1 we can determine the length of the semi-major axis of planet 2.

Solution: Kepler's third law tells us that

$$\left(\frac{T_1}{T_2}\right)^2 = \left(\frac{r_1}{r_2}\right)^3$$

This equation can be solved for r_2.

$$r_2 = \left(\left(\frac{T_2}{T_1}\right)^2 r_1^3\right)^{1/3} = \left(\frac{T_2}{T_1}\right)^{2/3} r_1 = \left(\frac{29.7}{3.87}\right)^{2/3}\left(2.00\times10^8\right) = 7.78\times10^8 \text{ m}$$

Section 6-6. Gravitational Field

The law of universal gravitation implies "action at a distance" since two objects not in contact with each other affect the motion of each other. The abstract notion of a **gravitational field** created by a massive body removes the "action at a distance problem." The gravitational field is a part of the massive body that extends through all space. The gravitational field, $\vec{g}$, at a particular point in space is defined as the ratio of the force acting on a test particle placed at that position and its mass:

$$\vec{g} = \frac{\vec{F}}{m}$$

The gravitational field of a point mass M is given by

$$\vec{g} = -\frac{GM}{r^2}\hat{\mathbf{r}}$$

where r is distance between the point mass M and the position at which the gravitational field is evaluated. The gravitational field of an extended body is the vector sum of the gravitational fields generated by the point masses that make up the extended body.

Section 6-7. Types of Forces in Nature

All of the interactions between objects that have been observed can be explained in terms of only four **fundamental forces**. The four fundamental forces, in order from weakest to strongest, are:
    The gravitational force (the weakest fundamental force)
    The weak nuclear force
    The electromagnetic force
    The strong nuclear force (the strongest fundamental force)
Physicists believe that these four fundamental forces are different manifestations of the same basic force. **Grand unified theories** (GUT) are theories that attempt to unify one or more of the fundamental forces.

Section 6-8. The Principle of Equivalence; Curvature of Space; Black Holes

Einstein's **principle of equivalence** states that it is experimentally impossible to detect whether you are in an inertial reference frame in a uniform gravitational field or in an accelerating reference frame with no gravitational field. When the principle of equivalence is applied to light, it implies that light should follow curved trajectories in space. This is contrary to what we expect light to do in empty space: follow the shortest distance between two points, which is a straight line.
An alternative method for dealing with gravitational fields is to give up the notion that space is described by Euclidean geometry and assume that space can be curved by the presence of mass. In **curved space-time**, the shortest distance between two points can be a curve. Einstein's general theory of relativity describes the behavior of curved space-time. If space has a strong enough curvature, the path of light may not allow it to escape from a region of space. This occurs around very massive bodies called **black holes**.

Practice Quiz

1. Which gravitational force has greater magnitude: the gravitational force of the Sun on the Earth or the gravitational force of the Earth on the Sun?
    a) The magnitude of the gravitational force of the Sun on the Earth is greater.
    b) The magnitude of the gravitational force of the Earth on the Sun is greater.
    c) The magnitude of the gravitational force of the Sun on the Earth is equal in magnitude to the gravitational force of the Earth on the Sun.
    d) Can't tell from the information given.

2. What happens to the magnitude of the gravitational force between two objects when the distance between them doubles?
    a) The magnitude of the force doubles.
    b) The magnitude of the force quadruples.
    c) The magnitude of the force decreases by a factor of two.
    d) The magnitude of the force decreases by a factor of four.

3. A planet has the same mass as the Earth, but its surface gravitational acceleration is $g/2$. What is the radius of the planet?
    a) $2 \times r_E$
    b) $\sqrt{2} \times r_E$
    c) $4 \times r_E$
    d) $r_E/2$

4. At which point is the gravitational force zero between two masses separated by a distance d when one mass has a mass m and the other has a mass $4m$?
    a) $(1/3)\, d$ away from the mass m
    b) $(1/3)\, d$ away from the mass $4m$
    c) $(1/4)\, d$ away from the mass m
    d) $(1/4)\, d$ away from the mass $4m$

5. A mass m is near a mass $3m$ as shown in the diagram. At which point in the diagram is the gravitational field approximately zero?
    a) A
    b) B
    c) C
    d) D

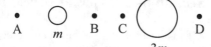

6. A satellite is in a circular orbit of radius r about a planet of mass m. Its orbital period is T. What is the period of an orbit with the same radius r about a planet of mass $2m$?
    a) $2T$
    b) $T/2$
    c) $T/\sqrt{2}$
    d) $T/4$

7. Consider this statement: The gravitational field on the surface of a spherical planet with uniform density is constant. What is wrong with that statement?
    a) Nothing is wrong with that statement.
    b) The magnitude of the gravitational field on the surface of the planet depends on the latitude.
    c) The magnitude of the gravitational field on the surface of the planet depends on the longitude.
    d) The magnitude of the gravitational field on the surface of the planet is constant, but the direction is not.

8. Why do we get correct results assuming that the force of gravity is constant near the surface of the Earth, but Newton's law of universal gravitation implies that the force of gravity should change with height?
   a) Newton's law of universal gravitation does not apply near the surface of the Earth.
   b) The Earth is not a point mass.
   c) The Earth is flat.
   d) The change in height is small compared to the radius of the Earth when we use that gravitational force is constant.

9. Kepler's third law implies that the kinetic energy of an object in a circular orbit decreases as the radius of the circular orbit increases, but positive work must be done on the object to increase the radius of the orbit. How can this be true?
   a) The increase in gravitational potential energy is greater than the decrease in kinetic energy.
   b) The statement is not true, only the gravitational potential energy changes as the orbital radius increases.
   c) The statement is not true, negative work must be done on the object to increase the orbital radius.
   d) The statement is not true, the kinetic energy increases as the orbital radius increases.

10. A spherical planet is made of uniform density material. How does the surface gravitational acceleration depend on the radius of the planet?
    a) $g \propto r$
    b) $g \propto r^2$
    c) $g \propto r^3$
    d) g is constant

11. Determine the orbital period of a satellite in a circular orbit that is just above the surface of the Sun. The mass of the Sun is 1.99×10^{30} kg and the radius of the Sun is 6.96×10^5 km.

12. Two planets are separated by a distance r. One planet has a mass that is three times the mass of the other planet. At what distance from the more massive planet is the sum of the gravitational forces of the two planets equal to zero along the line segment joining the two planets?

13. Tidal forces can tear a satellite apart if those forces are large enough. For a simple model to understand this, consider a satellite that consists of two spheres of mass m stacked on each other, held together by only their gravitational attraction for each other. The orientation of the stacking is shown in the diagram on the right. Since the centers of the two spheres are different distances from the planet, r and $r + d$, they have different gravitational forces acting on them. 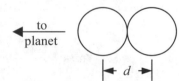 Because of the difference in gravitational forces from the planet, the spheres would separate if not held together by their gravitational attraction for each other. This difference can become large enough so that the gravitational force of attraction between the two spheres is not enough to hold the two spheres together. Calculate the force necessary to hold the two spheres together in orbit around a planet of mass M. Set this equal to gravitational force of attraction between the two spheres and solve for the distance r at which the planets would just barely be held together. At a distance to the planet smaller than this, the spheres would separate from each other. Assume $M \gg m$ and $r \gg d$.

14. Determine the orbital radius of a satellite in orbit around the Earth with an orbital period of two days.

15. Estimate the gravitational force between a 600-kg automobile and a 1200-kg truck parked 2 m apart.

Responses to Select End-of-Chapter Questions

1. Whether the apple is attached to a tree or falling, it exerts a gravitational force on the Earth equal to the force the Earth exerts on it, which is the weight of the apple (Newton's third law).

7. At the very center of the Earth, all of the gravitational forces would cancel, and the net force on the object would be zero.

13. Yes. At noon, the gravitational force on a person due to the Sun and the gravitational force due to the Earth are in the opposite directions. At midnight, the two forces point in the same direction. Therefore, your apparent weight at midnight is greater than your apparent weight at noon.

19. If you were in a satellite orbiting the Earth, you would have no apparent weight (no normal force). Walking, which depends on the normal force, would not be possible. Drinking would be possible, but only from a tube or pouch, from which liquid could be sucked. Scissors would not sit on a table (no apparent weight = no normal force).

25. If we treat $\vec{\mathbf{g}}$ as the *acceleration due to gravity*, it is the result of a force from one mass acting on another mass and causing it to accelerate. This implies action at a distance, since the two masses do not have to be in contact. If we view $\vec{\mathbf{g}}$ as a *gravitational field*, then we say that the presence of a mass changes the characteristics of the space around it by setting up a field, and the field then interacts with other masses that enter the space in which the field exists. Since the field is in contact with the mass, this conceptualization does not imply action at a distance.

Solutions to Select End-of-Chapter Problems

1. The spacecraft is at 3.00 Earth radii from the center of the Earth, or three times as far from the Earth's center as when at the surface of the Earth. Therefore, since the force of gravity decreases as the square of the distance, the force of gravity on the spacecraft will be one-ninth of its weight at the Earth's surface.

$$F_G = \tfrac{1}{9} mg_{\substack{\text{Earth's} \\ \text{surface}}} = \frac{(1480\,\text{kg})(9.80\,\text{m/s}^2)}{9} = \boxed{1610\,\text{N}}$$

This could also have been found using Eq. 6-1, Newton's law of universal gravitation.

7. The distance from the Earth's center is $r = R_{\text{Earth}} + 300\,\text{km} = 6.38 \times 10^6\,\text{m} + 3 \times 10^5\,\text{m} = 6.68 \times 10^6\,\text{m}$ (2 significant figures.) Calculate the acceleration due to gravity at that location.

$$g = G\frac{M_{\text{Earth}}}{r^2} = G\frac{M_{\text{Earth}}}{r^2} = \left(6.67 \times 10^{-11}\,\text{N·m}^2/\text{kg}^2\right)\frac{5.97 \times 10^{24}\,\text{kg}}{\left(6.68 \times 10^6\,\text{m}\right)^2} = 8.924\,\text{m/s}^2$$

$$= 8.924\,\text{m/s}^2\left(\frac{1\text{''}g\text{''}}{9.80\,\text{m/s}^2}\right) = \boxed{0.91g\text{'s}}$$

This is only about a 9% reduction from the value of g at the surface of the Earth.

13. To find the new weight of objects at the Earth's surface, the new value of g at the Earth's surface needs to be calculated. Since the spherical shape is being maintained, the Earth can be treated as a point mass. Find the density of the Earth using the actual values, and use that density to find g under the revised conditions.

$$g_{\text{original}} = G\frac{m_{\text{E}}}{r_{\text{E}}^2} \quad ; \quad \rho = \frac{m_{\text{E}}}{\frac{4}{3}\pi r_{\text{E}}^3} = \frac{3m_{\text{E}}}{4\pi r_{\text{E}}^3} \quad \rightarrow \quad r_{\text{E}} = \left(\frac{3m_{\text{E}}}{4\pi\rho}\right)^{1/3} \quad \rightarrow$$

$$g_{\text{original}} = G\frac{m_{\text{E}}}{\left(\dfrac{3m_{\text{E}}}{4\pi\rho}\right)^{2/3}} = G\frac{(m_{\text{E}})^{1/3}}{\left(\dfrac{3}{4\pi\rho}\right)^{2/3}} \quad ; \quad g_{\text{new}} = G\frac{(2m_{\text{E}})^{1/3}}{\left(\dfrac{3}{4\pi\rho}\right)^{2/3}} = 2^{1/3}G\frac{(m_{\text{E}})^{1/3}}{\left(\dfrac{3}{4\pi\rho}\right)^{2/3}} = 2^{1/3}g$$

Thus g is multiplied by $2^{1/3}$, and so the weight would be $\boxed{\text{multiplied by } 2^{1/3}}$.

19. The expression for g at the surface of the Earth is $g = G\, m_E/r_E^2$. Let $g + \Delta g$ be the value at a distance of $r_E + \Delta r$ from the center of Earth, which is Δr above the surface.

    (a) $g = G\dfrac{m_E}{r_E^2} \rightarrow g + \Delta g = G\dfrac{m_E}{\left(r_E + \Delta r\right)^2} = G\dfrac{m_E}{r_E^2\left(1 + \dfrac{\Delta r}{r_E}\right)^2} = G\dfrac{m_E}{r_E^2}\left(1 + \dfrac{\Delta r}{r_E}\right)^{-2} \approx g\left(1 - 2\dfrac{\Delta r}{r_E}\right) \rightarrow \boxed{\Delta g \approx -2g\dfrac{\Delta r}{r_E}}$

    (b) The minus sign indicated that the change in g is in the opposite direction as the change in r. So, if r increases, g decreases, and vice-versa.

    (c) Using this result:

    $$\Delta g \approx -2g\frac{\Delta r}{r_E} = -2\left(9.80\,\text{m/s}^2\right)\frac{1.25\times10^5\,\text{m}}{6.38\times10^6\,\text{m}} = -0.384\,\text{m/s}^2 \rightarrow g = \boxed{9.42\,\text{m/s}^2}$$

    Direct calculation:

    $$g = G\frac{m_E}{r^2} = \left(6.67\times10^{-11}\,\text{N}\bullet\text{m}^2/\text{kg}^2\right)\frac{\left(5.98\times10^{24}\,\text{kg}\right)}{\left(6.38\times10^6\,\text{m} + 1.25\times10^5\,\text{m}\right)^2} = 9.43\,\text{m/s}^2$$

    The difference is only about 0.1%.

25. Consider a free-body diagram of yourself in the elevator. $\vec{\mathbf{F}}_N$ is the force of the scale pushing up on you, and reads the normal force. Since the scale reads 76 kg, if it were calibrated in Newtons, the normal force would be $F_N = (76\text{ kg})(9.80\text{ m/s}^2) = 744.8\text{ N}$. Write Newton's second law in the vertical direction, with upward as positive.

    $$\sum F = F_N - mg = ma \rightarrow a = \frac{F_N - mg}{m} = \frac{744.8\,\text{N} - \left(65\,\text{kg}\right)\left(9.80\,\text{m/s}^2\right)}{65\,\text{kg}} = \boxed{1.7\,\text{m/s}^2 \text{ upward}}$$

    Since the acceleration is positive, the acceleration is upward.

31. Consider the free-body diagram for the astronaut in the space vehicle. The Moon is below the astronaut in the figure. We assume that the astronaut is touching the inside of the space vehicle, or in a seat, or strapped in somehow, and so a force will be exerted on the astronaut by the spacecraft. That force has been labeled $\vec{\mathbf{F}}_N$. The magnitude of that force is the apparent weight of the astronaut. Take down as the positive direction.

    (a) If the spacecraft is moving with a constant velocity, then the acceleration of the astronaut must be 0, and so the net force on the astronaut is 0.

    $$\sum F = mg - F_N = 0 \rightarrow$$

    $$F_N = mg = G\frac{mM_{\text{Moon}}}{r^2} = \left(6.67\times10^{-11}\,\text{N}\bullet\text{m}^2/\text{kg}^2\right)\frac{\left(75\text{ kg}\right)\left(7.4\times10^{22}\,\text{kg}\right)}{\left(2.5\times10^6\,\text{m}\right)^2} = 59.23\,\text{N}$$

    Since the value here is positive, the normal force points in the original direction as shown on the free-body diagram. The astronaut will be pushed "upward" by the floor or the seat. Thus the astronaut will perceive that he has a "weight" of $\boxed{59\text{ N, towards the Moon}}$.

    (b) Now the astronaut has an acceleration towards the Moon. Write Newton's second law for the astronaut, with down as the positive direction.

    $$\sum F = mg - F_N = ma \rightarrow F_N = mg - ma = 59.23\,\text{N} - \left(75\,\text{kg}\right)\left(2.3\,\text{m/s}^2\right) = -113.3\,\text{N}$$

    Because of the negative value, the normal force points in the opposite direction from what is shown on the free-body diagram – it is pointing towards the Moon. So perhaps the astronaut is pinned against the "ceiling" of the spacecraft, or safety belts are pulling down on the astronaut. The astronaut will perceive being "pushed downwards," and so has an upward apparent weight of $\boxed{110\text{ N, away from the Moon}}$.

37. Use Kepler's third law for objects orbiting the Earth. The following are given.

$$T_2 = \text{period of Moon} = (27.4 \text{ day})\left(\frac{86,400 \text{ s}}{1 \text{ day}}\right) = 2.367 \times 10^6 \text{ sec}$$

$$r_2 = \text{radius of Moon's orbit} = 3.84 \times 10^8 \text{ m}$$

$$r_1 = \text{radius of near-Earth orbit} = R_{\text{Earth}} = 6.38 \times 10^6 \text{ m}$$

$$\left(T_1/T_2\right)^2 = \left(r_1/r_2\right)^3 \rightarrow$$

$$T_1 = T_2 \left(r_1/r_2\right)^{3/2} = \left(2.367 \times 10^6 \text{ sec}\right)\left(\frac{6.38 \times 10^6 \text{ m}}{3.84 \times 10^8 \text{ m}}\right)^{3/2} = \boxed{5.07 \times 10^3 \text{ sec}} \left(= 84.5 \text{ min}\right)$$

43. Use Kepler's third law to find the radius of each moon of Jupiter, using Io's data for r_2 and T_2.

$$\left(r_1/r_2\right)^3 = \left(T_1/T_2\right)^2 \rightarrow r_1 = r_2 \left(T_1/T_2\right)^{2/3}$$

$$r_{\text{Europa}} = r_{\text{Io}} \left(T_{\text{Europa}}/T_{\text{Io}}\right)^{2/3} = \left(422 \times 10^3 \text{ km}\right)\left(3.55 \text{ d}/1.77 \text{ d}\right)^{2/3} = \boxed{671 \times 10^3 \text{ km}}$$

$$r_{\text{Ganymede}} = \left(422 \times 10^3 \text{ km}\right)\left(7.16 \text{ d}/1.77 \text{ d}\right)^{2/3} = \boxed{1070 \times 10^3 \text{ km}}$$

$$r_{\text{Callisto}} = \left(422 \times 10^3 \text{ km}\right)\left(16.7 \text{ d}/1.77 \text{ d}\right)^{2/3} = \boxed{1880 \times 10^3 \text{ km}}$$

The agreement with the data in the table is excellent.

49. (a) The gravitational field due to a spherical mass M, at a distance r from the center of the mass, is $g = GM/r^2$.

$$g_{\substack{\text{Sun at} \\ \text{Earth}}} = \frac{GM_{\text{Sun}}}{r^2_{\substack{\text{Sun to} \\ \text{Earth}}}} = \frac{\left(6.67 \times 10^{-11} \text{ N} \cdot \text{m}^2/\text{kg}^2\right)\left(1.99 \times 10^{30} \text{ kg}\right)}{\left(1.496 \times 10^{11} \text{ m}\right)^2} = \boxed{5.93 \times 10^{-3} \text{ m/s}^2}$$

(b) Compare this to the field caused by the Earth at the surface of the Earth.

$$\frac{g_{\substack{\text{Sun at} \\ \text{Earth}}}}{g_{\text{Earth}}} = \frac{5.93 \times 10^{-3} \text{ m/s}^2}{9.80 \text{ m/s}^2} = 6.05 \times 10^{-4}$$

$\boxed{\text{No}}$, this is not going to affect your weight significantly. The effect is less than 0.1 %.

55. The speed of an object in an orbit of radius r around a planet is given in Example 6-6 as $v = \sqrt{(GM_{\text{planet}}/r)}$ and is also given by $v = 2\pi r/T$, where T is the period of the object in orbit. Equate the two expressions for the speed and solve for T.

$$\sqrt{G\frac{M_{\text{Planet}}}{r}} = \frac{2\pi r}{T} \rightarrow T = 2\pi\sqrt{\frac{r^3}{GM_{\text{Planet}}}}$$

For this problem, the inner orbit has radius $r_{inner} = 7.3 \times 10^7$ m and the outer orbit has radius $r_{outer} = 1.7 \times 10^8$ m. Use these values to calculate the periods.

$$T_{inner} = 2\pi\sqrt{\frac{\left(7.3\times 10^7\,\text{m}\right)^3}{\left(6.67\times 10^{-11}\,\text{N•m}^2/\text{kg}^2\right)\left(5.7\times 10^{26}\,\text{kg}\right)}} = \boxed{2.0\times 10^4\,\text{s}}$$

$$T_{outer} = 2\pi\sqrt{\frac{\left(1.7\times 10^8\,\text{m}\right)^3}{\left(6.67\times 10^{-11}\,\text{N•m}^2/\text{kg}^2\right)\left(5.7\times 10^{26}\,\text{kg}\right)}} = \boxed{7.1\times 10^4\,\text{s}}$$

Saturn's rotation period (day) is 10 hr 39 min, which is about 3.8×10^4 s. Thus the inner ring will appear to move across the sky "faster" than the Sun (about twice per Saturn day), while the outer ring will appear to move across the sky "slower" than the Sun (about once every two Saturn days).

61. In the text, it says that Eq. 6-6 is valid if the radius r is replaced with the semi-major axis s. From Fig. 6-16, the distance of closest approach r_{min} is seen to be $r_{min} = s - es = s(1 - e)$, and so the semi-major axis is given by $s = r_{min}/(1 - e)$.

$$\frac{T^2}{s^3} = \frac{4\pi^2}{GM_{SgrA}} \rightarrow$$

$$M_{SgrA} = \frac{4\pi^2 s^3}{GT^2} = \frac{4\pi^2\left(\frac{r_{min}}{1-e}\right)^3}{GT^2} = \frac{4\pi^2\left(\frac{123\,\text{AU} \times \frac{1.5\times 10^{11}\,\text{m}}{1\,\text{AU}}}{1-0.87}\right)^3}{\left(6.67\times 10^{-11}\,\text{N•m}^2/\text{kg}^2\right)\left(15.2\,\text{y} \times \frac{3.156\times 10^7\,\text{s}}{1\,\text{y}}\right)^2} \approx \boxed{7.4\times 10^{36}\,\text{kg}}$$

or

$$\frac{M_{SgrA}}{M_{Sun}} = \frac{7.352\times 10^{36}\,\text{kg}}{1.99\times 10^{30}\,\text{kg}} = \boxed{3.7\times 10^6}$$

and so SgrA is almost 4 million times more massive than our Sun.

67. Since all of the masses (or mass holes) are spherical, and g is being measured outside of their boundaries, we can use the simple Newtonian gravitation expression. In the diagram, the distance $r = 2000$ m. The radius of the deposit is unknown.

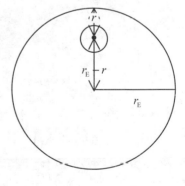

$$g_{actual} = g_{\substack{full \\ Earth}} - g_{\substack{missing \\ dirt\ mass}} + g_{oil} = g_{\substack{full \\ Earth}} - \frac{GM_{\substack{missing \\ dirt}}}{r^2} + \frac{GM_{oil}}{r^2}$$

$$= g_{\substack{full \\ Earth}} - \frac{G\left(M_{\substack{missing \\ dirt}} - M_{oil}\right)}{r^2}$$

$$\Delta g = g_{\substack{full \\ Earth}} - g_{actual} = \frac{G\left(M_{\substack{missing \\ dirt}} - M_{oil}\right)}{r^2} = \frac{G}{r^2}\left(V_{\substack{missing \\ dirt}}\rho_{\substack{missing \\ dirt}} - V_{oil}\rho_{oil}\right) = \frac{GV_{oil}}{r^2}\left(\rho_{\substack{missing \\ dirt}} - \rho_{oil}\right) = \frac{2}{10^7}g$$

$$V_{\text{oil}} = \frac{2}{10^7} g \frac{r^2}{G} \frac{1}{\left(\rho_{\substack{\text{missing} \\ \text{dirt}}} - \rho_{\text{oil}}\right)} = \frac{2}{10^7}\left(9.80\,\text{m/s}^2\right)\frac{\left(2000\,\text{m}\right)^2}{\left(6.67\times10^{-11}\,\text{N}\cdot\text{m}^2/\text{kg}^2\right)\left(3000-800\right)\text{kg/m}^2} \approx \boxed{5\times10^7\,\text{m}^3}$$

$$r_{\text{deposit}} = \left(\frac{3V_{\text{oil}}}{4\pi}\right)^{1/3} = 234\,\text{m} \approx \boxed{200\,\text{m}} \quad ; \quad m_{\text{deposit}} = V_{\text{oil}}\rho_{\text{oil}} = 4.27\times10^{10}\,\text{kg} \approx \boxed{4\times10^{10}\,\text{kg}}$$

$$\frac{M_{\text{Jupiter}}}{M_{\text{Sun}}} = \frac{\rho_{\text{Jupiter}} \frac{4}{3}\pi r^3_{\text{Jupiter}}}{\rho_{\text{Sun}} \frac{4}{3}\pi r^3_{\text{Sun}}} = \frac{\dfrac{4\pi^2 r^3_{\substack{\text{Callisto}\\\text{orbit}}}}{GT^2_{\text{Callisto}}}}{\dfrac{4\pi^2 r^3_{\substack{\text{Venus}\\\text{orbit}}}}{GT^2_{\text{Venus}}}} \rightarrow$$

$$\frac{\rho_{\text{Jupiter}}}{\rho_{\text{Sun}}} = \frac{r^3_{\substack{\text{Callisto}\\\text{orbit}}}}{T^2_{\text{Callisto}}}\frac{T^2_{\text{Venus}}}{r^3_{\substack{\text{Venus}\\\text{orbit}}}}\frac{r^3_{\text{Sun}}}{r^3_{\text{Jupiter}}} = \frac{\left(0.01253\right)^3}{\left(16.69\right)^2}\frac{\left(224.7\right)^2}{\left(0.724\right)^3}\frac{1}{\left(0.0997\right)^3} = \boxed{0.948}$$

And likewise for the Earth–Sun combination:

$$\frac{\rho_{\text{Earth}}}{\rho_{\text{Sun}}} = \frac{r^3_{\substack{\text{Moon}\\\text{orbit}}}}{T^2_{\text{Moon}}}\frac{T^2_{\text{Venus}}}{r^3_{\substack{\text{Venus}\\\text{orbit}}}}\frac{r^3_{\text{Sun}}}{r^3_{\text{Earth}}} = \frac{\left(0.003069\right)^3}{\left(27.32\right)^2}\frac{\left(224.7\right)^2}{\left(0.724\right)^3}\frac{1}{\left(0.0109\right)^3} = \boxed{3.98}$$

73. The initial force of 120 N can be represented as $F_{\text{grav}} = GM_{\text{planet}}/r^2 = 120$ N.
(a) The new radius is 1.5 times the original radius.

$$F_{\substack{\text{new}\\\text{radius}}} = \frac{GM_{\text{planet}}}{r^2_{\text{new}}} = \frac{GM_{\text{planet}}}{\left(1.5r\right)^2} = \frac{GM_{\text{planet}}}{2.25r^2} = \frac{1}{2.25}\left(120\,\text{N}\right) = \boxed{53\,\text{N}}$$

(b) With the larger radius, the period is $T = 7200$ seconds. As found in Example 6-6, orbit speed can be calculated by $v = \sqrt{(GM/r)}$

$$v = \sqrt{\frac{GM}{r}} = \frac{2\pi r}{T} \quad \rightarrow \quad M = \frac{4\pi^2 r^3}{GT^2} = \frac{4\pi^2\left(3.0\times10^7\,\text{m}\right)^3}{\left(6.67\times10^{-11}\,\text{N}\cdot\text{m}^2/\text{kg}^2\right)\left(7200\,\text{s}\right)^2} = \boxed{3.1\times10^{26}\,\text{kg}}$$

79. (a) The graph is shown.
(b) From the graph, we get this equation.

$$T^2 = 0.9999r^3 + 0.3412 \quad \rightarrow$$

$$r = \left(\frac{T^2 - 0.3412}{0.9999}\right)^{1/3}$$

$$r\left(T = 247.7\,\text{y}\right) =$$

$$\left(\frac{247.7^2 - 0.3412}{0.9999}\right)^{1/3} = \boxed{39.44\,\text{AU}}$$

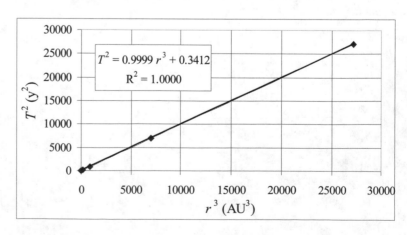

A quoted value for the mean distance of Pluto is 39.47 AU.

Chapter 7: Work and Energy

Chapter Overview and Objectives

In this chapter, the concepts of work and kinetic energy are introduced. The relation between these concepts and Newton's second law is discussed and the work-energy principle is derived.

After completing this chapter you should:
- Know the definitions of work done by both a constant force and a varying force.
- Know the definition of the scalar product of two vectors and the properties of the scalar product.
- Be able to calculate scalar products from knowledge of magnitudes and relative directions of the vectors.
- Be able to calculate scalar products from vectors expressed in terms of Cartesian unit vectors.
- Know the dimensions and SI units of work and energy.
- Know the work-energy principle and how to apply the work-energy principle to problems.

Summary of Equations

Definition of work done by a constant force:
$$W = \vec{\mathbf{F}} \cdot \vec{\mathbf{d}} = F_\| d = Fd\cos\theta \qquad \text{(Section 7-1)}$$

Definition of scalar product of two vectors:
$$\vec{\mathbf{A}} \cdot \vec{\mathbf{B}} = AB\cos\theta \qquad \text{(Section 7-2)}$$

Commutative property of scalar product:
$$\vec{\mathbf{A}} \cdot \vec{\mathbf{B}} = \vec{\mathbf{B}} \cdot \vec{\mathbf{A}} \qquad \text{(Section 7-2)}$$

Distributive property of scalar product:
$$\vec{\mathbf{A}} \cdot \left(\vec{\mathbf{B}} + \vec{\mathbf{C}}\right) = \vec{\mathbf{A}} \cdot \vec{\mathbf{B}} + \vec{\mathbf{A}} \cdot \vec{\mathbf{C}} \qquad \text{(Section 7-2)}$$

Scalar product in terms of Cartesian components:
$$\vec{\mathbf{A}} \cdot \vec{\mathbf{B}} = A_x B_x + A_y B_y + A_z B_z \qquad \text{(Section 7-2)}$$

Work done by varying force:
$$W = \int_a^b \vec{\mathbf{F}} \cdot d\vec{\mathbf{l}} \qquad \text{(Section 7-3)}$$

Definition of translational kinetic energy:
$$K = \frac{1}{2}mv^2 \qquad \text{(Section 7-4)}$$

Work-energy principle:
$$W_{net} = \Delta K = \frac{1}{2}mv_f^2 - \frac{1}{2}mv_i^2 \qquad \text{(Section 7-4)}$$

Chapter Summary

Section 7-1. Work Done by a Constant Force

The **work** done on an object by a force is defined to be the force on the object multiplied by the displacement of the object in the direction of the force:

$$W = F_\| d = Fd\cos\theta$$

where W is the work done on the object, $F_\|$ is the component of the force parallel to the displacement of the object, d is the magnitude of the displacement of the object, F is the magnitude of the force, and θ is the angle between the direction of the force and the direction of the displacement.

The dimensions of work are $[M][L]^2/[T]^2$ and the SI units of work are Newton · meter (Nm). A Newton · meter is defined to be a **Joule** of work: 1 J = 1 Nm. The British unit system unit of work is the foot-pound. One foot-pound is equal to 1.36 J.

Example 7-1-A. Work done while pushing a car. Your car of mass $m = 652$ kg runs out of gas. You need to push it up a hill to get to the gas station. The hill rises a height $h = 2.00$ m for every 100 m you travel horizontally. You push with a constant horizontal force $F = 538$ N on the car as you move the car a distance $d = 623$ m to the gas station. How much work is done by your force on the car, the force of gravity acting on the car, and the normal force of the road acting on the car?

Approach: To solve this problem we need to determine the three forces acting on the car, and use the definition of the work done by a force to calculate the work done by these three forces. In this type of calculation, it is critical to make sure that the sign of the work is determined properly.

Solution: We will consider the following three forces that act on the car: the gravitational force, the applied force, and the normal force. These three forces are schematically indicated in the following Figure

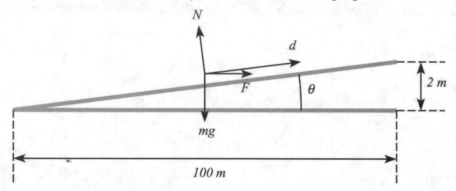

The force you apply on the car is directed in the horizontal direction and is thus not pointing in the same direction as the displacement of the car. The angle between the applied force and the displacement vector is equal to the angle of the incline:

$$\theta = \arctan\left(\frac{2.00\,\text{m}}{100\,\text{m}}\right) = 1.146°$$

The work done by the force you apply is thus equal to

$$W = Fd\cos\theta = (538\,\text{N})(623\,\text{m})\cos(1.146°) = 3.35 \times 10^5 \text{ J}$$

The work is positive since the angle between the force and the displacement is less than 90°.
The angle between the gravitational force and the displacement is $90° + \theta$ and the work done by the gravitational force will thus be negative. The work done by the gravitational force is equal to

$$W = mgd\cos(90° + \theta) = (6390\,\text{N})(623\,\text{m})\cos(91.146°) = -7.96 \times 10^4 \text{ J}$$

The work done by the normal force is zero since the normal force is directed perpendicular to the surface, and thus perpendicular to the displacement.

Section 7-2. Scalar Product of Two Vectors

Up to this point, the only multiplication operation involving vectors that has been introduced is multiplication of a vector by a scalar. It is useful to define two other vector multiplication operations involving vectors. The **scalar product** (also called the **dot product**) of two vectors is a scalar. The **vector product** of two vectors results in an **axial vector**. In this chapter, the scalar product is introduced.
The scalar product of vector $\vec{A}$ and vector $\vec{B}$ is denoted as $\vec{A}\cdot\vec{B}$ and is defined to be

$$\vec{A}\cdot\vec{B} = AB\cos\theta$$

where A and B are the magnitudes of vectors $\vec{A}$ and $\vec{B}$, respectively, and θ is the angle between the direction of vector $\vec{A}$ and the direction of vector $\vec{B}$.

The work W done on an object acted on by a constant force $\vec{F}$ while making a displacement $\vec{d}$ can be written as

$$W = \vec{F} \cdot \vec{d} = Fd\cos\theta$$

There are several properties of scalar multiplication that may be useful to know. The scalar product is **commutative** (the order of the multiplication does not matter):

$$\vec{A} \cdot \vec{B} = \vec{B} \cdot \vec{A}$$

Scalar multiplication distributes over vector addition:

$$\vec{A} \cdot \left(\vec{B} + \vec{C} \right) = \vec{A} \cdot \vec{B} + \vec{A} \cdot \vec{C}$$

The scalar product between vectors $\vec{A}$ and $\vec{B}$ can be expressed in terms of the Cartesian components of these vectors. If the vectors $\vec{A}$ and $\vec{B}$ are expressed in terms of Cartesian unit vectors as $\vec{A} = A_x\hat{i} + A_y\hat{j} + A_z\hat{k}$ and $\vec{B} = B_x\hat{i} + B_y\hat{j} + B_z\hat{k}$, respectively, then the scalar product of $\vec{A}$ and $\vec{B}$ is equal to

$$\vec{A} \cdot \vec{B} = A_x B_x + A_y B_y + A_z B_z$$

It is useful to remember the scalar products of the Cartesian unit vectors:

$$\hat{i} \cdot \hat{i} = \hat{j} \cdot \hat{j} = \hat{k} \cdot \hat{k} = 1$$

$$\hat{i} \cdot \hat{j} = \hat{j} \cdot \hat{k} = \hat{i} \cdot \hat{k} = 0$$

Example 7-2-A: Calculating the angle between two vectors. A vector $\vec{A}$ with magnitude $A = 5.34$ m lies in the xy plane. It makes an angle $\theta_A = +43°$ angle with respect to the $+x$ axis, toward the $+y$ axis. A second vector $\vec{B}$ with magnitude $B = 3.98$ m lies in the yz plane and makes an angle $\theta_B = +57°$ from the $+y$ axis, toward the $+z$ axis. Determine the scalar product of vectors $\vec{A}$ and $\vec{B}$. Determine the angle between the directions of vectors $\vec{A}$ and $\vec{B}$.

Approach: In order to calculate the angle between the two vectors we need to determine the components of the two vectors. Using the expression of the scalar product in terms of the components we can now determine the angle between these two vectors.

Solution: The vectors $\vec{A}$ and $\vec{B}$ can be expressed in terms of the Cartesian unit vectors:

$$\vec{A} = A\cos\theta_A\hat{i} + A\sin\theta_A\hat{j}$$

$$\vec{B} = B\cos\theta_B\hat{j} + B\sin\theta_B\hat{k}$$

We can directly calculate the scalar product in terms of the Cartesian components:

$$\vec{A} \cdot \vec{B} = A_x B_x + A_y B_y + A_z B_z = AB\sin\theta_A\cos\theta_B = 7.89$$

To determine the angle between vectors $\vec{A}$ and $\vec{B}$ we use the fact that the scalar product can be expressed in terms of the cosine of the angle between the two vectors:

$$\vec{A} \cdot \vec{B} = AB\cos\theta$$

The angle θ is thus equal to

$$\theta = \text{acos}\left(\frac{\vec{A} \cdot \vec{B}}{AB} \right) = \text{acos}\left(\frac{7.89}{(5.34)(3.98)} \right) = 68°$$

Section 7-3. Work Done by a Varying Force

When the force on an object is varying with position or the path the object follows is curved, calculus must be used to determine the work done on the object by the force between position a and position b:

$$W = \int_a^b \vec{\mathbf{F}} \cdot d\vec{\mathbf{l}}$$

where $d\vec{\mathbf{l}}$ is an infinitesimal path vector pointing along the direction of travel of the object. The integral is over the path from the initial position to the final position.

Example 7-3-A: Work done by the electrostatic force. The electrostatic force, $\vec{\mathbf{F}}$, between two charged particles has a magnitude $F = C/r^2$, where C is a constant and r is the distance between the charged particles. The force points along the line between the two particles and is repulsive. Determine the work done by this force on one particle as it directly approaches the other particle, moving from an initial distance r_1 to a final distance r_2.

Approach: We assume that the charged particles have the same charge and that the electrostatic force between them is repulsive. When we bring one of the particles from its initial position to its final position, the force will be pointing in a direction opposite to the direction of travel and the work done will thus be negative. Since the force will be a function of the distance between the particles, its magnitude will change when we move from the initial position to the final position. We will thus need to integrate the force in order to determine the work done.

Solution: The configuration is shown schematically in the Figure below. When the right charge moves, its displacement vector is pointing towards the left while the electrostatic force is pointing towards the right.

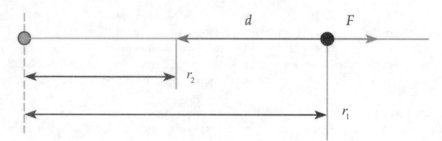

In order to determine the total work done on the charged particle, we break up its path into small segment of length dr and we assume that the force is constant along this small segment. The work done is equal to $-(C/r^2)dr$. The total work done can be found by integrating this expression with respect to r between the limits provided.

$$W = \int_{r_2}^{r_1} \left(-F \right) dr = \int_{r_2}^{r_1} \left(-\frac{C}{r^2} \right) dr = \left. \frac{C}{r} \right|_{r_2}^{r_1} = C \left(\frac{1}{r_1} - \frac{1}{r_2} \right)$$

Note that in the final integral we have reversed the limits in order to be consistent with our assumption that dr is positive. Since $r_1 > r_2$ we see that the work done is negative, which of course is no surprise since the displacement and the force are pointing in opposite directions.

Example 7-3-B: Work done by the spring force. A spring with spring constant k is attached to a mass that is constrained to slide along a straight rod as shown in the diagram. The natural length of the spring L is the perpendicular distance from the rod to the fixed end of the spring, i.e., the force is zero when the spring is perpendicular to the rod. How much work is done by the spring on the mass if the mass slides from position a to position b?

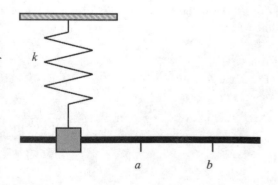

Approach: The spring force is a function of the elongation of the spring, and this force changes when we move the mass from position a to position b. In addition, the angle between the force and the rod will also change. In order to determine the work done we need to break up the path between a and b into small segments, and assume that the force is constant over these small segments. The total work done by the spring force can then be obtained from an integral over the contributions of each of these segments.

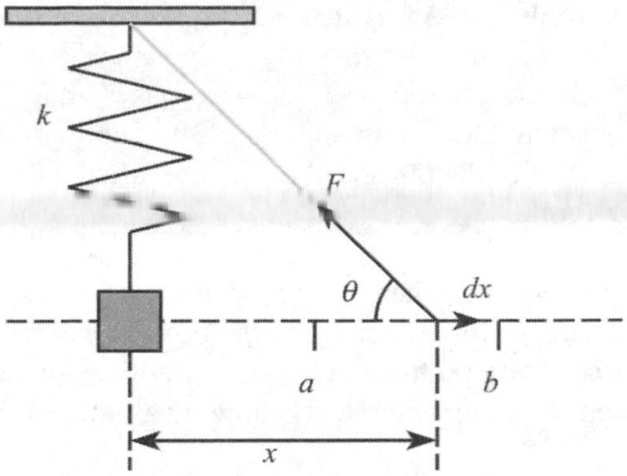

Solution: The position of the mass is specified in terms of an x coordinate. Our coordinate system is defined such that $x = 0$ m corresponds to the position of the mass in its equilibrium position. Consider the force that acts on the mass when it is located a distance x from its equilibrium position. At this position, the magnitude of the spring force will be equal to

$$F = k\left\{\sqrt{x^2 + L^2} - L\right\}$$

The force is directed as shown in the diagram on the right. The angle θ between the direction of the force and the horizontal plane is given by

$$\theta = \operatorname{atan}\left(\frac{L}{x}\right)$$

The angle between the force F and the displacement dx is $180° - \theta$. Since θ is less than $90°$, the angle between the force F and the displacement dx is larger than $90°$ and the work done by F will be negative:

$$\cos(180° - \theta) = -\cos\theta$$

The work dW done by the force F when we move the mass a distance dx to the right is equal to

$$dW = \vec{F}\cdot d\vec{x} = k\left\{\sqrt{x^2 + L^2} - L\right\}\cos(180° - \theta)\,dx = -k\left\{\sqrt{x^2 + L^2} - L\right\}\cos\theta\,dx = -k\left\{\sqrt{x^2 + L^2} - L\right\}\frac{x}{\sqrt{x^2 + L^2}}\,dx$$

The work done by the spring when we move the mass from position a to position b can be found by integrating the previous expression with respect to x:

$$W = \int dW = -k\int_{x_a}^{x_b}\left\{\sqrt{x^2 + L^2} - L\right\}\frac{x}{\sqrt{x^2 + L^2}}\,dx = -k\left\{\int_{x_a}^{x_b}x\,dx - L\int_{x_a}^{x_b}\frac{x}{\sqrt{x^2 + L^2}}\,dx\right\} =$$

$$= -k\left\{\frac{1}{2}\left(x_b^2 - x_a^2\right) - \frac{L}{2}\int_{x_a}^{x_b}\frac{1}{\sqrt{x^2 + L^2}}\,dx^2\right\} = -k\left\{\frac{1}{2}\left(x_b^2 - x_a^2\right) - L\left(\sqrt{x_b^2 + L^2} - \sqrt{x_a^2 + L^2}\right)\right\}$$

Section 7-4. Kinetic Energy and the Work-Energy Principle

The **translational kinetic energy** K of an object is defined as

$$K = \frac{1}{2}mv^2$$

where m is the mass of the object and v is its speed. Newton's second law can be used to relate the work done on an object by the net force and the change in its kinetic energy. This relationship is called the **work-energy principle**:

$$W_{net} = \Delta K = \frac{1}{2}mv_f^2 - \frac{1}{2}mv_i^2$$

where m is the mass of the object, and v_f and v_i are the initial and final speeds of the object, respectively.

Example 7-4-A: Applying the work-energy principle. A box of mass $m = 5.76$ kg is initially traveling with a speed $v_i = 3.43$ m/s as it starts up a ramp on which the kinetic friction force has a magnitude of $f = 37.6$ N. The ramp surface is tilted at an angle $\theta = 30°$ to horizontal, as shown in the Figure on the right. What is the speed of the box after it has moved a distance $d = 0.34$ m up the ramp?

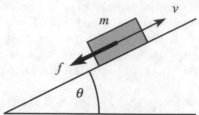

Approach: When we apply the work-energy principle we must be sure to include all forces acting on the object. The problem indicates the magnitude of the friction force, but this is not the only force acting on the block. The two other forces that act on the block are the gravitational force and the normal force. We need to calculate the work done by all these forces before we can apply the work-energy principle.

Solution: In order to solve this problem we need to examine all forces acting on the block and determine their direction with respect to the direction of motion. A free-body diagram of the block is shown in the following Figure. In this Figure we have also indicated the direction of the displacement of the block.

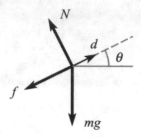

The work done by the normal force is 0 J since the normal force is perpendicular to the ramp and thus perpendicular to the displacement. The work done by the friction force is equal to

$$W_{friction} = \vec{f} \cdot \vec{d} = fd\cos(180°) = -(37.6\,\text{N})(0.34\,\text{m}) = -12.8\,\text{J}$$

The work done by the force of gravity is equal to

$$W_{gravity} = m\vec{g} \cdot \vec{d} = mgd\cos(90° + \theta) = -9.60\,\text{J}$$

The total work done by all the forces acting on the block can be found by adding the work done by each of the forces acting on the block:

$$W = W_{gravity} + W_{friction} + W_{normal} = (-9.6 - 12.8 + 0)\,\text{J} = -22.4\,\text{J}$$

According to the work-energy principle, the work done by all the forces is equal to the change in the kinetic energy of the object:

$$W = \Delta K = \tfrac{1}{2}mv_f^2 - \tfrac{1}{2}mv_i^2$$

The final velocity of the block is thus equal to

$$v_f = \sqrt{\frac{2}{m}\left(W + \frac{1}{2}mv_i^2\right)} = \sqrt{\frac{2W}{m} + v_i^2} = \sqrt{\frac{2(-22.4\,\text{J})}{(5.76\,\text{kg})} + (3.43\,\text{m/s})^2} = 2.00\,\text{m/s}$$

Practice Quiz

1. An object has a kinetic energy K. How much work do you have to do to double the speed of the object?
   a) K
   b) $2K$
   c) $3K$
   d) $4K$

2. You're working on a construction job and the boss asks you to hold a piece of plywood up at a height of 1.27 m above the floor. The weight of the plywood is 34.1 N. You hold the piece of plywood up for a time of 18.2 minutes. How much work have you done on the sheet of plywood while holding it at this height?
   a) 43.3 J
   b) -43.3 J
   c) 6.00×10^5 J
   d) 0 J

3. A force acting on an object accelerates it up to a speed v after starting from rest. Does accelerating to that speed in a shorter time increase or decrease the amount of work done by the force?
    a)   Shorter time means a bigger force so more work is done.
    b)   Shorter time means the car travels a shorter distance so less work is done.
    c)   Same change in kinetic energy regardless of time, so same amount of work is done.
    d)   More information is needed to answer the question.

4. What is the work done by the friction force around a closed path?
    a)   The work is positive.
    b)   The work is zero.
    c)   The work is negative.
    d)   More information is needed to answer the question.

5. What is the direction of the scalar product of two vectors?
    a)   Halfway between the direction of each of the vectors
    b) $\mathrm{acos}\left(\vec{A}\cdot\vec{B} \, / \, AB \right)$
    c) $\mathrm{acos}\left(AB \, / \, \vec{A}\cdot\vec{B} \right)$
    d)   The scalar product is a scalar; scalars don't have a direction.

6. Work is done on an object to triple its kinetic energy. By what factor does the object's speed change?
    a)   9
    b)   1/9
    c) $\sqrt{3}$
    d) $1/\sqrt{3}$

7. Determine which of the following is equivalent to $\vec{A}\cdot\vec{B} + \vec{C}\cdot\vec{B} - \vec{B}\cdot\vec{B} + \left(\vec{C}+\vec{B} \right)\cdot\vec{B}$.
    a) $\vec{A}\cdot\vec{B}$
    b) $\vec{C}\cdot\vec{B}$
    c) $\vec{A}\cdot\vec{B} + 2\vec{C}\cdot\vec{B}$
    d) $\vec{A}\cdot\vec{B} + 2\vec{C}\cdot\vec{B} + 2\vec{B}\cdot\vec{B}$

8. An object moves around a closed path so its displacement for the entire motion is zero. Can you conclude that the work done is zero?
    a)   Yes, no displacement implies no work.
    b)   No, the work done in going around a closed path is always negative.
    c)   No, the work done in going around a closed path is always positive.
    d)   Not enough information is given to determine whether the work was zero, positive, or negative.

9. Can an object have a net negative work done on it with a magnitude greater than its kinetic energy?
    a)   Yes, there is no limit to the amount of work done on the object.
    b)   No, once the kinetic energy is brought to zero, no more negative work can be done on the object.
    c)   Yes, if the object has relativistic kinetic energy.
    d)   Only by frictional forces.

10. An object on which a constant force F is acting starts from rest and reaches a speed v after a distance d. What force is required to increase its speed from zero to $2v$ in a distance $2d$?
    a) $F/2$
    b) $F/4$
    c) $2F$
    d) $4F$

11. Calculate the work done on an object that moves from $x = 1$ m to $x = 2$ m when a force

    $$F = 3.0 \, x^2$$

    acts on the object.

12. A box starts from rest and slides down a frictionless ramp that is tilted at 32° to the horizontal. How fast is the box traveling after it slides a distance 3.39 m down the ramp?

13. An object moves by a displacement $\vec{d} = 3.42\hat{i} + 2.31\hat{j} + 1.78\hat{k}$. While it is moving, it is acted upon by a constant force $\vec{F} = 56.3\hat{i} + 61.2\hat{j} - 49.2\hat{k}$. What is the work done by the force and what is the angle between the direction of the force and the direction of the displacement?

14. A child with a mass 38.6 kg slides down a playground slide that has a length of 2.82 m. The slide slopes downward at an angle of 44.1°. If the child starts from rest and reaches a speed of 4.34 m/s at the bottom of the slide, what is the frictional force acting on the child as the child goes down the slide?

15. A constant friction force of 50 N is acting on a car that is moving with an initial velocity of 10 m/s. The mass of the car is 1000 kg. How far will the block travel before it comes to rest?

Responses to Select End-of-Chapter Questions

1. "Work" as used in everyday language generally means "energy expended," which is similar to the way "work" is defined in physics. However, in everyday language, "work" can involve mental or physical energy expended, and is not necessarily connected with displacement, as it is in physics. So a student could say she "worked" hard carrying boxes up the stairs to her dorm room (similar in meaning to the physics usage), or that she "worked" hard on a problem set (different in meaning from the physics usage).

7. No. For instance, imagine $\vec{C}$ as a vector along the $+x$ axis. $\vec{A}$ and $\vec{B}$ could be two vectors with the same magnitude and the same x-component but with y-components in opposite directions, so that one is in quadrant I and the other in quadrant IV. Then $\vec{A} \cdot \vec{C} = \vec{B} \cdot \vec{C}$ even though $\vec{A}$ and $\vec{B}$ are different vectors.

13. The bullet with the smaller mass has a speed which is greater by a factor of $\sqrt{2} \approx 1.4$. Since their kinetic energies are equal, then $\frac{1}{2}m_1v_1^2 = \frac{1}{2}m_2v_2^2$. If $m_2 = 2m_1$, then $\frac{1}{2}m_1v_1^2 = \frac{1}{2} \cdot 2m_1v_2^2$, so $v_1 = \sqrt{2}v_2$. They can both do the same amount of work, however, since their kinetic energies are the same. (See the work-energy principle.)

Solutions to Select End-of-Chapter Problems

1. The force and the displacement are both downwards, so the angle between them is 0°. Use Eq. 7-1.

$$W_G = mgd\cos\theta = (280\,\text{kg})(9.80\,\text{m/s}^2)(2.80\,\text{m})\cos 0° = \boxed{7.7 \times 10^3\,\text{J}}$$

7. Draw a free-body diagram of the car on the incline. The minimum work will occur when the car is moved at a constant velocity. Write Newton's second law in the x direction, noting that the car is unaccelerated. Only the forces parallel to the plane do work.

$$\sum F_x = F_P - mg\sin\theta = 0 \ \rightarrow \ F_P = mg\sin\theta$$

The work done by $\vec{F}_P$ in moving the car a distance d along the plane (parallel to $\vec{F}_P$) is given by Eq. 7-1.

$$W_P = F_P d\cos 0° = mgd\sin\theta = (950\,\text{kg})(9.80\,\text{m/s}^2)(310\,\text{m})\sin 9.0° = \boxed{4.5 \times 10^5\,\text{J}}$$

13. (*a*) The gases exert a force on the jet in the same direction as the displacement of the jet. From the graph we see the displacement of the jet during launch is 85 m. Use Eq. 7-1 to find the work.

$$W_{gas} = F_{gas}d\cos 0° = (130 \times 10^3\,\text{N})(85\,\text{m}) = \boxed{1.1 \times 10^7\,\text{J}}$$

(b) The work done by catapult is the area underneath the graph in Figure 7-22. That area is a trapezoid.

$$W_{\text{catapult}} = \tfrac{1}{2}\left(1100 \times 10^3\,\text{N} + 65 \times 10^3\,\text{N}\right)\left(85\,\text{m}\right) = \boxed{5.0 \times 10^7\,\text{J}}$$

19. We utilize the fact that if $\vec{\mathbf{D}} = D_x\hat{\mathbf{i}} + D_y\hat{\mathbf{j}} + D_z\hat{\mathbf{k}}$, then $-\vec{\mathbf{D}} = \left(-D_x\right)\hat{\mathbf{i}} + \left(-D_y\right)\hat{\mathbf{j}} + \left(-D_z\right)\hat{\mathbf{k}}$.

$$\vec{\mathbf{A}}\cdot\left(-\vec{\mathbf{B}}\right) = A_x\left(-B_x\right) + A_y\left(-B_y\right) + A_z\left(-B_z\right)$$
$$= \left(-A_x\right)\left(B_x\right) + \left(-A_y\right)\left(B_y\right) + \left(-A_z\right)\left(B_z\right) = -\vec{\mathbf{A}}\cdot\vec{\mathbf{B}}$$

25. If $\vec{\mathbf{C}}$ is perpendicular to $\vec{\mathbf{B}}$, then $\vec{\mathbf{C}}\cdot\vec{\mathbf{B}} = 0$. Use this along with the value of $\vec{\mathbf{C}}\cdot\vec{\mathbf{A}}$ to find $\vec{\mathbf{C}}$. We also know that $\vec{\mathbf{C}}$ has no z-component.

$$\vec{\mathbf{C}} = C_x\hat{\mathbf{i}} + C_y\hat{\mathbf{j}} \; ; \; \vec{\mathbf{C}}\cdot\vec{\mathbf{B}} = C_xB_x + C_yB_y = 0 \; ; \; \vec{\mathbf{C}}\cdot\vec{\mathbf{A}} = C_xA_x + C_yA_y = 20.0 \; \rightarrow$$
$$9.6C_x + 6.7C_y = 0 \; ; \; -4.8C_x + 6.8C_y = 20.0$$

This set of two equations in two unknowns can be solved for the components of $\vec{\mathbf{C}}$.

$$9.6C_x + 6.7C_y = 0 \; ; \; -4.8C_x + 6.8C_y = 20.0 \; \rightarrow \; C_x = -1.4 \,, \, C_y = 2.0 \; \rightarrow$$
$$\boxed{\vec{\mathbf{C}} = -1.4\hat{\mathbf{i}} + 2.0\hat{\mathbf{j}}}$$

31. (a) Use the two expressions for dot product, Eqs. 7-2 and 7-4, to find the angle between the two vectors.

$$\vec{\mathbf{A}}\cdot\vec{\mathbf{B}} = AB\cos\theta = A_xB_x + A_yB_y + A_zB_z \; \rightarrow$$
$$\theta = \cos^{-1}\frac{A_xB_x + A_yB_y + A_zB_z}{AB}$$
$$= \cos^{-1}\frac{\left(1.0\right)\left(-1.0\right) + \left(1.0\right)\left(1.0\right) + \left(-2.0\right)\left(2.0\right)}{\left[\left(1.0\right)^2 + \left(1.0\right)^2 + \left(-2.0\right)^2\right]^{1/2}\left[\left(-1.0\right)^2 + \left(1.0\right)^2 + \left(2.0\right)^2\right]^{1/2}}$$
$$= \cos^{-1}\left(-\tfrac{2}{3}\right) = 132° \approx \boxed{130°}$$

(b) The negative sign in the argument of the inverse cosine means that the angle between the two vectors is obtuse.

37. See the graph of force vs. distance. The work done is the area under the graph. It can be found from the formula for a trapezoid.

$$W = \frac{1}{2}\left(12.0\,\text{m} + 4.0\,\text{m}\right)\left(380\,\text{N}\right)$$
$$= 3040\,\text{J} \approx \boxed{3.0 \times 10^3\,\text{J}}$$

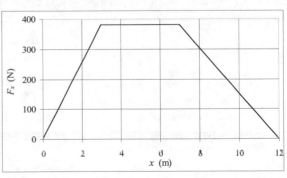

43. Since we are compressing the spring, the force and the displacement are in the same direction.

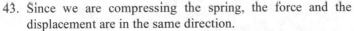
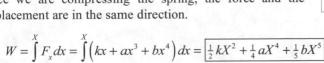

$$W = \int_0^X F_x\,dx = \int_0^X \left(kx + ax^3 + bx^4\right)dx = \boxed{\tfrac{1}{2}kX^2 + \tfrac{1}{4}aX^4 + \tfrac{1}{5}bX^5}$$

49. Let y represent the length of chain hanging over the table, and let λ represent the weight per unit length of the chain. Then the force of gravity (weight) of the hanging chain is $F_G = \lambda y$. As the next small length of chain dy comes over the table edge, gravity does an infinitesimal amount of work on the hanging chain given by the force times the distance, $F_G\, dy = \lambda y\, dy$. To find the total amount of work that gravity does on the chain, integrate that work expression, with the limits of integration representing the amount of chain hanging over the table.

$$W = \int_{y_{\text{initial}}}^{y_{\text{final}}} F_G\, dy = \int_{1.0\,\text{m}}^{3.0\,\text{m}} \lambda y\, dy = \tfrac{1}{2}\lambda y^2 \Big|_{1.0\,\text{m}}^{3.0\,\text{m}} = \tfrac{1}{2}\left(18\,\text{N/m}\right)\left(9.0\,\text{m}^2 - 1.0\,\text{m}^2\right) = \boxed{72\,\text{J}}$$

55. The force of the ball on the glove will be the opposite of the force of the glove on the ball, by Newton's third law. Both objects have the same displacement, and so the work done on the glove is opposite the work done on the ball. The work done on the ball is equal to the change in the kinetic energy of the ball.

$$W_{\text{on ball}} = \left(K_2 - K_1\right)_{\text{ball}} = \tfrac{1}{2}mv_2^2 - \tfrac{1}{2}mv_1^2 = 0 - \tfrac{1}{2}\left(0.145\,\text{kg}\right)\left(32\,\text{m/s}\right)^2 = -74.24\,\text{J}$$

So $W_{\text{on glove}} = 74.24$ J. But $W_{\text{on glove}} = F_{\text{on glove}}\, d \cos 0°$, because the force on the glove is in the same direction as the motion of the glove. Thus 74.24 J $= F_{\text{on glove}}\,(0.25$ m$)$ or $F_{\text{on glove}} = (74.24/0.25)$ N $= \boxed{300\ \text{N}}$, in the direction of the original velocity of the ball.

61. The work done by the net force is the change in kinetic energy.

$$W = \Delta K = \frac{1}{2}mv_2^2 - \frac{1}{2}mv_1^2$$
$$= \frac{1}{2}\left(4.5\,\text{kg}\right)\left[\left(15.0\,\text{m/s}\right)^2 + \left(30.0\,\text{m/s}\right)^2\right] - \frac{1}{2}\left(4.5\,\text{kg}\right)\left[\left(10.0\,\text{m/s}\right)^2 + \left(20.0\,\text{m/s}\right)^2\right] = \boxed{1400\,\text{J}}$$

67. (*a*) In the Earth frame of reference, the ball changes from a speed of v_1 to a speed of $v_1 + v_2$.

$$\Delta K_{\text{Earth}} = \frac{1}{2}m\left(v_1 + v_2\right)^2 - \frac{1}{2}mv_1^2 = \frac{1}{2}m\left(v_1^2 + 2v_1 v_2 + v_2^2\right) - \frac{1}{2}mv_1^2 = mv_1 v_2 + \tfrac{1}{2}mv_2^2 = \boxed{\frac{1}{2}mv_2^2\left(1 + 2\frac{v_1}{v_2}\right)}$$

(*b*) In the train frame of reference, the ball changes from a speed of 0 to a speed of v_2.

$$\Delta K_{\text{train}} = \frac{1}{2}mv_2^2 - 0 = \boxed{\frac{1}{2}mv_2^2}$$

(*c*) The work done is the change of kinetic energy, in each case.

$$W_{\text{Earth}} = \boxed{\frac{1}{2}mv_2^2\left(1 + 2\frac{v_1}{v_2}\right)} \ ; \ W_{\text{train}} = \boxed{\frac{1}{2}mv_2^2}$$

(*d*) The difference can be seen as due to the definition of work as force exerted through a distance. In both cases, the force on the ball is the same, but relative to the Earth, the ball moves further during the throwing process than it does relative to the train. Thus more work is done in the Earth frame of reference. Another way to say it is that kinetic energy is very dependent on reference frame, and so since work is the change in kinetic energy, the amount of work done will be very dependent on reference frame as well.

73. Consider the free-body diagram for the block as it moves up the plane.

(*a*) $K_1 = \dfrac{1}{2}mv_1^2 = \dfrac{1}{2}\left(6.10\,\text{kg}\right)\left(3.25\,\text{m/s}\right)^2 = 32.22\text{ J} \approx \boxed{32.2\,\text{J}}$

(*b*) $W_P = F_P d \cos 37° = \left(75.0\,\text{N}\right)\left(9.25\,\text{m}\right)\cos 37° \approx \boxed{554\,\text{J}}$

(*c*) $W_G = mgd \cos 127° = \left(6.10\,\text{kg}\right)\left(9.80\,\text{m/s}^2\right)\left(9.25\,\text{m}\right)\cos 127° \approx \boxed{-333\,\text{J}}$

(*d*) $W_N = F_N d \cos 90° = \boxed{0\,\text{J}}$

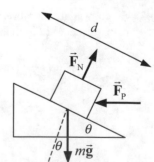

(*e*) Apply the work-energy theorem.

$$W_{\text{total}} = K_2 - K_1 \rightarrow$$

$$KE_2 = W_{\text{total}} + K_1 = W_P + W_G + W_N + K_1 = (554.05 - 332.78 + 0 + 32.22)\,\text{J} \approx \boxed{253\,\text{J}}$$

79. The force is constant, and so we may calculate the force by Eq. 7-3. We may also use that to calculate the angle between the two vectors.

$$W = \vec{\mathbf{F}} \cdot \vec{\mathbf{d}} = \left[\left(10.0\hat{\mathbf{i}} + 9.0\hat{\mathbf{j}} + 12.0\hat{\mathbf{k}} \right)\text{kN} \right] \cdot \left[\left(5.0\hat{\mathbf{i}} + 4.0\hat{\mathbf{j}} \right)\text{m} \right] = \boxed{86\,\text{kJ}}$$

$$F = \left[(10.0)^2 + (9.0)^2 + (12.0)^2 \right]^{1/2}\,\text{kN} = 18.0\,\text{kN} \; ; \; d = \left[(5.0)^2 + (4.0)^2 \right]^{1/2}\,\text{m} = 6.40\,\text{m}$$

$$W = Fd\cos\theta \rightarrow \theta = \cos^{-1}\frac{W}{Fd} = \cos^{-1}\frac{8.6 \times 10^4\,\text{J}}{\left(1.80 \times 10^4\,\text{N}\right)\left(6.40\,\text{m}\right)} = \boxed{42°}$$

85. If the rider is riding at a constant speed, then the positive work input by the rider to the (bicycle + rider) combination must be equal to the negative work done by gravity as he moves up the incline. The net work must be 0 if there is no change in kinetic energy.

(*a*) If the rider's force is directed downwards, then the rider will do an amount of work equal to the force times the distance parallel to the force. The distance parallel to the downward force would be the diameter of the circle in which the pedals move. Then consider that by using 2 feet, the rider does twice that amount of work when the pedals make one complete revolution. So in one revolution of the pedals, the rider does the work calculated below.

$$W_{\text{rider}} = 2\left(0.90\,m_{\text{rider}}\,g \right)d_{\substack{\text{pedal} \\ \text{motion}}}$$

In one revolution of the front sprocket, the rear sprocket will make 42/19 revolutions, and so the back wheel (and the entire bicycle and rider as well) will move a distance of $(42/19)(2\pi r_{\text{wheel}})$. That is a distance along the plane, and so the height that the bicycle and rider will move is $h = (42/19)(2\pi r_{\text{wheel}})\sin\theta$. Finally, the work done by gravity in moving that height is calculated.

$$W_G = \left(m_{\text{rider}} + m_{\text{bike}} \right)gh\cos 180° = -\left(m_{\text{rider}} + m_{\text{bike}} \right)gh = -\left(m_{\text{rider}} + m_{\text{bike}} \right)g\left(42/19 \right)\left(2\pi r_{\text{wheel}} \right)\sin\theta$$

Set the total work equal to 0, and solve for the angle of the incline.

$$W_{\text{rider}} + W_G = 0 \rightarrow 2\left[0.90\,m_{\text{rider}}\,g \right]d_{\substack{\text{pedal} \\ \text{motion}}} - \left(m_{\text{rider}} + m_{\text{bike}} \right)g\left(42/19 \right)\left(2\pi r_{\text{wheel}} \right)\sin\theta = 0 \rightarrow$$

$$\theta = \sin^{-1}\frac{\left(0.90\,m_{\text{rider}} \right)d_{\substack{\text{pedal} \\ \text{motion}}}}{\left(m_{\text{rider}} + m_{\text{bike}} \right)\left(42/19 \right)\left(\pi r_{\text{wheel}} \right)} = \sin^{-1}\frac{0.90(65\,\text{kg})(0.36\,\text{m})}{(77\,\text{kg})(42/19)\pi(0.34\,\text{m})} = \boxed{6.7°}$$

(*b*) If the force is tangential to the pedal motion, then the distance that one foot moves while exerting a force is now half of the circumference of the circle in which the pedals move. The rest of the analysis is the same.

$$W_{\text{rider}} = 2\left(0.90\,m_{\text{rider}}\,g \right)\left(\pi r_{\substack{\text{pedal} \\ \text{motion}}} \right) \; ; \; W_{\text{rider}} + W_G = 0 \rightarrow$$

$$\theta = \sin^{-1}\frac{\left(0.90\,m_{\text{rider}} \right)\pi r_{\substack{\text{pedal} \\ \text{motion}}}}{\left(m_{\text{rider}} + m_{\text{bike}} \right)\left(42/19 \right)\left(\pi r_{\text{wheel}} \right)} = \sin^{-1}\frac{0.90(65\,\text{kg})(0.18\,\text{m})}{(77\,\text{kg})(42/19)(0.34\,\text{m})} = 10.5° \approx \boxed{10°}$$

91. Refer to the free body diagram. The coordinates are defined simply to help analyze the components of the force. At any angle θ, since the mass is not accelerating, we have the following.

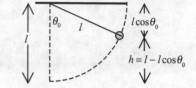

$$\sum F_x = F - mg\sin\theta = 0 \quad\rightarrow\quad F = mg\sin\theta$$

Find the work done in moving the mass from $\theta = 0$ to $\theta = \theta_0$.

$$W_F = \int \vec{\mathbf{F}}\cdot d\vec{\mathbf{s}} = \int_{\theta=0}^{\theta=\theta_0} F\cos\theta\, ld\theta = mgl \int_{\theta=0}^{\theta=\theta_0} \sin\theta\, d\theta =$$

$$= -mgl\cos\theta\Big|_0^{\theta_0} = mgl\left(1-\cos\theta_0\right)$$

See the second diagram to find the height that the mass has risen. We see that

$$h = l - l\cos\theta_0 = l\left(1-\cos\theta_0\right)$$

and so

$$\boxed{W_F = mgl\left(1-\cos\theta_0\right) = mgh}$$

Chapter 8: Conservation of Energy

Chapter Overview and Objectives

In this chapter, the concept of potential energy for conservative forces is defined and the principle of conservation of energy is introduced. The concepts of power, potential energy diagrams, equilibrium, and stability are discussed.

After completing this chapter you should:
- Know the difference between conservative and nonconservative forces.
- Know how to calculate the potential energy function associated with a conservative force.
- Know how to determine the force from the potential energy function.
- Know how to apply the law of conservation of energy to solve problems.
- Know how to apply the concept of mechanical power.
- Know the conditions for mechanical equilibrium and how to determine the stability of the equilibrium.

Summary of Equations

Definition of change in potential energy:	$\Delta U = U_2 - U_1 = -\int_1^2 \vec{\mathbf{F}} \cdot d\vec{\mathbf{l}}$	(Section 8-2)
Gravitational potential energy near the Earth's surface:	$U(y) = mgy$	(Section 8-2)
Potential energy of a spring:	$U(x) = \frac{1}{2} kx^2$	(Section 8-2)
Calculation of the force from the potential energy function:	$\vec{\mathbf{F}} = -\dfrac{\partial U}{\partial x} \hat{\mathbf{i}} - \dfrac{\partial U}{\partial y} \hat{\mathbf{j}} - \dfrac{\partial U}{\partial z} \hat{\mathbf{k}}$	(Section 8-2)
Conservation of mechanical energy:	$\Delta K + \Delta U = 0$	(Section 8-3)
	$K_1 + U_1 = K_2 + U_2$	(Section 8-3)
Definition of total mechanical energy:	$E = K + U$	(Section 8-3)
Conservation of energy:	$\Delta K + \Delta U = W_{NC}$	(Section 8-6)
General gravitational potential energy:	$U(r) = -\dfrac{GMm}{r}$	(Section 8-7)
Definition of instantaneous power:	$P = \dfrac{dW}{dt} = \dfrac{dE}{dt}$	(Section 8-8)
Definition of average power:	$\overline{P} = \dfrac{W}{t} = \dfrac{\Delta E}{t}$	(Section 8-8)
Power due to a force acting on a moving object:	$P = \vec{\mathbf{F}} \cdot \vec{\mathbf{v}}$	(Section 8-8)
Definition of efficiency of a machine:	$e = \dfrac{P_{out}}{P_{in}}$	(Section 8-8)

Condition for equilibrium:	$\dfrac{dU}{dx} = 0$	(Section 8-9)
Condition for stable equilibrium:	$\dfrac{d^2U}{dx^2} > 0$	(Section 8-9)
Condition for unstable equilibrium:	$\dfrac{d^2U}{dx^2} < 0$	(Section 8-9)
Condition for neutral equilibrium:	$\dfrac{d^2U}{dx^2} = 0$	(Section 8-9)

Chapter Summary

Section 8-1. Conservative and Nonconservative Forces

Forces will be categorized into two types:

> *Conservative forces*: The work done by a conservative force depends only on the initial and the final position, and not on the path taken. This is equivalent to stating that the work done by a conservative force on the object moving around any closed path is zero.

> *Nonconservative force*: The work done by a nonconservative force depends not only on the initial and final position, but also on the path taken.

Section 8-2. Potential Energy

Potential energy functions can be defined for conservative forces. The change in the potential energy of a system is defined to be the negative of the work done by the conservative force:

$$\Delta U = U_2 - U_1 = -\int_1^2 \vec{\mathbf{F}} \cdot d\vec{\mathbf{l}}$$

In this expression, ΔU is the change in the potential energy between position 1 and position 2, U_1 and U_2 are the values of the potential energy at position 1 and position 2, respectively, $\vec{\mathbf{F}}$ is the conservative force acting on the object, $d\vec{\mathbf{l}}$ is a infinitesimal displacement along the path followed by the object, and the limits of the integral specify the start and end points of the path. Because only differences in potential energy are physically meaningful, the potential energy is not uniquely defined by the force. The choice of where $U = 0$ J is arbitrary and can be chosen wherever it is most convenient. Any potential energy function can have a constant added to it and it remains a potential energy function for a given force.

Given a potential energy function U we can calculate the corresponding force by calculating the opposite of the gradient of the potential energy:

$$\vec{\mathbf{F}} = -\frac{\partial U}{\partial x}\hat{\mathbf{i}} - \frac{\partial U}{\partial y}\hat{\mathbf{j}} - \frac{\partial U}{\partial z}\hat{\mathbf{k}}$$

The gravitational potential energy associated with the gravitational force near the surface of the Earth is given by

$$U(y) = mgy$$

where m is the mass of the object. This form of the potential energy assumes that the potential energy is zero on the surface of the Earth.

The potential energy of a spring that follows Hooke's Law is given by

$$U\left(x\right) = \frac{1}{2}kx^2$$

where x is the displacement of the end of the spring from its natural, unstretched, length and k is the spring constant of the spring.

Example 8-2-A: Calculating the potential energy. Given a force as a function of position

$$\vec{F}(x) = \left(-\frac{C}{x^2} + B\sin sx\right)\hat{i}$$

Determine the change in the potential energy between x_1 and x_2.

Approach: The force is a function of x. The force is directed along the x axis and will thus be parallel to the displacement $d\vec{l}$ when we move from x_1 to x_2 (assuming that $x_1 < x_2$). We can determine the change in the potential energy between x_1 and x_2 by integrating the force integral.

Solution: The change in the potential energy between x_1 and x_2 can be obtained by integrating $\vec{F} \cdot d\vec{l}$ between x_1 and x_2:

$$\Delta U = -\int_1^2 \vec{F} \cdot d\vec{l} = -\int_1^2 \left(-\frac{C}{x^2} + B\sin sx\right)dx = -\left(\frac{C}{x} - \frac{B}{s}\cos sx\right)\Bigg|_{x_1}^{x_2} = C\left(\frac{1}{x_1} - \frac{1}{x_2}\right) + \frac{B}{s}\left(\cos sx_2 - \cos sx_1\right)$$

Example 8-2-B: Calculating the force. Given a potential energy function $U\left(x,y\right) = Ax^3 + Bxy^2 + Cy^3$, determine the force as a function of position.

Approach: We can use the relation between the force and potential energy to determine the force that will produce the potential energy $U(x, y)$.

Solution: The potential energy U only depends on x and y and the force will only have components along the x and y axes:

$$\vec{F} = -\frac{\partial U}{\partial x}\hat{i} - \frac{\partial U}{\partial y}\hat{j} - \frac{\partial U}{\partial z}\hat{k} = -\frac{\partial U}{\partial x}\hat{i} - \frac{\partial U}{\partial y}\hat{j} = -\left(3Ax^2 + By^2\right)\hat{i} - \left(2Bxy + 3Cy^2\right)\hat{j}$$

Example 8-2-C: Compressing a spring. A spring with a spring constant $k = 120$ N/m has a natural length of $L = 24$ cm. The spring is compressed until the potential energy stored in the spring is $U = 1.8$ J. How long is the spring when it is compressed?

Approach: The potential energy of a compressed spring depends on the compression or elongation of the spring. The problem provides us with information about the potential energy of the spring and the spring constant; using this information we can determine the compression or elongation of the spring. The potential energy of the spring is 0 K when it has its natural length. In order to calculate the length of the spring when it is compressed we must make sure we include this natural length in our calculation.

Solution: The potential energy of a compressed spring is equal to

$$U = \frac{1}{2}kx^2$$

The amount of compression of the spring is thus equal to

$$x = \sqrt{\frac{2U}{k}}$$

Since the spring has a natural length L and since the spring is compressed, we find that its length l must be equal to

$$l = L - x = L - \sqrt{\frac{2U}{k}} = (0.24 \text{ m}) - \sqrt{\frac{2(1.8 \text{ J})}{(120 \text{ N/m})}} = 0.067 \text{ m} = 6.7 \text{ cm}$$

Section 8-3. Mechanical Energy and Its Conservation

If we consider a system with only conservative forces acting, the work-energy theorem can be written in a new form:

$$\Delta K + \Delta U = 0$$

where ΔK is the change in the total kinetic energy of the system and ΔU is the change in the total potential energy of the system. This relationship is an expression of the **principle of conservation of energy**. The relationship is often written as

$$K_1 + U_1 = K_2 + U_2$$

where K_1 and U_1 are the total kinetic and potential energy of the system in configuration 1 and K_2 and U_2 are the total kinetic and potential energy of the system in configuration 2.

We define the total mechanical energy E as the sum of the kinetic and potential energy:

$$E = K + U$$

Section 8-4. Problem Solving Using Conservation of Mechanical Energy

Problems that relate the initial and final positions to the initial and final speeds of an object often can be solved by applying the principle of conservation of mechanical energy. When we use this principle to solve problems, we identify what types of potential energies may be changing during the process and include a term in the total energy for those particular types of potential energy. Remember that the form of conservation of energy above only applies if conservative forces act on the system.

Example 8-4-A: Conservation of mechanical energy for a spring system. A box of mass $m = 2.98$ kg slides along a frictionless surface with a speed of $v = 12.5$ m/s. It moves toward a wall with a spring of spring constant $k = 1{,}280$ N/m as shown in the Figure on the right. What is the compression of the spring when the box comes to rest?

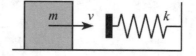

Approach: Since there are no nonconservative forces acting on the box, mechanical energy will be conserved. The initial mechanical energy is entirely in the form of kinetic energy while the final mechanical energy is in the form of potential energy associated with the compression of the spring. Conservation of mechanical energy will allow us to determine the compression of the spring.

Solution: The initial mechanical energy of the system is equal to the kinetic energy of the box:

$$E_i = K_i + U_i = \frac{1}{2} m v^2$$

The final mechanical energy of the system is equal to

$$E_f = K_f + U_f = \frac{1}{2} k x^2$$

where x is the compression of the spring. Applying conservation of mechanical energy tells us that

$$\frac{1}{2} m v^2 = \frac{1}{2} k x^2 \quad \text{or} \quad x = \sqrt{\frac{m}{k} v^2} = \sqrt{\frac{m}{k}} v = \sqrt{\frac{(2.98 \text{ kg})}{(1280 \text{ N/m})}} \times (12.5 \text{ m/s}) = 0.603 \text{ m}$$

Example 8-4-B: Conservation of mechanical energy for a spring system. A pendulum bob of mass $m = 0.47$ kg hangs vertically and rests against a spring compressed a distance of $x = 1.7$ cm. The length of the pendulum from the pivot to the center of mass of the bob is $L = 83$ cm. The spring constant of the spring is $k = 340$ N/m. If the pendulum bob is released, to what maximum angle θ will the pendulum bob swing upward?

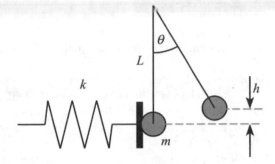

Approach: Since there are no nonconservative forces acting on the system, the mechanical energy will be conserved. The initial mechanical energy is entirely in the form of potential energy, associated with the compression of the spring. The potential energy of the spring is converted into kinetic energy of the bob. The bob will swing upward, thereby converting its kinetic energy into potential energy. At the maximum angle, the mechanical energy of the system will entirely be in the form of gravitational potential energy of the bob. Conservation of mechanical energy will allow us to determine the maximum height of the bob, and thus the maximum angle.

Solution: The initial mechanical energy of the system is equal to the potential energy of the compressed spring:

$$E_i = U_i + K_i = \frac{1}{2}kx^2$$

We have assumed that the initial potential energy of the bob is zero. This is equivalent to saying that we define the gravitational potential energy to be zero at the initial height of the bob. The final mechanical energy of the system is equal to the potential energy of the bob:

$$E_f = U_f + K_f = mgh$$

Since mechanical energy is conserved we can calculate the height of the bob:

$$h = \frac{E_f}{mg} = \frac{1}{2}\frac{kx^2}{mg}$$

Once we know the height h we can calculate the angle θ:

$$\cos\theta = \frac{L-h}{L} = 1 - \frac{h}{L} = 1 - \frac{1}{2}\frac{kx^2}{mgL}$$

or

$$\theta = \text{acos}\left(1 - \frac{1}{2}\frac{kx^2}{mgL}\right) = 9.2°$$

Section 8-5. The Law of Conservation of Energy

Nonconservative forces can be accounted for in the law of conservation of energy. The most common nonconservative force encountered in daily life is friction. The dissipative work done by frictional forces is always negative and acts to decrease the total macroscopic mechanical energy of a system. It has been recognized that the energy does not disappear, but acts to increase the microscopic energy of the system. This microscopic energy cannot be "seen" in the same way we can see that macroscopic objects have kinetic and potential energy. This **internal** or **thermal energy** is associated with the kinetic and potential energy of the individual atoms of the object. We can recognize some changes in the internal energy as changes in phase, such as the change from solid to liquid (melting), or changes in temperature. The generalization of the principle of conservation of energy is:

The total energy in a closed system is neither increased nor decreased in any physical process. The energy can be changed from one form into another, but the total amount of energy remains constant.

The principle of conservation of energy in the presence of conservative and nonconservative forces can be written as $\Delta K + \Delta U = W_{NC}$ where W_{NC} is the total work done by the nonconservative forces.

Example 8-6-A. Work done by the friction force. A car of mass $m = 733$ kg is going down an incline with an angle of inclination $\theta = 4.78°$ and an initial speed $v_i = 26.5$ m/s. To avoid an accident, the car must slow down to $v_f = 18.3$ m/s by the time it has gone another distance $d = 123$ m down the incline. What amount of work must be done on the car by nonconservative forces to slow the car down? What amount of work must be done on the car by nonconservative forces to slow the car down if it is headed up the hill instead of down the hill?

Approach: The problem provides us with the information required to calculate the total work done by all forces acting on the car. The work done by these forces will contain contributions from conservative forces, such as the gravitational force, and nonconservative forces, such as the friction force. The work done by the conservative forces can also be determined by calculating the change in the potential energy associated with these conservative forces.

Solution: As the car slows down, there are changes in its kinetic energy and its gravitational potential energy. In the absence of nonconservative forces, the reduction in potential energy when the car moves down the incline will result in an increase in its kinetic energy. However, in this situation, the kinetic energy actually decreases and there must be some work done by nonconservative forces. The work done by the nonconservative forces can be determined by applying the law of conservation of energy

$$\Delta K + \Delta U = W_{NC}$$

The change in the kinetic energy of the car can be determined from the information provided in the problem:

$$\Delta K = K_f - K_i = \frac{1}{2}mv_f^2 - \frac{1}{2}mv_i^2 = \frac{1}{2}(733 \text{ kg})\left\{(18.3 \text{ m/s})^2 - (26.5 \text{ m/s})^2\right\} = -1.35 \times 10^5 \text{ J}$$

The change in gravitational potential energy of the car can be determined from the known angle of inclination and the distance traveled along the incline:

$$\Delta U = mgy_f - mgy_i = mg\Delta y = mgd \sin\theta = -(733 \text{ kg}) \times (9.80 \text{ m/s}^2) \times (123 \text{ m}) \times \sin(4.78°) = -7.36 \times 10^4 \text{ J}$$

Note that the minus indicates that the potential energy is decreasing ($y_f < y_i$).
Using the calculated changes in the potential and kinetic energy we can now determine the work that must be done by the nonconservative forces acting on the car:

$$W_{NC} = \Delta K + \Delta U = -1.35 \times 10^5 - 7.36 \times 10^4 = -2.09 \times 10^5 \text{ J}$$

If the car is headed uphill, the change in gravitational potential energy becomes positive, and the work that must be done by the nonconservative force acting on the car must be equal to:

$$W_{NC} = \Delta K + \Delta U = -1.35 \times 10^5 + 7.36 \times 10^4 = -6.1 \times 10^4 \text{ J}$$

Section 8-7. Gravitational Potential Energy and Escape Velocity

The gravitational potential energy associated with the universal gravitational force is given by:

$$U(r) = -\frac{GMm}{r}$$

where $U(r)$ is the gravitational potential energy as a function of r, the distance between the two bodies of mass M and m. Note that the potential energy is zero when the distance is infinite. This is the usual position where the potential energy is set to zero when the force increases without bound as the distance decreases.
For a body with a very small mass m interacting with a body with a very large mass M, such as man-made size objects of mass m interacting with the Earth of mass M_E, the change in the kinetic energy of the object of mass M is often negligible. In these circumstances we can write the principle of conservation of energy as:

$$\frac{1}{2}mv^2 - \frac{GMm}{r} = \text{constant}$$

where v is the speed of the object of mass m and r is the distance between the object of mass m and the object of mass M. If the total energy of the system is zero, it is easy to see that the speed of the object will go to zero as the distance between the objects goes to infinity. The speed necessary to obtain this condition at a given separation between the objects is called the **escape speed**:

$$v_{esc} = \sqrt{\frac{2GM}{r}}$$

Example 8-7-A: Applying conservation of energy to study the motion of astronomical objects. An asteroid undergoes a collision far from the Earth that essentially brings it to rest relative to the Earth. Assuming gravitational forces from other bodies can be ignored, what would the speed of the asteroid be when it collides with the Earth? If the mass of the asteroid is $m = 10^6$ kg, what will its kinetic energy be when it reaches the surface of the Earth?

Approach: The problem tells us that right after the collision, the total energy of the asteroid is 0 J (its kinetic energy is 0 J since it is at rest and its potential energy is 0 J since it is far away from the Earth and it is assumed that the gravitational force from other bodies can be ignored). Since the only force that acts on the system is the gravitational force, mechanical energy is conserved and the mechanical energy of the asteroid when it hits the Earth must therefore be equal to 0 J.

Solution: The mechanical energy of the asteroid when it hits the Earth is the sum of its kinetic energy and its potential energy:

$$E = \frac{1}{2}mv^2 - \frac{GM_E m}{r_E}$$

Since the mechanical energy is 0 J when the asteroid is far from the Earth and since mechanical energy is conserved, the mechanical energy when the asteroid hits the Earth will still be 0 J. The previous equation can thus be used to determine the velocity of the asteroid when it hits the Earth:

$$v = \sqrt{\frac{\left(\frac{GM_E m}{r_E}\right)}{\frac{1}{2}m}} = \sqrt{\frac{2GM_E}{r_E}} = \sqrt{\frac{2\left(6.67\times10^{-11}\ \text{Nm}^2/\text{kg}^2\right)\left(5.97\times10^{24}\ \text{kg}\right)}{\left(6.38\times10^6\ \text{m}\right)}} = 1.12\times10^4\ \text{m/s}$$

or, in terms of the more familiar miles per hour

$$v = 1.12\times10^4\ \text{m/s} = \left(1.12\times10^4\ \text{m/s}\right)\left(\frac{1\ \text{mi/h}}{0.447\ \text{m/s}}\right) = 25,000\ \text{mi/h}$$

The kinetic energy of the asteroid is equal to

$$K = \frac{1}{2}mv^2 = \frac{1}{2}m\left(\frac{GM_E}{r_E}\right) = \frac{GM_E m}{2r_E} = 6.24\times10^{13}\ \text{J}$$

This energy can be converted into kilotons of TNT (1 kt is the energy released when 1,000 tons of TNT explodes):

$$K = 6.24\times10^{13}\ \text{J} = \left(6.24\times10^{13}\ \text{J}\right)\left(\frac{1\ \text{kt}}{4.186\times10^{12}\ \text{J}}\right) = 14.5\ \text{kt}$$

The atomic bomb that was dropped on Hiroshima released an energy of 12.5 kt. Clearly the damage that can be caused by the asteroid will be very significant.

Example 8-7-B: Calculating the escape velocity. The mass of the sun is $M_{sun} = 1.99 \times 10^{30}$ kg and its radius is $r_{sun} = 6.96 \times 10^5$ km. What is the escape speed at the surface of the sun?

Approach: The escape velocity can be calculated directly using the expression derived in this Section. Note that the escape velocity does not depend on the mass of the object that is escaping. This is a result of the fact that both the gravitational potential energy and the kinetic energy are proportional to the mass of the escaping object.

Solution: The escape speed is given by

$$v_{esc} = \sqrt{\frac{2GM_{sun}}{r_{sun}}} = \sqrt{\frac{2\left(6.67 \times 10^{-11} \text{ Nm}^2/\text{kg}^2\right)\left(1.99 \times 10^{30} \text{ kg}\right)}{\left(6.96 \times 10^8 \text{ m}\right)}} = 6.18 \times 10^5 \text{ m/s}$$

Section 8-8. Power

Instantaneous power is defined as the rate at which work is done:

$$P = \frac{dW}{dt}$$

or, more generally, the rate at which energy is transformed from one form into another:

$$P = \frac{dE}{dt}$$

The time average of the power is:

$$\bar{P} = \frac{W}{t} = \frac{\Delta E}{t}$$

The SI unit of power is the **watt** (W). One watt is equal to one Joule per second. A common unit of power used from the British unit system is the **horsepower**. One horsepower is equal to 746 watts.

A force $\vec{F}$ acting on an object with velocity $\vec{v}$ generates a power:

$$P = \vec{F} \cdot \vec{v}$$

The **efficiency** of a machine is usually defined as the ratio of the useful power output and the power input:

$$e = \frac{P_{out}}{P_{in}}$$

Example 8-8-A: Power required to lift bags. A motor runs a conveyor system that raises bags of rice that each weigh $W = 50.0$ pounds. The bags are raised a distance $d = 10.0$ ft per minute. What average mechanical power is required from the motor to lift each bag of rice?

Approach: The problem provides us with information that allows us to calculate the work that is done on each bag during a time interval of 1 minute. In order to calculate the work that must be done by the motor we will assume that the bags are raised with constant velocity (the net force on the bags is zero). We must be very careful to use the correct units in this problem and should start with converting all units to the SI system.

Solution: Since the net force on each bag must be zero, the magnitude of the force applied by the motor on the bag must be mg (the weight of each bag). The direction of the force applied by the motor is pointing in the same direction as the displacement of the bags, and the work done by this force during 1 minute is thus equal to mgd. To calculate the work done, we need to convert the variables provided in the problem to the proper SI units:

$$d = 10 \text{ ft} = \left(10 \text{ ft}\right)\left(\frac{1 \text{ m}}{3.28 \text{ ft}}\right) = 3.05 \text{ m}$$

and

$$mg = 50 \text{ lb} = \left(50 \text{ lb}\right)\left(\frac{1 \text{ N}}{0.225 \text{ lb}}\right) = 222 \text{ N}$$

The power required to lift one bag is thus equal to

$$P = \frac{W}{\Delta t} = \frac{mgd}{\Delta t} = \frac{(222 \text{ N})(3.05 \text{ m})}{(60 \text{ s})} = 11.3 \text{ W}$$

Example 8-8-B: Power and friction. A car has a maximum speed $v_{max} = 94.2$ mph. It has an engine that can provide a maximum power $P_{max} = 235$ hp to the drive system. What is the magnitude of the total friction force acting against the vehicle's motion?

Approach: The problem tells us that the car has a maximum speed. This implies that even though the engine does work on the car, there is no change in its kinetic energy. The work done by the engine must thus be used entirely to cancel the work done by the friction forces that act on the car.

Solution: Consider the motion of the car during a time interval Δt. During this interval the car travels a distance d where

$$d = v_{max} \Delta t$$

The work done by the engine during this period is equal to

$$W_{engine} = P_{max} \Delta t$$

Since the total work done on the car must be zero, the work done by the friction force must be opposite to the work done by the engine:

$$W_{friction} = -P_{max} \Delta t$$

We will assume that the friction force is directed in a direction opposite to the displacement of the car. The work done by the friction force during the time interval Δt is thus equal to

$$W_{friction} = \vec{\mathbf{F}}_{friction} \cdot \vec{\mathbf{d}} = -F_{friction} d = -F_{friction} v_{max} \Delta t$$

The friction force is thus equal to

$$F_{friction} = -\frac{W_{friction}}{v_{max} \Delta t} = \frac{P_{max} \Delta t}{v_{max} \Delta t} = \frac{P_{max}}{v_{max}} = \frac{(235 \text{ hp})\left(\dfrac{746 \text{ W}}{1 \text{ hp}}\right)}{(94.2 \text{ mph})\left(\dfrac{0.447 \text{ m/s}}{1 \text{ mph}}\right)} = 4.16 \times 10^3 \text{ N}$$

Example 8-8-C: Efficiency. Determine the efficiency of the conveyor system in Example 8-8-A if the power used by the motor that runs the system is one-half horsepower. Assume that the system raises 12 bags of rice per minute.

Approach: In problem 8-8-A we calculated the power required to lift one bag of rice. This information can be used to determine the useful power generated by the system. Since the problem provides us with information about the power required to run the system we can determine its efficiency.

Solution: The useful power generated by the conveyor system is equal to

$$P_{out} = 12 \times 11.3 = 136 \text{ W}$$

The efficiency can be easily calculated by taking the ratio of the useful power output to power input:

$$e = \frac{P_{out}}{P_{in}} = \frac{136 \text{ W}}{(0.5 \text{ hp}) \times \left(\dfrac{746 \text{ W}}{1 \text{ hp}}\right)} = 0.365 = 36.5\%$$

Section 8-9. Potential Energy Diagrams; Stable and Unstable Equilibrium

A graph of the potential energy as a function of position is useful in determining the qualitative nature of the motion of a system. Consider the diagram of the potential energy U of a system as a function of position shown in the Figure on the right.

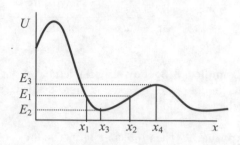

If the system has the energy E_1, the system can evolve between positions x_1 and x_2. We say that the motion of the system is bound. Any position at which dU/dx is zero is an **equilibrium** position. If the system has zero kinetic energy when located at the equilibrium position it will not move. If the system has an energy exactly equal to E_2, no motion will occur because $F = -dU/dx$ is zero at x_3. If the system has an energy exactly equal to E_3 and the object is located at x_4, no motion will occur because $F = -dU/dx$ is zero at x_4. However, there is something very different about the two equilibrium positions at x_3 and x_4. The equilibrium position at x_3 is a **stable equilibrium**, while the equilibrium position at x_4 is an **unstable equilibrium**. If the system is disturbed slightly from equilibrium at position x_3, the forces act in a direction to return the system toward the equilibrium position. If the system at x_4 is disturbed slightly from equilibrium, the forces act in a direction to move the system away from equilibrium. A mathematical test for stability of equilibrium is to determine the sign of the second derivative of the potential energy function. If the second derivative d^2U/dx^2 is positive, the equilibrium is stable. If the second derivative is negative, the equilibrium is unstable. If the second derivative is zero, the equilibrium is neutral.

Example 8-9-A: Determining equilibrium positions. Consider the following potential energy distribution $U(x)$

$$U(x) = Ax^3 - Bx$$

where A and B are both positive numbers and x is the position. Determine the equilibrium positions and their stability (stable, unstable, or neutral).

Approach: In order to determine the location of the equilibrium positions we determine where $dU/dx = 0$. The type of equilibrium can be determined by evaluating d^2U/dx^2 at those locations.

Solution: We start the solution by determining at which position $dU/dx = 0$:

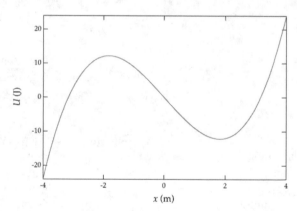

$$\frac{dU}{dx} = \frac{d}{dx}\left(Ax^3 - Bx\right) = 3Ax^2 - B = 0$$

The equilibrium positions are thus

$$x = \pm\sqrt{\frac{B}{3A}}$$

To determine the nature of the equilibrium we calculate d^2U/dx^2:

$$\frac{d^2U}{dx^2} = \frac{d}{dx}\left(3Ax^2 - B\right) = 6Ax$$

We see that $d^2U/dx^2 < 0$ if $x < 0$ and $d^2U/dx^2 > 0$ if $x > 0$. The nature of the equilibrium positions at $x = \pm\sqrt{\dfrac{B}{3A}}$ is thus as follows:

$$x = +\sqrt{\frac{B}{3A}} : \qquad \text{stable equilibrium}$$

$$x = -\sqrt{\frac{B}{3A}} : \qquad \text{unstable equilibrium}$$

An example of the potential energy as function of x is shown in the graph above for $A = 1$ J/m^2 and $B = 10$ J/m. You can clearly see that the equilibrium point at $x < 0$ m is a point of unstable equilibrium while the equilibrium point at $x > 0$ m is a point of stable equilibrium.

Practice Quiz

1. What evidence is there that nonconservative forces act on an automobile while it is in motion? Identify as many of the nonconservative forces as you can.

2. Is the change in gravitational potential energy when you walk up a hill larger or smaller If one walks directly up the steep side of a hill compared to when one walks around the hill on a gently sloping path?

3. If the potential energy and spring constant of a spring are known, does that imply that the displacement of the end of the spring from its equilibrium position is known?

4. What is the form of the potential energy for a force that is constant in magnitude and direction?

5. If energy is always conserved, what do you think environmentally conscious people mean when they remind us to conserve energy when possible?

6. Consider frictional forces that do not depend on speed. Car A travels from point X to point Y at a constant speed v. Car B travels from point X to point Y at a constant speed $2v$. Does the fact that power is proportional to velocity imply that more energy is used by car B than by car A?

7. In our discussion of the conditions for equilibrium and stability we have only considered one-dimensional systems. What do you expect the conditions for equilibrium to be in a three-dimensional system? What do you expect the conditions for stability to be in a three-dimensional system?

8. The conditions for stable equilibrium were discussed in this chapter, but nothing was stated concerning why some equilibria are more stable than others. A house of cards must be stable if it is to stand, but why does it collapse much more readily than a textbook standing on edge? In other words, what determines how stable a configuration is?

9. There have often been claims of perpetual motion, machines that effectively provide greater output power than input power. Explain how a machine of this type violates the principle of conservation of energy.

10. If the shape of a potential well can be described by a parabola, that is $V(x)$ is proportional to x^2, what can you conclude about the shape of the force that is responsible for this potential well?

11. Einstein's general theory of relativity includes a description of how light is affected by gravity. Suppose we simply assume that light cannot leave a place where the escape speed is greater than or equal to the speed of light, 3.00×10^8 m/s. If a body with the mass of the Earth had this escape speed, what would the radius of that body be?

12. The force exerted by a nonlinear spring is given by $F(x) = -kx - qx^3$, where x is the displacement of the end of the spring from its natural length. Determine the potential energy of the spring as a function of x.

13. A ball of mass 0.0343 kg rests atop a spring that is compressed 3.67 cm. The spring has a spring constant 27.34 N/m. When the spring is released, how high will the ball be projected into the air?

14. An object with a mass of 4.76 kg slides down a 30.0° slope. It starts with a speed of 2.54 m/s and after sliding a distance of 3.29 m down the slope the object has a speed of 4.84 m/s. What is the magnitude of the frictional force acting on the object?

15. Determine the equilibrium positions and the stability of those equilibrium positions for a potential
    $U(x) = 2x + 3x^2 - x^3$.

Responses to Select End-of-Chapter Questions

1. Friction is not conservative; it dissipates energy in the form of heat, sound, and light. Air resistance is not conservative; it dissipates energy in the form of heat and the kinetic energy of fluids. "Human" forces, for example, the forces produced by your muscles, are also not conservative. They dissipate energy in the form of heat and also through chemical processes.

7. At the top of the pendulum's swing, all of its energy is gravitational potential energy; at the bottom of the swing, all of the energy is kinetic.
(*a*) If we can ignore friction, then energy is transformed back and forth between potential and kinetic as the pendulum swings.
(*b*) If friction is present, then during each swing energy is lost to friction at the pivot point and also to air resistance. During each swing, the kinetic energy and the potential energy decrease, and the pendulum's amplitude decreases. When a grandfather clock is wound up, the energy lost to friction and air resistance is replaced by energy stored as potential energy (either elastic or gravitational, depending on the clock mechanism).

13. (*a*) As a car accelerates uniformly from rest, the potential energy stored in the fuel is converted into kinetic energy in the engine and transmitted through the transmission into the turning of the wheels, which causes the car to accelerate (if friction is present between the road and the tires).
(*b*) If there is a friction force present between the road and the tires, then when the wheels turn, the car moves forward and gains kinetic energy. If the static friction force is large enough, then the point of contact between the tire and the road is instantaneously at rest – it serves as an instantaneous axis of rotation. If the static friction force is not large enough, the tire will begin to slip, or skid, and the wheel will turn without the car moving forward as fast. If the static friction force is very small, the wheel may spin without moving the car forward at all, and the car will not gain any kinetic energy (except the kinetic energy of the spinning tires).

19. When the ball is released, its potential energy will be converted into kinetic energy and then back into potential energy as the ball swings. If the ball is not pushed, it will lose a little energy to friction and air resistance, and so will return almost to the initial position, but will not hit the instructor. If the ball is pushed, it will have an initial kinetic energy, and will, when it returns, still have some kinetic energy when it reaches the initial position, so it will hit the instructor in the nose. (Ouch!)

25. When you climb a mountain by going straight up, the force needed is large (and the distance traveled is small), and the power needed (work per unit time) is also large. If you take a zigzag trail, you will use a smaller force (over a longer distance, so that the work done is the same) and less power, since the time to climb the mountain will be longer. A smaller force and smaller power output make the climb seem easier.

Solutions to Select End-of-Chapter Problems

1. The potential energy of the spring is given by $U_{el} = \frac{1}{2}kx^2$ where x is the distance of stretching or compressing of the spring from its natural length.

$$x = \sqrt{\frac{2U_{el}}{k}} = \sqrt{\frac{2(35.0\,\text{J})}{82.0\,\text{N/m}}} = \boxed{0.924\,\text{m}}$$

7. (*a*) This force is conservative, because the work done by the force on an object moving from an initial position x_1 to a final position x_2 depends only on the endpoints.

$$W = \int_{x_1}^{x_2} \vec{\mathbf{F}} \cdot d\vec{\mathbf{l}} = \int_{x_1}^{x_2} F_x\,dx = \int_{x_1}^{x_2} \left(-kx + ax^3 + bx^4\right)dx = \left(-\frac{1}{2}kx^2 + \frac{1}{4}ax^4 + \frac{1}{5}bx^5\right)\Big|_{x_1}^{x_2}$$

$$= \left(-\frac{1}{2}kx_2^2 + \frac{1}{4}ax_2^4 + \frac{1}{5}bx_2^5\right) - \left(-\frac{1}{2}kx_1^2 + \frac{1}{4}ax_1^4 + \frac{1}{5}bx_1^5\right)$$

The expression for the work only depends on the endpoints.

(b) Since the force is conservative, there is a potential energy function U such that $F_x = -\partial U/\partial x$.

$$F_x = \left(-kx + ax^3 + bx^4\right) = -\frac{\partial U}{\partial x} \quad \rightarrow \quad \boxed{U(x) = \frac{1}{2}kx^2 - \frac{1}{4}ax^4 - \frac{1}{5}bx^5 + C}$$

13. We assume that all the forces on the jumper are conservative, so that the mechanical energy of the jumper is conserved. Subscript 1 represents the jumper at the bottom of the jump, and subscript 2 represents the jumper at the top of the jump. Call the ground the zero location for gravitational potential energy ($y = 0$). We have $y_1 = 0$, $v_2 = 0.70$ m/s, and $y_2 = 2.10$ m. Solve for v_1, the speed at the bottom.

$$\frac{1}{2}mv_1^2 + mgy_1 = \frac{1}{2}mv_2^2 + mgy_2 \quad \rightarrow \quad \frac{1}{2}mv_1^2 + 0 = \frac{1}{2}mv_2^2 + mgy_2 \quad \rightarrow$$

$$v_1 = \sqrt{v_2^2 + 2gy_2} = \sqrt{(0.70\,\text{m/s})^2 + 2(9.80\,\text{m/s}^2)(2.10\text{ m})} = 6.454\,\text{m/s} \approx \boxed{6.5\,\text{m/s}}$$

19. Use conservation of energy. The level of the ball on the uncompressed spring is taken as the zero location for both gravitational potential energy ($y = 0$) and elastic potential energy ($x = 0$). It is diagram 2 in the figure. Take "up" to be positive for both x and y.

(a) Subscript 1 represents the ball at the launch point, and subscript 2 represents the ball at the location where it just leaves the spring, at the uncompressed length. We have $v_1 = 0$, $x_1 = y_1 = -0.160$ m, and $x_2 = y_2 = 0$. Solve for v_2.

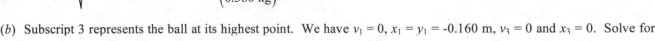

$$E_1 = E_2 \quad \rightarrow \quad \frac{1}{2}mv_1^2 + mgy_1 + \frac{1}{2}kx_1^2 = \frac{1}{2}mv_2^2 + mgy_2 + \frac{1}{2}kx_2^2 \quad \rightarrow$$

$$0 + mgy_1 + \frac{1}{2}kx_1^2 = \frac{1}{2}mv_2^2 + 0 + 0 \quad \rightarrow \quad v_2 = \sqrt{\frac{kx_1^2 + 2mgy_1}{m}}$$

$$v_2 = \sqrt{\frac{(875\,\text{N/m})(0.160\text{ m})^2 + 2(0.380\text{ kg})(9.80\,\text{m/s}^2)(-0.160\text{ m})}{(0.380\text{ kg})}} = \boxed{7.47\,\text{m/s}}$$

(b) Subscript 3 represents the ball at its highest point. We have $v_1 = 0$, $x_1 = y_1 = -0.160$ m, $v_3 = 0$ and $x_3 = 0$. Solve for y_3.

$$E_1 = E_3 \quad \rightarrow \quad \frac{1}{2}mv_1^2 + mgy_1 + \frac{1}{2}kx_1^2 = \frac{1}{2}mv_3^2 + mgy_3 + \frac{1}{2}kx_3^2 \quad \rightarrow$$

$$0 + mgy_1 + \frac{1}{2}kx_1^2 = 0 + mgy_2 + 0 \quad \rightarrow \quad y_2 - y_1 = \frac{kx_1^2}{2mg} = \frac{(875\,\text{N/m})(0.160\text{ m})^2}{2(0.380\text{ kg})(9.80\,\text{m/s}^2)} = \boxed{3.01\text{m}}$$

25. Since there are no dissipative forces in the problem, the mechanical energy of the pendulum bob is conserved. Subscript 1 represents the bob at the release point, and subscript 2 represents the ball at some subsequent position. The lowest point in the swing of the pendulum is the zero location for potential energy ($y = 0$). We have $v_1 = 0$ and $y_1 = l(1 - \cos\theta)$. The "second" point for the energy conservation will vary from part to part of the problem.

(a) The second point is at the bottom of the swing, so $y_2 = 0$.

$$E_1 = E_2 \quad \rightarrow \quad \frac{1}{2}mv_1^2 + mgy_1 = \frac{1}{2}mv_2^2 + mgy_2 \quad \rightarrow \quad mgl(1 - \cos 30.0°) = \frac{1}{2}mv_2^2 \quad \rightarrow$$

$$v_2 = \sqrt{2gl(1 - \cos 30.0°)} = \sqrt{2(9.80\,\text{m/s}^2)(2.00\text{ m})(1 - \cos 30.0°)} = \boxed{2.29\,\text{m/s}}$$

(b)   The second point is displaced from equilibrium by 15.0°, so $y_2 = l\,(1 - \cos 15.0°)$.

$$E_1 = E_2 \;\; \rightarrow \;\; \frac{1}{2}mv_1^2 + mgy_1 = \frac{1}{2}mv_2^2 + mgy_2 \;\; \rightarrow$$

$$mgl\left(1 - \cos 30.0°\right) = \frac{1}{2}mv_2^2 + mgl\left(1 - \cos 15.0°\right) \;\; \rightarrow$$

$$v_2 = \sqrt{2gl\left(\cos 15.0° - \cos 30.0°\right)} = \sqrt{2\left(9.80\,\text{m/s}^2\right)\left(2.00\,\text{m}\right)\left(\cos 15.0° - \cos 30.0°\right)} = \boxed{1.98\,\text{m/s}}$$

(c)   The second point is displaced from equilibrium by -15.0°. The pendulum bob is at the same height at -15.0° as it was at 15.0°, and so the speed is the same. Also, since $\cos(-\theta) = \cos(\theta)$ the mathematics is identical. Thus $v_2 = \boxed{1.98\text{ m/s}}$.

(d)   The tension always pulls radially on the pendulum bob, and so is related to the centripetal force on the bob. The net centripetal force is always mv^2/r. Consider the free body diagram for the pendulum bob at each position.

$(a)\;\; F_{\text{T}} - mg = \dfrac{mv^2}{r} \;\; \rightarrow \;\; F_{\text{T}} = m\left(g + \dfrac{v^2}{l}\right) = m\left(g + \dfrac{2gl\left(1 - \cos 30.0°\right)}{l}\right)$

$$= mg\left(3 - 2\cos 30.0°\right) = \left(0.0700\,\text{kg}\right)\left(9.80\,\text{m/s}^2\right)\left(3 - 2\cos 30.0°\right) = \boxed{0.870\,\text{N}}$$

$(b)\;\; F_{\text{T}} - mg\cos\theta = \dfrac{mv^2}{r} \;\; \rightarrow \;\; F_{\text{T}} = m\left(g\cos\theta + \dfrac{v^2}{l}\right)$

$$= m\left(g\cos 15.0° + \dfrac{2gl\left(\cos 15.0° - \cos 30.0°\right)}{l}\right) =$$

$$= mg\left(3\cos 15.0° - 2\cos 30.0°\right) =$$

$$= \left(0.0700\,\text{kg}\right)\left(9.80\,\text{m/s}^2\right)\left(3\cos 15.0° - 2\cos 30.0°\right) = \boxed{0.800\,\text{N}}$$

(c)   Again, as earlier, since the cosine and the speed are the same for -15.0° as for 15.0°, the tension will be the same, $\boxed{0.800\text{ N}}$.

(e)   Again use conservation of energy, but now we have $v_1 = v_0 = 1.20$ m/s.
(a)   The second point is at the bottom of the swing, so $y_2 = 0$.

$$\frac{1}{2}mv_1^2 + mgl\left(1 - \cos 30.0°\right) = \frac{1}{2}mv_2^2 \;\; \rightarrow$$

$$v_2 = \sqrt{v_1^2 + 2gl\left(1 - \cos 30.0°\right)} = \sqrt{\left(1.20\,\text{m/s}\right)^2 + 2\left(9.80\,\text{m/s}^2\right)\left(2.00\,\text{m}\right)\left(1 - \cos 30.0°\right)} = \boxed{2.59\,\text{m/s}}$$

(b)   The second point is displaced from equilibrium by 15.0°, so $y_2 = l\,(1 - \cos 15.0°)$.

$$\frac{1}{2}mv_1^2 + mgl\left(1 - \cos 30.0°\right) = \frac{1}{2}mv_2^2 + mgl\left(1 - \cos 15.0°\right) \;\; \rightarrow$$

$$v_2 = \sqrt{v_1^2 + 2gl\left(\cos 15.0° - \cos 30.0°\right)}$$

$$= \sqrt{\left(1.20\,\text{m/s}\right)^2 + 2\left(9.80\,\text{m/s}^2\right)\left(2.00\,\text{m}\right)\left(\cos 15.0° - \cos 30.0°\right)} = \boxed{2.31\,\text{m/s}}$$

(c)   As before, the pendulum bob is at the same height at -15.0° as it was at 15.0°, and so the speed is the same. Thus $v_2 = \boxed{2.31\text{ m/s}}$.

31. (*a*) See the free-body diagram for the ski. Write Newton's second law for forces perpendicular to the direction of motion, noting that there is no acceleration perpendicular to the plane.

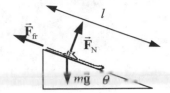

$$\sum F_\perp = F_N - mg\cos\theta \;\rightarrow$$

$$F_N = mg\cos\theta \;\rightarrow$$

$$F_{fr} = \mu_k F_N = \mu_k mg\cos\theta$$

Now use conservation of energy, including the nonconservative friction force. Subscript 1 represents the ski at the top of the slope, and subscript 2 represents the ski at the bottom of the slope. The location of the ski at the bottom of the incline is the zero location for gravitational potential energy ($y = 0$). We have $v_1 = 0$, $y_1 = l\sin\theta$, and $y_2 = 0$. Write the conservation of energy condition, and solve for the final speed. Note that $F_{fr} = \mu_k F_N = \mu_k mg\cos\theta$.

$$\frac{1}{2}mv_1^2 + mgy_1 = \frac{1}{2}mv_2^2 + mgy_2 + F_{fr}l \;\rightarrow$$

$$mgl\sin\theta = \frac{1}{2}mv_2^2 + \mu_k mgl\cos\theta \;\rightarrow$$

$$v_2 = \sqrt{2gl\left(\sin\theta - \mu_k\cos\theta\right)} = \sqrt{2\left(9.80\,\text{m/s}^2\right)\left(85\ \text{m}\right)\left(\sin 28^\circ - 0.090\cos 28^\circ\right)} = 25.49\,\text{m/s} \approx \boxed{25\,\text{m/s}}$$

(*b*) Now, on the level ground, $F_{fr} = \mu_k mg$, and there is no change in potential energy. We again use conservation of energy, including the nonconservative friction force, to relate position 2 with position 3. Subscript 3 represents the ski at the end of the travel on the level, having traveled a distance l_3 on the level. We have $v_2 = 25.49$ m/s, $y_2 = 0$, $v_3 = 0$, and $y_3 = 0$.

$$\frac{1}{2}mv_2^2 + mgy_2 = \frac{1}{2}mv_3^2 + mgy_3 + F_{fr}l_3 \;\rightarrow$$

$$\frac{1}{2}mv_2^2 = \mu_k mgl_3 \;\rightarrow$$

$$l_3 = \frac{v_2^2}{2g\mu_k} = \frac{\left(25.49\,\text{m/s}\right)^2}{2\left(9.80\,\text{m/s}^2\right)\left(0.090\right)} = 368.3\,\text{m} \approx \boxed{370\,\text{m}}$$

37. Use conservation of energy, including the nonconservative frictional force, as developed in Eq. 8-15. The block is on a level surface, so there is no gravitational potential energy change to consider. The frictional force is given by $F_{fr} = \mu_k F_N = \mu_k mg$ since the normal force is equal to the weight. Subscript 1 represents the block at the compressed location, and subscript 2 represents the block at the maximum stretched position. The location of the block when the spring is neither stretched nor compressed is the zero location for elastic potential energy ($x = 0$). Take right to be the positive direction. We have $v_1 = 0$, $x_1 = -0.050$ m, $v_2 = 0$, and $x_2 = 0.023$ m.

$$E_1 = E_2 + F_{fr}l \;\rightarrow$$

$$\frac{1}{2}mv_1^2 + \frac{1}{2}kx_1^2 = \frac{1}{2}mv_2^2 + \frac{1}{2}kx_2^2 + F_{fr}\left(x_2 - x_1\right) \;\rightarrow$$

$$\frac{1}{2}kx_1^2 = \frac{1}{2}kx_2^2 + \mu_k mg\left(x_2 - x_1\right) \;\rightarrow$$

$$\mu_k = \frac{k\left(x_1^2 - x_2^2\right)}{2mg\left(x_2 - x_1\right)} = \frac{-k\left(x_2 + x_1\right)}{2mg} = \frac{-\left(180\,\text{N/m}\right)\left[\left(-0.050\text{m}\right) + \left(0.023\text{m}\right)\right]}{2\left(0.620\ \text{kg}\right)\left(9.80\,\text{m/s}^2\right)} = \boxed{0.40}$$

43. Because friction does work, Eq. 8-15 applies.

    (a) The spring is initially uncompressed, so $x_0 = 0$. The block is stopped at the maximum compression, so $v_f = 0$.

    $$\Delta K + \Delta U + F_{fr} l = \frac{1}{2} m \left(v_f^2 - v_0^2 \right) + \frac{1}{2} k \left(x_f^2 - x_0^2 \right) + m g \mu_k \left(x_f - x_0 \right) = 0 \quad \rightarrow$$

    $$\frac{1}{2} k x_f^2 + m g \mu_k x_f - \frac{1}{2} m v_0^2 = 0 \quad \rightarrow$$

    $$x_f = \frac{-m g \mu_k \pm \sqrt{\left(m g \mu_k \right)^2 - 4 \left(\frac{1}{2} k \right) \left(-\frac{1}{2} m v_0^2 \right)}}{2 \left(\frac{1}{2} k \right)} = \frac{-m g \mu_k \pm \sqrt{\left(m g \mu_k \right)^2 + k m v_0^2}}{k}$$

    $$= \frac{m g \mu_k}{k} \left(-1 \pm \sqrt{1 + \frac{k m v_0^2}{\left(m g \mu_k \right)^2}} \right)$$

    $$= \frac{\left(2.0 \, \text{kg} \right) \left(9.80 \, \text{m/s}^2 \right) \left(0.30 \right)}{\left(120 \, \text{N/m} \right)} \left(-1 \pm \sqrt{1 + \frac{\left(120 \, \text{N/m} \right) \left(2.0 \, \text{kg} \right) \left(1.3 \, \text{m/s} \right)^2}{\left(2.0 \, \text{kg} \right)^2 \left(9.80 \, \text{m/s}^2 \right)^2 \left(0.30 \right)^2}} \right)$$

    $$= 0.1258 \, \text{m} \approx \boxed{0.13 \, \text{m}}$$

    (b) To remain at the compressed position with the minimum coefficient of static friction, the magnitude of the force exerted by the spring must be the same as the magnitude of the maximum force of static friction.

    $$k x_f = \mu_s m g \quad \rightarrow \quad \mu_s = \frac{k x_f}{m g} = \frac{\left(120 \, \text{N/m} \right) \left(0.1258 \, \text{m} \right)}{\left(2.0 \, \text{kg} \right) \left(9.80 \, \text{m/s}^2 \right)} = 0.7702 \approx \boxed{0.77}$$

    (c) If static friction is not large enough to hold the block in place, the spring will push the block back towards the equilibrium position. The block will detach from the decompressing spring at the equilibrium position because at that point the spring will begin to slow down while the block continues moving. Use Eq. 8-15 to relate the block at the maximum compression position to the equilibrium position. The block is initially at rest, so $v_0 = 0$. The spring is relaxed at the equilibrium position, so $x_f = 0$.

    $$\Delta K + \Delta U + F_{fr} l = \frac{1}{2} m \left(v_f^2 - v_0^2 \right) + \frac{1}{2} k \left(x_f^2 - x_0^2 \right) + m g \mu_k \left(x_f - x_0 \right) = 0 \quad \rightarrow$$

    $$\frac{1}{2} m v_f^2 - \frac{1}{2} k x_0^2 + m g \mu_k x_0 = 0 \quad \rightarrow$$

    $$v_f = \sqrt{\frac{k}{m} x_0^2 - 2 g \mu_k x_0} = \sqrt{\frac{\left(120 \, \text{N/m} \right)}{\left(2.0 \, \text{kg} \right)} \left(0.1258 \, \text{m} \right)^2 - 2 \left(9.80 \, \text{m/s}^2 \right) \left(0.30 \right) \left(0.1258 \, \text{m} \right)}$$

    $$= 0.458 \, \text{m/s} \approx \boxed{0.5 \, \text{m/s}}$$

49. The escape velocity for an object located a distance r from a mass M is given by Eq. 8-19, $v_{esc} = \sqrt{\dfrac{2 M G}{r}}$. The orbit speed for an object located a distance r from a mass M is $v_{orb} = \sqrt{\dfrac{M G}{r}}$.

    (a) $v_{\substack{esc \, at \\ Sun's \\ surface}} = \sqrt{\dfrac{2 M_{Sun} G}{r_{Sun}}} = \sqrt{\dfrac{2 \left(2.0 \times 10^{30} \, \text{kg} \right) \left(6.67 \times 10^{-11} \, \text{N} \bullet \text{m}^2 / \text{kg}^2 \right)}{7.0 \times 10^8 \, \text{m}}} = \boxed{6.2 \times 10^5 \, \text{m/s}}$

(b) $v_{\substack{\text{esc at} \\ \text{Earth} \\ \text{orbit}}} = \sqrt{\dfrac{2M_{\text{Sun}}G}{r_{\text{Earth orbit}}}} = \sqrt{\dfrac{2(2.0\times10^{30}\,\text{kg})(6.67\times10^{-11}\,\text{N}\bullet\text{m}^2/\text{kg}^2)}{1.50\times10^{-11}\,\text{m}}} = \boxed{4.2\times10^4\,\text{m/s}}$

$\dfrac{v_{\substack{\text{esc at} \\ \text{Earth} \\ \text{orbit}}}}{v_{\substack{\text{Earth} \\ \text{orbit}}}} = \dfrac{\sqrt{\dfrac{2M_{\text{Sun}}G}{r_{\text{Earth orbit}}}}}{\sqrt{\dfrac{M_{\text{Sun}}G}{r_{\text{Earth orbit}}}}} = \sqrt{2} \quad\to\quad \boxed{v_{\substack{\text{esc at} \\ \text{Earth} \\ \text{orbit}}} = \sqrt{2}\,v_{\substack{\text{Earth} \\ \text{orbit}}}}$

Since $v_{\substack{\text{esc at} \\ \text{Earth} \\ \text{orbit}}} \approx 1.4 v_{\substack{\text{Earth} \\ \text{orbit}}}$, the orbiting object will not escape the orbit.

55. (a) From Eq. 8-19, the escape velocity at a distance $r \geq r_{\text{E}}$ from the center of the Earth is

$$v_{\text{esc}} = \sqrt{\dfrac{2GM_{\text{E}}}{r}} = r^{-1/2}\sqrt{2GM_{\text{E}}} \quad\to\quad \dfrac{dv_{\text{esc}}}{dr} = -\tfrac{1}{2}r^{-3/2}\sqrt{2GM_{\text{E}}} = \boxed{-\sqrt{\dfrac{GM_{\text{E}}}{2r^3}}}$$

(b) $\Delta v_{\text{esc}} \approx \dfrac{dv_{\text{esc}}}{dr}\Delta r = -\sqrt{\dfrac{GM_{\text{E}}}{2r^3}}\,\Delta r = -\sqrt{\dfrac{(6.67\times10^{-11}\,\text{N}\bullet\text{m}^2/\text{kg}^2)(5.98\times10^{24}\,\text{kg})}{2(6.38\times10^6\,\text{m})^3}}\,(3.2\times10^5\,\text{m}) = -280\,\text{m/s}$

The escape velocity has decreased by 280 m/s, and so is $v_{\text{esc}} = 1.12 \times 10^4$ m/s - 280 m/s = $\boxed{1.09 \times 10^4 \text{ m/s}}$.

61. (a) The escape speed from the surface of the Earth is $v_{\text{E}} = \sqrt{(2GM_{\text{E}}/r_{\text{E}})}$. The escape velocity from the gravitational field of the sun, is $v_{\text{S}} = \sqrt{(2GM_{\text{S}}/r_{\text{ES}})}$. In the reference frame of the Earth, if the spacecraft leaves the surface of the Earth with speed v (assumed to be greater than the escape velocity of Earth), then the speed v' at a distance far from Earth, relative to the Earth, is found from energy conservation.

$$\tfrac{1}{2}mv^2 - \dfrac{GM_{\text{E}}m}{r_{\text{E}}^2} = \tfrac{1}{2}mv'^2 \quad\to\quad v'^2 = v^2 - \dfrac{2GM_{\text{E}}}{r_{\text{E}}^2} = v^2 - v_{\text{E}}^2 \quad\to\quad v^2 = v'^2 + v_{\text{E}}^2$$

The reference frame of the Earth is orbiting the sun with speed v_0. If the rocket is moving with speed v' relative to the Earth, and the Earth is moving with speed v_0 relative to the Sun, then the speed of the rocket relative to the Sun is $v' + v_0$ (assuming that both speeds are in the same direction). This is to be the escape velocity from the Sun, and so $v_{\text{S}} = v' + v_0$, or $v' = v_{\text{S}} - v_0$. Combine this with the relationship from above.

$$v^2 = v'^2 + v_{\text{E}}^2 = (v_{\text{S}} - v_0)^2 + v_{\text{E}}^2 \quad\to\quad \boxed{v = \sqrt{(v_{\text{S}} - v_0)^2 + v_{\text{E}}^2}}$$

$$v_{\text{E}} = \sqrt{\dfrac{2GM_{\text{E}}}{r_{\text{E}}}} = \sqrt{\dfrac{2(6.67\times10^{-11}\,\text{N}\bullet\text{m}^2/\text{kg}^2)(5.98\times10^{24}\,\text{kg})}{6.38\times10^6\,\text{m}}} = 1.118\times10^4\,\text{m/s}$$

$$v_{\text{S}} = \sqrt{\dfrac{2GM_{\text{S}}}{r_{\text{ES}}}} = \sqrt{\dfrac{2(6.67\times10^{-11}\,\text{N}\bullet\text{m}^2/\text{kg}^2)(1.99\times10^{30}\,\text{kg})}{1,496\times10^{11}\,\text{m}}} = 4.212\times10^4\,\text{m/s}$$

$$v_0 = \dfrac{2\pi r_{\text{SE}}}{T_{\text{SE}}} = \dfrac{2\pi(1.496\times10^{11}\,\text{m})}{(3.156\times10^7\,\text{s})} = 2.978\times10^4\,\text{m/s}$$

$$v = \sqrt{(v_{\text{S}} - v_0)^2 + v_{\text{E}}^2} = \sqrt{(4.212\times10^4\,\text{m/s} - 2.978\times10^4\,\text{m/s})^2 + (1.118\times10^4\,\text{m/s})^2}$$

$$= 1.665\times10^4\,\text{m/s} \approx \boxed{16.7\,\text{km/s}}$$

(*b*) Calculate the kinetic energy for a 1.00 kg mass moving with a speed of 1.665 x 10^4 m/s. This is the energy required per kilogram of spacecraft mass.

$$K = \frac{1}{2}mv^2 = \frac{1}{2}(1.00\,\text{kg})(1.665\times10^4\,\text{m/s})^2 = \boxed{1.39\times10^8\,\text{J}}$$

67. The power is the force that the motor can provide times the velocity, as given in Eq. 8-21. The force provided by the motor is parallel to the velocity of the boat. The force resisting the boat will be the same magnitude as the force provided by the motor, since the boat is not accelerating, but in the opposite direction to the velocity.

$$P = \vec{\mathbf{F}}\cdot\vec{\mathbf{v}} = Fv \rightarrow F = \frac{P}{v} = \frac{(55\,\text{hp})(746\,\text{W/1 hp})}{(35\,\text{km/h})\left(\dfrac{1\,\text{m/s}}{3.6\,\text{km/h}}\right)} = 4220\,\text{N} \approx 4200\,\text{N}$$

So the force resisting the boat is $\boxed{4200\,\text{N, opposing the velocity}}$.

73. The net rate of work done is the power, which can be found by $P = Fv = mav$. The velocity is given by

$$v = \frac{dx}{dt} = 15.0t^2 - 16.0t - 44 \text{ and } a = \frac{dv}{dt} = 30.0t - 16.0$$

(*a*) $P = mav = (0.28\,\text{kg})\left(\left[30.0(2.0) - 16.0\right]\text{m/s}^2\right)\left[15.0(2.0)^2 - 16.0(2.0) - 44\right]\text{m/s}$

$$= -197.1\,\text{W} \approx \boxed{-2.0\times10^2\,\text{W}}$$

(*b*) $P = mav = (0.28\,\text{kg})\left(\left[30.0(4.0) - 16.0\right]\text{m/s}^2\right)\left[15.0(4.0)^2 - 16.0(4.0) - 44\right]\text{m/s}$

$$= 3844\,\text{W} \approx \boxed{3800\,\text{W}}$$

The average net power input is the work done divided by the elapsed time. The work done is the change in kinetic energy.

Note: $\quad v(0) = -44\,\text{m/s}$

$$v(2.0) = 15.0(2.0)^2 - 16.0(2.0) - 44 = -16\,\text{m/s}$$

$$v(4.0) = 15.0(4.0)^2 - 16.0(4.0) - 44 = 132\,\text{m/s}$$

(*c*) $P_{\substack{\text{avg}\\0\text{ to }2.0}} = \dfrac{\Delta K}{\Delta t} = \dfrac{\frac{1}{2}m(v_f^2 - v_i^2)}{\Delta t} = \dfrac{\frac{1}{2}(0.28\,\text{kg})\left[(-16\,\text{m/s})^2 - (-44\,\text{m/s})^2\right]}{2.0\,\text{s}} = \boxed{-120\,\text{W}}$

(*d*) $P_{\substack{\text{avg}\\2.0\text{ to }4.0}} = \dfrac{\Delta K}{\Delta t} = \dfrac{\frac{1}{2}m(v_f^2 - v_i^2)}{\Delta t} = \dfrac{\frac{1}{2}(0.28\,\text{kg})\left[(132\,\text{m/s})^2 - (16\,\text{m/s})^2\right]}{2.0\,\text{s}} = \boxed{1200\,\text{W}}$

79. The power must exert a force equal to the weight of the elevator, through the vertical height, in the given time.

$$P = \frac{mgh}{t} = \frac{(885\,\text{kg})(9.80\,\text{m/s}^2)(32.0\,\text{m})}{(11.0\,\text{s})} = \boxed{2.52\times10^4\,\text{W}}$$

85. (*a*) The tension in the cord is perpendicular to the path at all times, and so the tension in the cord does not do any work on the ball. Thus only gravity does work on the ball, and so the mechanical energy of the ball is conserved. Subscript 1 represents the ball when it is horizontal, and subscript 2 represents the ball at the lowest point on its path. The lowest point on the path is the zero location for potential energy ($y = 0$). We have $v_1 = 0$, $y_1 = l$, and $y_2 = 0$. Solve for v_2.

$$E_1 = E_2 \;\; \rightarrow \;\; \frac{1}{2}mv_1^2 + mgy_1 = \frac{1}{2}mv_2^2 + mgy_2 \;\; \rightarrow$$

$$mgl = \frac{1}{2}mv_2^2 \;\; \rightarrow \;\; v_2 = \boxed{\sqrt{2gl}}$$

(*b*) Use conservation of energy, to relate points 2 and 3. Point 2 is as described above. Subscript 3 represents the ball at the top of its circular path around the peg. The lowest point on the path is the zero location for potential energy ($y = 0$). We have $v_2 = \sqrt{(2gl)}$, $y_2 = 0$, and $y_3 = 2(l - h) = 2(l - 0.80l) = 0.40l$. Solve for v_3.

$$E_2 = E_3 \;\; \rightarrow \;\; \frac{1}{2}mv_2^2 + mgy_2 = \frac{1}{2}mv_3^2 + mgy_3 \;\; \rightarrow$$

$$\frac{1}{2}m(2gl) = \frac{1}{2}mv_3^2 + mg(0.40l) \;\; \rightarrow \;\; \boxed{v_3 = \sqrt{1.2gl}}$$

91. (*a*) Use conservation of energy for the swinging motion. Subscript 1 represents the student initially grabbing the rope, and subscript 2 represents the student at the top of the swing. The location where the student initially grabs the rope is the zero location for potential energy ($y = 0$). We have $v_1 = 5.0$ m/s, $y_1 = 0$, and $v_2 = 0$. Solve for y_2.

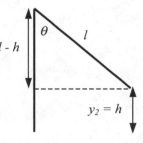

$$E_1 = E_2 \;\; \rightarrow$$

$$\frac{1}{2}mv_1^2 + mgy_1 = \frac{1}{2}mv_2^2 + mgy_2 \;\; \rightarrow$$

$$\frac{1}{2}mv_1^2 = mgy_2 \;\; \rightarrow$$

$$y_2 = \frac{v_1^2}{2g} = h$$

Calculate the angle from the relationship in the diagram.

$$\cos\theta = \frac{l-h}{l} = 1 - \frac{h}{l} = 1 - \frac{v_1^2}{2gl} \;\; \rightarrow$$

$$\theta = \cos^{-1}\left(1 - \frac{v_1^2}{2gl}\right) = \cos^{-1}\left(1 - \frac{(5.0\,\text{m/s})^2}{2(9.80\,\text{m/s}^2)(10.0\,\text{m})}\right) = \boxed{29°}$$

(*b*) At the release point, the speed is 0, and so there is no radial acceleration, since $a_R = v^2/r$. Thus the centripetal force must be 0. Use the free-body diagram to write Newton's second law for the radial direction.

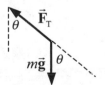

$$\sum F_R = F_T - mg\cos\theta = 0 \;\; \rightarrow \;\; F_T = mg\cos\theta = (56\,\text{kg})(9.80\,\text{m/s}^2)\cos 29° = \boxed{480\,\text{N}}$$

(*c*) Write Newton's second law for the radial direction for any angle, and solve for the tension.

$$\sum F_R = F_T - mg\cos\theta = mv^2/r \;\; \rightarrow \;\; F_T = mg\cos\theta + mv^2/r$$

As the angle decreases, the tension increases, and as the speed increases, the tension increases. Both effects are greatest at the bottom of the swing, and so that is where the tension will be at its maximum.

$$F_{T\,\text{max}} = mg\cos 0 + mv_1^2/r = (56\,\text{kg})(9.80\,\text{m/s}^2) + \frac{(56\,\text{kg})(5.0\,\text{m/s})^2}{10.0\,\text{m}} = \boxed{690\,\text{N}}$$

97. The energy to be stored is the power multiplied by the time: $E = Pt$. The energy will be stored as the gravitational potential energy increase in the water: $E = \Delta U = mg\Delta y = \rho Vg\Delta y$, where ρ is the density of the water, and V is the volume of the water.

$$Pt = \rho Vg\Delta y \rightarrow$$

$$V = \frac{Pt}{\rho g\Delta y} = \frac{\left(180\times10^{6}\,\text{W}\right)\left(3600\,\text{s}\right)}{\left(1.00\times10^{3}\,\text{kg}/\text{m}^{3}\right)\left(9.80\,\text{m}/\text{s}^{2}\right)\left(380\,\text{m}\right)} = \boxed{1.7\times10^{5}\,\text{m}^{3}}$$

103. The only forces acting on the bungee jumper are gravity and the elastic force from the bungee cord, so the jumper's mechanical energy is conserved. Subscript 1 represents the jumper at the bridge, and subscript 2 represents the jumper at the bottom of the jump. Let the lowest point of the jumper's motion be the zero location for gravitational potential energy ($y = 0$). The zero location for elastic potential energy is the point at which the bungee cord begins to stretch. See the diagram in the textbook. We have $v_1 = v_2 = 0$, $y_1 = h$, $y_2 = 0$, and the amount of stretch of the cord $x_2 = h$ - 15. Solve for h.

$$E_1 = E_2 \rightarrow \frac{1}{2}mv_1^2 + mgy_1 + \frac{1}{2}kx_1^2 = \frac{1}{2}mv_2^2 + mgy_2 + \frac{1}{2}kx_2^2 \rightarrow mgh = \frac{1}{2}k\left(h-15\right)^2 \rightarrow$$

$$h^2 - \left(30 + 2\frac{mg}{k}\right)h + 225 = 0 \rightarrow h^2 - 59.4h + 225 = 0 \rightarrow$$

$$h = \frac{59.4 \pm \sqrt{59.4^2 - 4\left(225\right)}}{2} = 55\,\text{m}, 4\,\text{m} \rightarrow h = \boxed{60\,\text{m}}$$

The larger answer must be taken because $h > 15$ m. And only 1 significant figure is justified.

109. A point of stable equilibrium will have $\frac{dU}{dx} = 0$ and $\frac{d^2U}{dx^2} > 0$, indicating a minimum in the potential equilibrium function.

$$U\left(x\right) = \frac{a}{x} + bx \qquad \frac{dU}{dx} = -\frac{a}{x^2} + b = 0 \rightarrow x^2 = \frac{a}{b} \rightarrow x = \pm\sqrt{a/b}$$

But since the problem restricts us to $x > 0$, the point of equilibrium must be $x = \boxed{\sqrt{a/b}}$.

$$\left.\frac{d^2U}{dx^2}\right|_{x=\sqrt{a/b}} = \left.\frac{2a}{x^3}\right|_{x=\sqrt{a/b}} = \frac{2a}{\left(a/b\right)^{3/2}} = \frac{2}{ab^{3/2}} > 0 \text{ and so the point } x = \sqrt{a/b} \text{ gives a minimum in the potential energy}$$

function.

Chapter 9: Linear Momentum and Collisions

Chapter Overview and Objectives

In this chapter, the concept of linear momentum is introduced. The law of conservation of linear momentum and related topics are discussed.

After completing this chapter you should:
- Know the definition of momentum.
- Know the definition of impulse.
- Know how Newton's laws can be expressed as the conservation of momentum.
- Know how to use the law of conservation of momentum to solve collision problems.
- Know how to determine the center of mass of an extended body and a system of particles.
- Know how to apply the law of conservation of momentum to a body that is changing mass.
- Know the SI units of momentum and impulse and their abbreviations.

Summary of Equations

Definition of momentum:
$$\vec{\mathbf{p}} = m\vec{\mathbf{v}}$$
(Section 9-1)

Newton's second law in terms of momentum:
$$\sum \vec{\mathbf{F}} = \frac{d\vec{\mathbf{p}}}{dt}$$
(Section 9-1)

Conservation of momentum:
$$\sum \frac{d\vec{\mathbf{p}}}{dt} = \sum \vec{\mathbf{F}}_{ext}$$
(Section 9-2)

Integral form of Newton's second law:
$$\int_{t_1}^{t_2} \vec{\mathbf{F}} dt = \int_{\vec{\mathbf{p}}_1}^{\vec{\mathbf{p}}_2} d\vec{\mathbf{p}} = \vec{\mathbf{p}}_2 - \vec{\mathbf{p}}_1 = \Delta\vec{\mathbf{p}}$$
(Section 9-3)

Definition of impulse:
$$\vec{\mathbf{J}} = \int_{t_1}^{t_2} \vec{\mathbf{F}} dt$$
(Section 9-3)

Expression for average force in terms of impulse:
$$\vec{\mathbf{F}}_{average} = \frac{\vec{\mathbf{J}}}{t_2 - t_1}$$
(Section 9-3)

Definition of center of mass of a system of particles:
$$\vec{\mathbf{r}}_{CM} = \frac{\sum m_i \vec{\mathbf{r}}_i}{\sum m_i}$$
(Section 9-8)

Definition of center of mass of an extended body:
$$\vec{\mathbf{r}}_{CM} = \frac{\int \vec{\mathbf{r}}\, dm}{\int dm} = \frac{1}{M}\int \vec{\mathbf{r}}\, dm$$
(Section 9-8)

Newton's second law applied to center-of-mass:
$$\sum \vec{\mathbf{F}}_{ext} = M\vec{\mathbf{a}}_{CM}$$
(Section 9-9)

Principle of conservation of momentum applied to system with changing mass:

$$d\vec{\mathbf{P}}_{system} = \left(\vec{\mathbf{v}} - \vec{\mathbf{u}}\right) dM + M\, d\vec{\mathbf{v}}$$
(Section 9-10)

Chapter Summary

Section 9-1. Momentum and Its Relation to Force

The **linear momentum** of an object is defined to be the product of its mass and its velocity:

$$\vec{\mathbf{p}} = m\vec{\mathbf{v}}$$

As the velocity of an object is a vector and its mass is a scalar, this product is the product of a scalar with a vector and results in a vector quantity. Thus, linear momentum is a vector quantity. The SI units of momentum are kg m/s. Newton's second law of motion can be written in terms of momentum as:

$$\sum \vec{\mathbf{F}} = \frac{d\vec{\mathbf{p}}}{dt}$$

The above relationship can be multiplied on both sides by dt and integrated

$$\int_{t_1}^{t_2} \sum \vec{\mathbf{F}}\, dt = \int_{\vec{\mathbf{p}}_1}^{\vec{\mathbf{p}}_2} d\vec{\mathbf{p}} = \vec{\mathbf{p}}_2 - \vec{\mathbf{p}}_1 = \Delta\vec{\mathbf{p}}$$

Example 9-1-A. Lift generated by a wing. As an airplane flies through the air, it accelerates air downward at a rate of 1060 kg/s. The air starts at rest and is deflected downward with a velocity of 140 m/s. What is the force on the airplane?

Approach: In this problem we will only consider the motion of the air in the vertical direction. When the air is deflected by the wing, its linear momentum in the vertical direction changes. The change in the linear momentum of the air is a result of a force exerted by the wing on the air. Newton's third law tells us that this force is equal in magnitude but directed in an opposite direction to the force the air exerts on the wing.

Solution: The change of the linear momentum of the air is

$$dp = \left(v_f - v_i\right)dm \quad \text{or} \quad \frac{dp}{dt} = \left(v_f - v_i\right)\frac{dm}{dt} = \left(v_f - 0\right)\frac{dm}{dt} = -\left(140\,\text{m/s}\right)\left(1060\,\text{kg/s}\right) = -1.48 \times 10^5\,\text{N}$$

The force the wing exerts on the air is equal to the rate of change of its linear momentum. The minus sign in the equation indicates that the force is directed downwards. The force exerted by the air on the wing has a magnitude 1.48×10^5 N and is directed in the upward direction. This force is called lift.

Section 9-2. Conservation of Momentum

The relationship above can be applied to a system of interacting objects and summed over the momentum changes for all the objects in the system. Applying Newton's third law to the sum results in a total of zero momentum change for interactions between objects in the system. This results in the law of conservation of momentum. This law states that the sum of the changes in momentum of the objects in an isolated system is equal to zero.

$$\sum \Delta\vec{\mathbf{p}} = 0 \quad \text{or} \quad \sum_{\text{intial}} \vec{\mathbf{p}} = \sum_{\text{final}} \vec{\mathbf{p}} \quad \text{or} \quad \sum \frac{d\vec{\mathbf{p}}}{dt} = 0$$

An isolated system is a system with no forces acting on any component of the system arising from outside the system. If external forces act on the system, this can be generalized to:

$$\sum \frac{d\vec{\mathbf{p}}}{dt} = \sum \vec{\mathbf{F}}_{ext}$$

where $\vec{\mathbf{F}}_{ext}$ represents all of the forces acting on the system that arise from sources external to the system.

Example 9-2-A. Conservation of linear momentum for a two-body system in one dimension. Two people run toward each other on ice, collide, and embrace each other after the collision. One person has a mass of 106 kg and runs with a speed of 2.67 m/s. The other person has a mass of 78.4 kg and runs with a speed of 6.42 m/s. What is the motion of the two people after they collide?

Approach: In this problem we apply the principle for conservation of linear momentum. The problem provides us with information about the linear momentum before the collision. Since the people embrace each other after the collision, they travel with the same velocity, and this velocity can be calculated by requiring that the final linear momentum is equal to the initial linear momentum, which can be determined from the information provided.

Solution. The initial velocities of the people are in opposite directions. We will choose our coordinate system such that the 106-kg person moves in the positive direction, $v_1 = +2.67$ m/s and the 78.4-kg person moves in the negative direction, $v_2 = -6.42$ m/s. The initial linear momentum of the two people is equal to

$$\sum p_i = m_1 v_1 + m_2 v_2$$

The linear momentum after the collision is equal to the total mass of the two people multiplied by their common velocity:

$$\sum p_f = m_1 v_f + m_2 v_f = \left(m_1 + m_2 \right) v_f$$

Since linear momentum is conserved we can calculate the final velocity of the two people:

$$v_f = \frac{m_1 v_1 + m_2 v_2}{m_1 + m_2} = -1.19 \, \text{m/s}$$

The two people after the collision will thus move in the negative direction.

Example 9-2-B. Conservation of linear momentum for a two-body system in three dimensions. During a football game, a thrown ball with a mass of 0.214 kg hits a duck with a mass of 4.52 kg flying over the field. Before the collision, the football is traveling with a velocity $\vec{v}_{ball} = 4.5\hat{i} + 6.3\hat{j} + 1.4\hat{k}$, where the velocity is expressed in units of m/s, $\hat{i}$ points to the east, $\hat{j}$ to the north, and $\hat{k}$ upward. The duck's velocity before the collision is 8.2 m/s, horizontal directly toward the west before the collision ($\vec{v}_{duck} = -8.2\hat{i}$). If the football is deflected so that its velocity is straight downward at 5.2 m/s after the collision, what is the velocity of the duck after the collision?

Approach: This problem can be solved by applying conservation of linear momentum. Since the motion of the system is three-dimensional motion, linear momentum must be conserved along the three coordinate axes.

Solution: The initial linear momentum of the system is equal to

$$\sum \vec{p}_i = m_{ball} \vec{v}_{ball} + m_{duck} \vec{v}_{duck} = m_{ball} \left(4.5\,\hat{i} + 6.3\,\hat{j} + 1.4\,\hat{k} \right) + m_{duck} \left(-8.2\,\hat{i} \right) =$$
$$= \left(4.5 m_{ball} - 8.2 m_{duck} \right)\hat{i} + \left(6.3 m_{ball} \right)\hat{j} + \left(1.4 m_{ball} \right)\hat{k}$$

The final linear momentum of the system is equal to

$$\sum \vec{p}_f = m_{ball} \vec{v}'_{ball} + m_{duck} \vec{v}'_{duck} = m_{ball} \left(-5.2\,\hat{k} \right) + m_{duck} \left(v'_{x,duck}\,\hat{i} + v'_{y,duck}\,\hat{j} + v'_{z,duck}\,\hat{k} \right) =$$
$$= m_{duck} v'_{x,duck}\,\hat{i} + m_{duck} v'_{y,duck}\,\hat{j} + \left(m_{duck} v'_{z,duck} - 5.2 m_{ball} \right)\hat{k}$$

Conservation of linear momentum requires that

$$\left(4.5 m_{ball} - 8.2 m_{duck} \right)\hat{i} + \left(6.3 m_{ball} \right)\hat{j} + \left(1.4 m_{ball} \right)\hat{k} = m_{duck} v'_{x,duck}\,\hat{i} + m_{duck} v'_{y,duck}\,\hat{j} + \left(m_{duck} v'_{z,duck} - 5.2 m_{ball} \right)\hat{k}$$

The left-hand side and the right-hand side are equal if the components of the vectors on either side are the same. This requires that

$$4.5 m_{ball} - 8.2 m_{duck} = m_{duck} v'_{x,duck}$$
$$6.3 m_{ball} = m_{duck} v'_{y,duck}$$
$$1.4 m_{ball} = m_{duck} v'_{z,duck} - 5.2 m_{ball}$$

These equations can be used to determine the three components of the velocity of the duck after the collision:

$$v'_{x,duck} = \frac{4.5m_{ball} - 8.2m_{duck}}{m_{duck}} = 4.5\frac{m_{ball}}{m_{duck}} - 8.2 = -7.99 \text{ m/s}$$

$$v'_{y,duck} = 6.3\frac{m_{ball}}{m_{duck}} = 0.30 \text{ m/s}$$

$$v'_{z,duck} = \frac{5.2m_{ball} + 1.4m_{ball}}{m_{duck}} = 6.6\frac{m_{ball}}{m_{duck}} = 0.31 \text{ m/s}$$

The final velocity of the duck is equal to:

$$\vec{v}'_{duck} = 7.99\hat{i} + 0.30\hat{j} + 0.31\hat{k}$$

Section 9-3. Collisions and Impulse

The integral of force over time is called the **impulse**, $\vec{J}$:

$$\vec{J} = \int_{t_1}^{t_2} \vec{F}\, dt$$

Impulse is a vector quantity with the same direction as the force. The change in the linear momentum of an object is equal to the total impulse that acts on the object:

$$\Delta\vec{p} = \vec{J}$$

There are times when the details of the force as a function of time are unknown. In that situation it makes sense to talk about the average force $\vec{F}_{average}$. The average force is defined as the impulse divided by the time interval over which the force acts:

$$\vec{F}_{average} = \frac{\vec{J}}{t}$$

Example 9-3-A. **Using the impulse of a force to determine the change in the linear momentum.** A force F has a time dependence given by $1.00 - 6.37\, t^2$. Determine the change in the linear momentum of a particle on which this force is acting, assuming the force acts during the time interval between $t = 0$ s and $t = 3.42$ s.

Approach: Since the problem provides information about the time dependence of the force, we can use this information to calculate the impulse associated with the force. The impulse of the force, over the time period specified, is equal to the change in the linear momentum of the particle over this time period.

Solution: Using the relationship between impulse and change of momentum

$$\Delta p = J = \int_0^{3.42} F\, dt = \int_0^{3.42} \left(1.00 - 6.37\, t^2\right) dt = \left(1.00\, t - \frac{1}{2}6.37\, t^3\right)\Big|_0^{3.42} = -81.5 \text{kg m/s}$$

Example 9-3-B. **Calculating the average force by observing the change in linear momentum.** The function of an airbag is to reduce the average force exerted on a person involved in a collision. Consider a 100-kg driver in a car, driving at 55 mph, who collides with the pillars of a bridge. As a result of the collision, the car comes to an immediate stop. The inflated airbag, with a thickness of 12.5 cm, ensures that it takes 10 ms for the driver to come to rest. What is the average force exerted by the airbag on the drive?

Approach: In this problem the change in the linear momentum, and thus the impulse associated with the airbag, can be determined from the initial and the final velocity of the driver. The impulse will be independent of the design of the airbag, but the time over which the linear momentum changes will be a function of the design; the 10 ms quoted in this problem corresponds to an average velocity of 27.5 mph during the time the driver comes to rest. The average force exerted by the airbag on the driver is equal to the ratio of the impulse and the time during which the force acts.

Solution: The change in the linear momentum of the driver is equal to

$$\Delta p = p_f - p_i = 0 - (100)(55 \cdot 0.447) = -2,500 \text{ kg m/s}$$

The average force exerted by the airbag on the driver is

$$F_{average} = \frac{J}{t} = \frac{-2,500}{10 \times 10^{-3}} = 250,000 \text{ N} \approx 25,000g$$

Section 9-4. Conservation of Energy and Momentum in Collisions

In collisions between objects, the details of the forces that act on the objects are usually unknown. Newton's second law cannot be used to determine the motion resulting from the collision without this knowledge. However, using the laws of conservation of momentum and conservation of energy, the motions before and after the collision can be related to each other. In any collision in which external forces acting on the colliding bodies can be neglected, the law of conservation of momentum can be applied. In some collisions, the total kinetic energy of the colliding bodies remains constant. These collisions are called elastic collisions. For **elastic collisions**, a law of conservation of kinetic energy can be written down to relate the motion before the collision to the motion after the collision. For a system of two objects, 1 and 2, this can be written

$$\tfrac{1}{2}m_1v_1^2 + \tfrac{1}{2}m_2v_2^2 = \tfrac{1}{2}m_1v_1'^2 + \tfrac{1}{2}m_2v_2'^2$$

where the unprimed speeds are before the collision and the primed speeds are after the collision. The subscripts on the quantities refer to which particle, 1 or 2, they correspond to.
If the total kinetic energy does not remain constant during the collision, the collision is said to be an **inelastic collision**. During inelastic collisions, some of the initial kinetic energy is transformed into other forms of energy:

$$\tfrac{1}{2}m_1v_1^2 + \tfrac{1}{2}m_2v_2^2 = \tfrac{1}{2}m_1v_1'^2 + \tfrac{1}{2}m_2v_2'^2 + \text{other energy}$$

Example 9-4-A. Kinetic energies in collisions. A collision between two objects occurs. Initially, object A, with a mass of $m_A = 12$ kg, is traveling with a speed of $v_A = 3.4$ m/s and object B, with a mass of $m_B = 16$ kg, is traveling with a speed of $v_B = 2.8$ m/s. After the collision, object A is traveling with a speed of $v'_A = 2.2$ m/s and object B is traveling with a speed of $v'_B = 3.0$ m/s. Determine whether this is an elastic or inelastic collision. If the collision is an inelastic collision, how much kinetic energy was transformed into other forms of energy during the collision?

Approach: Although the problem does not provide us with information about the dimensional nature of the collision, we do not need this type of information in order to study the change in the kinetic energy of the system. By comparing the total kinetic energy before and after the collision we can determine if the collision is elastic or inelastic.

Solution: The kinetic energy before the collision is equal to

$$K_i = \frac{1}{2}m_A v_A^2 + \frac{1}{2}m_B v_B^2 = \frac{1}{2}(12)(3.4)^2 + \frac{1}{2}(16)(2.8)^2 = 1.3 \times 10^2 \text{ J}$$

The kinetic energy after the collision is

$$K_f = \frac{1}{2}m_A v_A'^2 + \frac{1}{2}m_B v_B'^2 = \frac{1}{2}(12)(2.2)^2 + \frac{1}{2}(16)(3.0)^2 = 1.0 \times 10^2 \text{ J}$$

The change in the kinetic energy of the system is 0.3×10^2 J and the collision is thus inelastic.

Section 9-5. Elastic Collisions in One Dimension

Collisions in one dimension have all velocities, both initial and final, lying along the same line. If the collision is between two particles and is elastic, we can apply conservation of momentum and conservation of kinetic energy:

$$m_1 v_1 + m_2 v_2 = m_1 v_1' + m_2 v_2'$$

$$\tfrac{1}{2} m_1 v_1^2 + \tfrac{1}{2} m_2 v_2^2 = \tfrac{1}{2} m_1 v_1'^2 + \tfrac{1}{2} m_2 v_2'^2$$

Eliminating m_1 and m_2 from these two equations results in:

$$v_1 - v_2 = -\left(v_1' - v_2' \right)$$

The result of the collision is to reverse the relative velocity of the two objects.

Example 9-5-A. The center-of-mass reference frame. Many collision problems are solved in a reference frame in which the center of mass of the system is at rest. What are the final velocities of particle 1 of mass m_1 and particle 2 of mass m_2 if they are involved in an elastic collision and have initial velocities equal to v_1 and v_2, respectively, in their center-of-mass frame of reference?

Approach: If the collision force is the only force acting on the system, the total external system is equal to 0 N and the center of mass of the system will remain at rest. This requirement imposes a stringent requirement on the velocities of the particles.

Solution: The position of the center of mass of the two particles is given by

$$r_{CM} = \sum m_i x_i = m_1 x_1 + m_2 x_2$$

If the motion of particles is described in their center-of-mass frame of reference, the position of the center of mass will remain at rest. This requires that

$$0 = \frac{dr_{CM}}{dt} = \frac{d}{dt}\left(m_1 x_1 + m_2 x_2 \right) = m_1 v_1 + m_2 v_2$$

The same requirement must be satisfied by the velocities after the collision. We must require that

$$0 = \frac{dr'_{CM}}{dt} = \frac{d}{dt}\left(m_1 x'_1 + m_2 x'_2 \right) = m_1 v'_1 + m_2 v'_2$$

This applies both before and after the collision. Writing down this condition along with the fact that an elastic collision reverses the relative velocity of the two objects

$$v_2 - v_1 = v'_1 - v'_2 \qquad m_1 v_1 + m_2 v_2 = 0 \qquad m_1 v'_1 + m_2 v'_2 = 0$$

Multiplying the first equation by m_1 and adding these three equations together we obtain

$$m_1 \left(v_2 - v_1 \right) + \left(m_1 v_1 + m_2 v_2 \right) + \left(m_1 v'_1 + m_2 v'_2 \right) = m_1 \left(v'_1 - v'_2 \right)$$

This equation can be rewritten as

$$\left(m_1 + m_2 \right) v_2 + m_1 v'_1 + m_2 v'_2 = m_1 v'_1 - m_2 v'_2 \quad \Rightarrow \quad v_2 = -v'_2$$

Similarly, multiplying the first equation by m_2 and adding the three equations we obtain

$$v_1 = -v'_1$$

In the center of mass reference frame, an elastic collision simply reverses the directions of the colliding particles' velocities!

Example 9-5-B. Using elastic collisions to determine mass differences. One way in which a pool hustler can beat you is by substituting a cue ball whose mass is different from the other billiard balls. The directions of the balls after the collisions are not the same as for a cue ball with identical mass as the other balls, thus throwing off the opponent. The hustler has practiced with the heavy cue ball and knows what to expect; the other player doesn't. You have been playing pool and are having difficulty making shots. Your suspicions are confirmed when you see an elastic head-on collision with a ball at rest in which the spin of the balls had no effect, but the cue ball continues to move after the collision with a velocity equal to one tenth its initial velocity. By what percentage is the cue ball bigger in mass than the ball it struck?

Approach: In this problem, we consider a head-on collision between a moving billiard ball and one that is at rest. Although collisions between billiard balls in general require two dimensions to be described, this special type of collision can be described in term of one-dimensional motion. Using the theory of elastic collisions in one dimension and making the assumption that the balls have equal mass, we expect that after this type of collision, the ball that moves initially comes to rest, and the ball that was initially at rest continues to move with the same velocity is that of the incoming ball. The observed motion is different, immediately indicating that the two balls have different masses. The observation that the incoming ball continues to move in its original direction implies that the mass of the incoming ball is larger than the mass of the ball that is initially at rest. Based on knowing the final velocity of the incoming ball, we can use the equations we derived for elastic one-dimensional collisions to determine the mass difference between the balls.

Solution: In this Section we have seen that the relative velocities in a one-dimensional elastic collision is reversed:

$$v'_2 - v'_1 = -(v_2 - v_1) = v_1 - v_2$$

The initial velocity of ball 2, assumed to be the ball that is initially at rest, is 0 m/s. The final velocity of ball 1 is 1/10 of its initial velocity:

$$v'_1 = \frac{1}{10}v_1$$

Combining these two equations we can determine the final velocity of ball 2:

$$v'_2 = v_1 - v_2 + v'_1 = v_1 + 0 + \frac{1}{10}v_1 = \frac{11}{10}v_1$$

Since linear momentum must be conserved in the collision, we must require that

$$\sum p = m_1 v_1 = \sum p' = m_1 v'_1 + m_2 v'_2 = \frac{1}{10}m_1 v_1 + \frac{11}{10}m_2 v_1$$

This equation can be rewritten as

$$m_1 = \frac{1}{10}m_1 + \frac{11}{10}m_2 \quad \Rightarrow \quad \frac{9}{10}m_1 = \frac{11}{10}m_2 \quad \Rightarrow \quad m_1 = \frac{11}{9}m_2 = 1.22\,m_2$$

The cue ball is 22% more massive than the other ball.

Section 9-6. Inelastic Collisions

In inelastic collisions, the total kinetic energy of the objects is not conserved. To solve problems in which inelastic collisions are involved, usually some additional information will be necessary to solve the collision problem. One common type of problem encountered is the **completely inelastic** collision problem. In a completely inelastic collision, the colliding objects move with identical velocities after the collision.

$$v'_1 = v'_2$$

Example 9-6-A. Head-on Inelastic Collisions. Two cars collide head on and lock together as they do so. The first car had a mass of $m_1 = 800$ kg and was traveling north at $v_1 = 18$ m/s. The second car had a mass of $m_2 = 600$ kg and was traveling south. Evidence at the scene of the accident is consistent with the cars moving at velocity of $v' = 1.12$ m/s north immediately after the collision before friction slowed them to a rest. The police believe the second car was exceeding the speed limit of 20 m/s just prior to the collision. Determine the speed of the second car prior to the collision to help the police determine whether the second car was speeding or not.

Approach: This collision is an example of a completely inelastic collision. Given the fact that we know the masses of the cars, their velocities after the collision, and the velocity of one of the cars before the collision, we have sufficient information to determine the velocity of the other car before the collision.

Solution: Consider a system consisting of the two cars. In this system, the collision force is an internal force, and if no other external forces are present along the direction of motion, the linear momentum in this direction is conserved. The coordinate system used to describe the motion of the cars is defined such that traveling north corresponds to a positive velocity. Conservation of linear momentum requires that

$$\sum p_i = \sum p_f \quad \Rightarrow \quad m_1 v_1 + m_2 v_2 = \left(m_1 + m_2\right)v'$$

The first car's initial velocity, the final velocity of both cars, and their masses are given. This information can be used to determine the initial velocity of the second car

$$v_2 = \frac{\left(m_1 + m_2\right)v' - m_1 v_1}{m_2} = \frac{\left(800\,\text{kg} + 600\,\text{kg}\right)\left(1.12\,\text{m/s}\right) - \left(800\,\text{kg}\right)\left(18\,\text{m/s}\right)}{600\,\text{kg}} = -21\,\text{m/s}$$

The minus sign for the velocity of car two is consistent with this car initially traveling toward the south. The car was exceeding the speed limit.

Note: the velocity of the cars after the collision can usually be determined from a measurement of the distance required to bring them to a stop and the coefficient of friction between the tires and the road.

Section 9-7. Collisions in Two or Three Dimensions

The law of conservation of momentum as written covers problems of any dimension because it is written in vector form. The mathematical details of the problem can be more complex in two or three dimensions as compared to one dimension. There will be one conservation of momentum equation for each component of the motion, two components in two dimensions and three components in three dimensions. If the final velocities of the particles are considered unknowns, then in a two-particle collision there will be two unknown final velocities for each dimension. With one conservation of momentum equation for each dimension and two unknown velocities, there are not enough equations to solve for the unknown velocities in two or three dimensions even if the condition of an elastic collision is included. In two dimensions, one more parameter must be included and in three dimensions two more parameters are needed to make the elastic collision problem solvable. A completely inelastic problem will be solvable because the complete inelasticity condition reduces the number of unknown velocity components to the number of dimensions.

As an example, in two dimensions, the additional parameter given in an elastic collision problem is often one of the angles of one of the final velocities. Consider how this additional information provides enough information to make the problem solvable in the following example.

Example 9-7-A. A two-dimensional elastic collision. Two identical balls collide elastically. Initially, the two balls are traveling at right angles to one another; the first with a speed of 8.73 m/s and the second with a speed of 2.97 m/s. After the collision, the first ball is deflected by an angle of 24.6° from its initial path. What are the speeds of the two balls after the collision and what is the direction of the final velocity of the second ball?

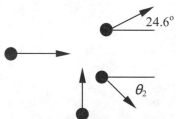

Approach: This is an example of a two-dimensional collision. Since there are no external forces acting on the balls in the plane of the collision, linear momentum will be conserved. Since the collision is elastic, the kinetic energy of the system will also be conserved. We thus have three equations (two for conservation of linear momentum and one for conservation of kinetic energy) with three unknown (the final velocities of the balls and the direction of motion of the second ball). This set of equations can be solved.

Solution: In order to solve this problem, we define our coordinate system such that ball 1 initially moves along the *x* axis with a positive velocity and ball 2 moves along the *y* axis, also with a positive velocity. With this choice of coordinate system, we will measure the angles with respect to the *x* axis. With this convention, the angle of ball 1 will positive, while the angle of ball 2 will be negative.

Conservation of linear momentum in the x direction, conservation of linear momentum in the y direction, and conservation of kinetic energy require

$$mv_1 + 0 = mv_1' \cos 24.6° + mv_2' \cos \theta_2$$
$$0 + mv_2 = mv_1' \sin 24.6° + mv_2' \sin \theta_2$$
$$\frac{1}{2}mv_1^2 + \frac{1}{2}mv_2^2 = \frac{1}{2}mv_1'^2 + \frac{1}{2}mv_2'^2$$

This set of equations can be simplified by dividing each equation by the mass m, and multiplying the equation for conservation of kinetic energy by a factor of 2.

$$v_1 + 0 = v_1' \cos 24.6° + v_2' \cos \theta_2$$
$$0 + v_2 = v_1' \sin 24.6° + v_2' \sin \theta_2$$
$$v_1^2 + v_2^2 = v_1'^2 + v_2'^2$$

To solve this set of equations we start by rewriting the equations for conservation of linear momentum in the following way:

$$v_1 - v_1' \cos 24.6° = v_2' \cos \theta_2 \Rightarrow \left(v_1 - v_1' \cos 24.6°\right)^2 = \left(v_2' \cos \theta_2\right)^2$$
$$v_2 - v_1' \sin 24.6° = v_2' \sin \theta_2 \Rightarrow \left(v_2 - v_1' \sin 24.6°\right)^2 = \left(v_2' \sin \theta_2\right)^2$$

Adding these two equations will eliminate θ_2.

$$\left(v_1 - v_1' \cos 24.6°\right)^2 + \left(v_2 - v_1' \sin 24.6°\right)^2 = v_1^2 + v_2^2 + v_1'^2 - 2\left(v_1 \cos 24.6° + v_2 \sin 24.6°\right)v_1' = v_2'^2$$

Substituting this expression into the equation obtained for conservation of kinetic energy we obtain:

$$v_1^2 + v_2^2 = v_1'^2 + v_2'^2 = v_1^2 + v_2^2 + 2v_1'^2 - 2\left(v_1 \cos 24.6° + v_2 \sin 24.6°\right)v_1' \Rightarrow 2\left(v_1 \cos 24.6° + v_2 \sin 24.6°\right)v_1' = 2v_1'^2$$

Assuming that the final velocity of ball 1 is not equal to 0, we can use this expression to calculate this velocity:

$$v_1' = \left(v_1 \cos 24.6° + v_2 \sin 24.6°\right) = 9.17 \text{ m/s}$$

Using the previously derived equation for the final velocity of ball 2 and the final velocity of ball 1, we can determine the final velocity of ball 2:

$$v_2'^2 = v_1^2 + v_2^2 + v_1'^2 - 2\left(v_1 \cos 24.6° + v_2 \sin 24.6°\right)v_1' = v_1^2 + v_2^2 - v_1'^2 \Rightarrow v_2' = \sqrt{v_1^2 + v_2^2 - v_1'^2} = 0.93 \text{ m/s}$$

At this point we can use any of the equations associated with conservation of linear momentum to determine the angle of motion of ball 2.

$$v_1 + 0 = v_1' \cos 24.6° + v_2' \cos \theta_2 \Rightarrow \cos \theta_2 = \frac{v_1 - v_1' \cos 24.6°}{v_2'} \Rightarrow \theta_2 = a\cos\left(\frac{v_1 - v_1' \cos 24.6°}{v_2'}\right) = \pm 65°$$

Given our choice of coordinate systems, the angle specifying the direction of ball 2 should be negative: $\theta_2 = -65°$.

Section 9 8. Center of Mass (CM)

General motion of rigid objects includes rotational motion as well as translational motion. Collections of particles can have very complex motions. However, the center of mass of a rigid body or collection of objects moves translationally according to Newton's second law of motion the same as a body with the total mass of the rigid object or collection of objects subject to the total external force on the rigid body or collection of objects.

The location of the center of mass of a system of point-like particles is given by

$$\vec{r}_{CM} = \frac{\sum m_i \vec{r}_i}{\sum m_i} = \frac{1}{M} \sum m_i \vec{r}_i$$

where $\vec{r}_i$ is the position of the ith particle of mass m_i and M is the total mass of the system of particles. To calculate the location of the center of mass of extended objects, we replace the summation over point-like particles with a volume integral over the extended object:

$$\vec{r}_{CM} = \frac{\int \vec{r}\, dm}{\int dm} = \frac{1}{M} \int \vec{r}\, dm$$

The mass dm in this equation is equal to the product of the appropriate mass density and the corresponding infinitesimal volume at position $\vec{r}$.

For a one-dimensional object, the mass dm of a segment of infinitesimal length dl is equal to the product of the linear mass density (mass per unit length) λ and length dl: $dm = \lambda\, dl$. The location of the center of mass of the one-dimensional object is given by

$$\vec{r}_{CM} = \frac{\int \vec{r}\, \lambda\, dl}{\int \lambda\, dl} = \frac{1}{M} \int \vec{r}\, \lambda\, dl$$

where M is the total mass of the object.

For a two-dimensional object, the mass dm of a segment of infinitesimal area dA is equal to the product of the area mass density (mass per unit area) σ and area dA: $dm = \sigma\, dA$. The location of the center of mass of the two-dimensional object is given by

$$\vec{r}_{CM} = \frac{\int \vec{r}\, \sigma\, dA}{\int \sigma\, dA} = \frac{1}{M} \int \vec{r}\, \sigma\, dA$$

where M is the total mass of the object.

For a three-dimensional object the mass dm of a segment of infinitesimal volume dV is equal to the product of the volume mass density (mass per unit volume) ρ and area dV: $dm = \rho\, dV$. The location of the center of mass of the two-dimensional object is given by

$$\vec{r}_{CM} = \frac{\int \vec{r}\, \rho\, dV}{\int \rho\, dV} = \frac{1}{M} \int \vec{r}\, \rho\, dV$$

where M is the total mass of the object.

Example 9-8-A. Position of the center of mass of a set of discrete masses. Four masses are located at the corners of a square of side a, as shown in the Figure. The masses, starting in the lower left-hand corner and proceeding counterclockwise around the square are 1.00 kg, 2.00 kg, 3.00 kg, and 4.00 kg. What is the location of the center of mass?

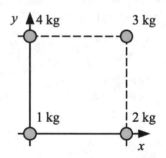

Approach: The coordinate system that will be used to solve this problem has its origin located at the position of the mass located in the lower left-hand corner of the square. With this choice of coordinate system we can define the position of each mass in terms of its coordinates, and use these position vectors to determine the location of the center of mass in this coordinate system. Note: a different choice of coordinate system will result in a different location of the center of mass when expressed in terms of this coordinate system, but the position of the center of mass relative to the position of the masses is unchanged.

Solution: Using the coordinate system with the origin located on the lower left-hand corner of the square, the position vectors of the four masses are

$$\vec{r}_1 = 0\,\hat{i} + 0\,\hat{j} \qquad \vec{r}_2 = a\,\hat{i} + 0\,\hat{j} \qquad \vec{r}_3 = a\,\hat{i} + a\,\hat{j} \qquad \vec{r}_4 = 0\,\hat{i} + a\,\hat{j}$$

We determine the center of mass by

$$\vec{r}_{CM} = \frac{\sum m_i \vec{r}_i}{\sum m_i} = \frac{1(0\,\hat{i}+0\,\hat{j}) + 2(a\,\hat{i}+0\,\hat{j}) + 3(a\,\hat{i}+a\,\hat{j}) + 4(0\,\hat{i}+a\,\hat{j})}{1+2+3+4} = 0.5a\,\hat{i} + 0.7a\,\hat{j}$$

For a coordinate system for which the origin is located on the upper right-hand corner of the square, the position vectors of the four masses are

$$\vec{r}'_1 = -a\,\hat{i}' - a\,\hat{j}' \qquad \vec{r}'_2 = 0\,\hat{i}' - a\,\hat{j}' \qquad \vec{r}'_3 = 0\,\hat{i}' + 0\,\hat{j}' \qquad \vec{r}'_4 = -a\,\hat{i}' + 0\,\hat{j}'$$

The location of the center of mass, in terms of this coordinate system, is

$$\vec{r}'_{CM} = \frac{\sum m_i \vec{r}'_i}{\sum m_i} = \frac{1(-a\,\hat{i}' - a\,\hat{j}') + 2(0\,\hat{i}' - a\,\hat{j}') + 3(0\,\hat{i}' + 0\,\hat{j}') + 4(-a\,\hat{i}' + 0\,\hat{j}')}{1+2+3+4} = -0.5a\,\hat{i}' - 0.3a\,\hat{j}'$$

Although the coordinates of the center of mass differ for the two coordinate systems, it is easy to see that the position of the center of mass with respect to the location of the other mass is the same for both coordinate systems.

Example 9-8-B. Calculating the position of the center of mass of a non-uniform object. A flat square piece of material with sides of length s has an area density given by $\sigma(x, y) = Axy^2$. The coordinate system used to describe the mass density has an origin that coincides with the lower left-hand corner of the square (see Figure). Determine the position of the center of mass.

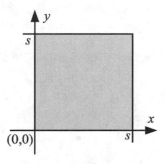

Approach: The mass density of the problem is provided in Cartesian coordinates and it is most convenient to use Cartesian coordinates to determine the integrals that must be evaluated to determine the position of the center of mass. The limits of x and y in these integrals are 0 and s.

Solution: Since the object we are considering is a two-dimensional object, we use the two-dimensional version of the expression of the position of the center of mass:

$$\vec{r}_{CM} = \frac{\int \vec{r}\,\sigma\,dA}{\int \sigma\,dA} = \frac{\int_{x=0}^{s}\int_{y=0}^{s}\left\{(x\,\hat{i} + y\,\hat{j})(Axy^2)\right\}dx\,dy}{\int_{x=0}^{s}\int_{y=0}^{s}(Axy^2)\,dx\,dy} = \frac{\int_{x=0}^{s}\int_{y=0}^{s}\left\{(x^2y^2)\,\hat{i} + (xy^3)\,\hat{j}\right\}dx\,dy}{\int_{x=0}^{s}\int_{y=0}^{s}(xy^2)\,dx\,dy}$$

The integral in the denominator can be evaluated as follows:

$$\int_{x=0}^{s}\int_{y=0}^{s}\left\{(x^2y^2)\,\hat{i} + (xy^3)\,\hat{j}\right\}dy\,dx = \int_{x=0}^{s}\left\{\left(\frac{1}{3}x^2y^3\right)\hat{i} + \left(\frac{1}{4}xy^4\right)\hat{j}\bigg|_{y=0}^{s}\right\}dx = \int_{x=0}^{s}\left\{\left(\frac{1}{3}x^2s^3\right)\hat{i} + \left(\frac{1}{4}xs^4\right)\hat{j}\right\}dx =$$

$$= \left\{\left(\frac{1}{9}x^3s^3\right)\hat{i} + \left(\frac{1}{8}x^2s^4\right)\hat{j}\bigg|_{x=0}^{s}\right\} = \left\{\left(\frac{1}{9}s^6\right)\hat{i} + \left(\frac{1}{8}s^6\right)\hat{j}\right\}$$

The integral in the numerator can be evaluated as follows:

$$\int_{x=0}^{s}\int_{y=0}^{s}(xy^2)\,dy\,dx = \int_{x=0}^{s}\left\{\frac{1}{3}xy^3\bigg|_{y=0}^{s}\right\}dx = \int_{x=0}^{s}\left\{\frac{1}{3}xs^3\right\}dx = \frac{1}{6}x^2s^3\bigg|_{x=0}^{s} = \frac{1}{6}s^5$$

The position of the center of mass of the square is thus located at

$$\hat{r}_{CM} = \frac{\frac{1}{9}s^6\,\hat{i} + \frac{1}{8}s^6\,\hat{j}}{\frac{1}{6}s^5} = \frac{2}{3}s\,\hat{i} + \frac{3}{4}s\,\hat{j}$$

Section 9-9. Center of Mass and Translational Motion

As stated in the previous section, Newton's second law describes the motion of the center of mass of an extended object or system of particles if the mass in Newton's second law is the total mass and the force is the sum of all the external forces on the system:

$$\sum F_{ext} = Ma_{CM}$$

where F_{ext} are the external forces acting on the system, M is the total mass of the system, and a_{CM} is the acceleration of the center of mass of the system.

Example 9-9-A. Oscillatory motion. Two blocks of masses $m_1 = 1$ kg and $m_2 = 2$ kg are connected by a spring, as shown in the Figure. The blocks are initially at rest and when they are released, they move towards each other. At a later time, the left block is found at a position $x = 0$ m. What is the position of the right block at this time?

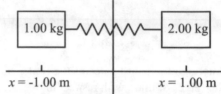

$x = -1.00$ m $x = 1.00$ m

Approach: To solve this problem we take as our system the two blocks. For this system, the spring force is considered an internal force, and there are thus no external forces acting on the system. As a result, the center of mass of the blocks, initially at rest, will remain at rest.

Solution: The initial position of the center of mass is given by

$$x_{CM} = \frac{\sum m_i x_i}{\sum m_i} = \frac{(1.00)(-1.00)+(2.00)(+1.00)}{(1.00)+(2.00)} = 0.33\,\text{m}$$

The final position of the center of mass must be at the same location. This requires that

$$x'_{CM} = \frac{(1.00)(0.00)+(2.00)x'_2}{(1.00)+(2.00)} = 0.33\,\text{m}$$

This equation has one unknown, the final position of block 2, which can now be determined.

$$x'_2 = \frac{(0.33)(1.00+2.00)}{2.00} = 0.50\,\text{m}$$

Section 9-10. Systems of Variable Mass; Rocket Propulsion

The law of conservation of momentum can be applied to systems that change mass. To see how changing mass of a system enters into the momentum, we use the product rule to differentiate the momentum expression of a given object in terms of M and v:

$$d\vec{P}_{object} = d\left(M\vec{v}\right) = \vec{v}\,dM + M\,d\vec{v}$$

where $d\vec{P}_{object}$ is the change in the object's linear momentum, $\vec{v}$ is the velocity of the object, $d\vec{v}$ is the infinitesimal change in the object's velocity, M is the mass of the object, and dM is the infinitesimal mass added to the object. This is only the change in momentum of the original object, but the object has now changed. To determine the change in momentum of the entire system, the original momentum of the infinitesimal mass element dM before it was added to the object must be considered. If the infinitesimal mass element dM had a velocity $\vec{u}$ before being added to the object, its original momentum was $\vec{u}dM$ so the total change in momentum of the system is

$$d\vec{P}_{system} = \left(\vec{v} - \vec{u}\right)dM + M\,d\vec{v}$$

Applying Newton's second law to this system:

$$\sum \vec{F}_{ext} = \frac{d\vec{P}_{system}}{dt} = \left(\vec{v} - \vec{u}\right)\frac{dM}{dt} + M\frac{d\vec{v}}{dt}$$

Example 9-10-A. A system of variable mass. A beaker of mass M rests on a scale. A liquid of density ρ leaves a tap with negligible velocity and a volume rate Q from a distance L above the beaker. The liquid is caught in the beaker. What is the scale reading as a function of time?

Approach: This system is an example of a variable-mass system. Although the current problem appears to be very different from the rocket equations discussed in this Section, the same principles apply. When applying the variable-mass equations we need to make sure we use the proper system and identify the variables correctly. Since all forces and velocities/momenta are directed along the vertical direction our current problem is a one-dimensional problem.

Solution: The system that we will use to solve the current problem consists of the beaker and the liquid coming from the tap. The external forces acting on this system are the gravitational force exerted on the beaker and the liquid and the force exerted by the scale on the beaker. The net external force acting on the beaker and the liquid can be obtained from the rocket equation derived in this Section:

$$\sum F_{ext} = \frac{dP_{system}}{dt} = \left(v - u\right)\frac{dm}{dt} + m\frac{dv}{dt} = -F_{grav} + F_{scale}$$

For the current system, the variables used in this equation have the following meaning:
- v is the velocity of the beaker before a drop of liquid of mass dm hits the beaker. Since the beaker remains at rest on top the scale, $v = 0$ and $dv/dt = 0$.
- u is the velocity of the liquid just before it hits the beaker. Since the liquid starts from rest and falls a distance L. When it reaches the beaker it has a velocity u where

    $$u = -\sqrt{2gL}$$

    Note: we have taken the upward direction as positive and the velocity of the liquid when it reaches the beaker is thus negative.
- m is the mass of the beaker and the liquid already collected in it. At a time t, the mass of the liquid collected is $\rho Q t$ and $m = M + \rho Q t$. We also conclude that $dm/dt = \rho Q$.

The original equation for the external force acting on the system can now be rewritten as

$$-u\frac{dm}{dt} + m\frac{dv}{dt} = \sqrt{2gL}\rho Q = -F_{grav} + F_{scale} = -\left(M + \rho Q t\right)g + F_{scale}$$

The force exerted by the scale on the beaker is thus equal to

$$F_{scale} = \sqrt{2gL}\rho Q + \left(M + \rho Q t\right)g = Mg + \rho Q\left(\sqrt{2gL} + g t\right)$$

Practice Quiz

1. A car is traveling down a road at a high speed that implies that it has momentum in the direction it is traveling. The car applies its brakes and comes to a rest. The momentum is now zero. If momentum is conserved, what happened to the original momentum of the car?

2. A bomb is sitting at rest. When it explodes it breaks into two pieces, one being more massive than the other. Which piece will fly out at the highest speed?

3. What can you say about the relative direction of the two pieces of bomb in the previous question?

4. A rail transportation system makes many stops and starts. To get the train up to a given speed requires a certain impulse. Explain the trade-offs in creating the impulse over a relatively short time or over a relatively long time in terms of passenger comfort and total travel time between stops.

5. In a one-dimensional collision, applying conservation of energy and conservation of momentum completely determines the motion of two objects after the collision if the motion before the collision is known. Why isn't this true in two dimensions and three dimensions?

6. Is it possible for the center of mass of an object to lie outside of the object? If so, give an example.

7. In most of the examples of this chapter, we have been somewhat lax about enforcing the condition that momentum is conserved in a system *only if no net external force acts on the system*. What property of many collisions allows approximate momentum conservation even though a significant force may be acting during the collision?

8. During a collision with an immovable object, the change in momentum of the colliding object is the same whether the surface is padded or not. Why does less damage occur if the surface is padded?

9. There are occasionally accidents in Indy style car racing in which a driver strikes the wall at over 200 mph and the car is destroyed, but the driver receives only minor injuries. This is often explained in terms of energy considerations. Explain how the collapse of the car decreases the force on the driver in terms of impulse.

10. Rubber bullets tend to bounce off of objects they are fired at rather than penetrating into them. Besides the obvious advantage that this may cause less damage than a bullet that penetrates the object that it hits, there is an additional advantage. Explain why a rubber bullet is more effective at knocking down what it hits than a bullet of equal mass and speed that penetrates the object it hits.

11. A force depends on time as $F(t) = 2 \text{ N} + (3 \text{ N/s}) \, t$. What impulse does this force create during the time interval 0 s to 10 s?

12. You jump off the surface of the Earth with enough velocity to reach a height of 0.732 m above the surface of the Earth. What was the velocity of the Earth at the moment you stopped being in contact with it?

13. A bowling ball has a weight of 16 lbs. It is traveling at a speed of 8.42 m/s when it strikes a bowling pin with a weight of 30 ounces. Assuming the collision between the bowling ball and the pin is elastic and the direction of the pin after the collision is the same as the initial direction of the bowling ball, determine the velocity of the bowling pin after the collision.

14. Determine the location of the center of mass of an object consisting of a uniform bar of length 1.08 m and mass 1.86 kg with a small sphere attached to one end with a mass of 2.95 kg.

15. A 1000-kg car, heading north at a speed of 18.5 m/s, collides completely inelastically with an 800-kg car heading west at a speed of 13.8 m/s. What is the resulting velocity of the two cars immediately after the collision?

Responses to Select End-of-Chapter Questions

1. Momentum is conserved if the sum of the external forces acting on an object is zero. In the case of moving objects sliding to a stop, the sum of the external forces is not zero; friction is an unbalanced force. Momentum will not be conserved in that case.

7. (*d*) The truck and the car will have the same change in the magnitude of momentum because momentum is conserved. (The sum of the changes in momentum must be zero.)

13. If the force is non-constant, and reverses itself over time, it can give a zero impulse. For example, the spring force would give a zero impulse over one period of oscillation.

19. When a ball is thrown into the air, it has only a vertical component of velocity. When the batter hits the ball, usually in or close to the horizontal direction, the ball acquires a component of velocity in the horizontal direction from the bat. If the ball is pitched, then when it is hit by the bat it reverses its horizontal component of velocity (as it would if it bounced off of a stationary wall) and acquires an additional contribution to its horizontal component of velocity from the bat. Therefore, a pitched ball can be hit farther than one tossed into the air.

25.

Lying down

Sitting up

CM is within the body, approximately halfway between the head and feet.

CM is outside the body.

31. If there were only two particles involved in the decay, then by conservation of momentum, the momenta of the particles would have to be equal in magnitude and opposite in direction, so that the momenta would be required to lie along a line. If the momenta of the recoil nucleus and the electron do not lie along a line, then some other particle must be carrying off some of the momentum.

Solutions to Select End-of-Chapter Problems

1. The force on the gas can be found from its change in momentum. The speed of 1300 kg of the gas changes from rest to 4.5 x 10^4 m/s, over the course of one second.

$$F = \frac{\Delta p}{\Delta t} = \frac{m\Delta v}{\Delta t} = \Delta v \frac{m}{\Delta t} = \left(4.5 \times 10^4 \text{ m/s}\right)\left(1300 \text{ kg/s}\right) = 5.9 \times 10^7 \text{ N, opposite to the velocity}$$

The force on the rocket is the Newton's third law pair (equal and opposite) to the force on the gas, and so is $\boxed{5.9 \times 10^7 \text{ N in the direction of the velocity}}$.

7. To alter the course by 35.0°, a velocity perpendicular to the original velocity must be added. Call the direction of the added velocity, $\vec{v}_{add}$, the positive direction. From the diagram, we see that $v_{add} = v_{orig} \tan\theta$. The momentum in the perpendicular direction will be conserved, considering that the gases are given perpendicular momentum in the opposite direction of $\vec{v}_{add}$. The gas is expelled oppositely to $\vec{v}_{add}$, and so a negative value is used for $v_{\perp gas}$.

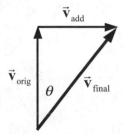

$$p_{\perp_{before}} = p_{\perp_{after}} \rightarrow 0 = m_{gas} v_{\perp gas} + \left(m_{rocket} - m_{gas}\right)v_{add} \rightarrow$$

$$m_{gas} = \frac{m_{rocket} v_{add}}{\left(v_{add} - v_{\perp gas}\right)} = \frac{\left(3180 \text{ kg}\right)\left(115 \text{ m/s}\right)\tan 35.0°}{\left[\left(115 \text{ m/s}\right)\tan 35.0° - \left(-1750 \text{ m/s}\right)\right]} = \boxed{1.40 \times 10^2 \text{ kg}}$$

13. The throwing of the package is a momentum-conserving action, if the water resistance is ignored. Let A represent the boat and child together, and let B represent the package. Choose the direction that the package is thrown as the positive direction. Apply conservation of momentum, with the initial velocity of both objects being 0.

$$p_{initial} = p_{final} \rightarrow \left(m_A + m_B\right)v = m_A v'_A + m_B v'_B \rightarrow$$

$$v'_A = -\frac{m_B v'_B}{m_A} = -\frac{\left(5.70 \text{ kg}\right)\left(10.0 \text{ m/s}\right)}{\left(24.0 \text{ kg} + 35.0 \text{ kg}\right)} = \boxed{-0.966 \text{ m/s}}$$

The boat and child move in the opposite direction as the thrown package, as indicated by the negative velocity.

19. Since no outside force acts on the two masses, their total momentum is conserved.

$$m_1 \vec{v}_1 = m_1 \vec{v}_1' + m_2 \vec{v}_2' \quad \rightarrow$$

$$\vec{v}_2' = \frac{m_1}{m_2}\left(\vec{v}_1 - \vec{v}_1'\right) = \frac{2.0\,\text{kg}}{3.0\,\text{kg}}\left[\left(4.0\hat{i} + 5.0\hat{j} - 2.0\hat{k}\right)\text{m/s} - \left(-2.0\hat{i} + 3.0\hat{k}\right)\text{m/s}\right]$$

$$= \frac{2.0\,\text{kg}}{3.0\,\text{kg}}\left[\left(6.0\hat{i} + 5.0\hat{j} - 5.0\hat{k}\right)\text{m/s}\right]$$

$$= \boxed{\left(4.0\hat{i} + 3.3\hat{j} - 3.3\hat{k}\right)\text{m/s}}$$

25. The impulse given the ball is the change in the ball's momentum. From the symmetry of the problem, the vertical momentum of the ball does not change, and so there is no vertical impulse. Call the direction AWAY from the wall the positive direction for momentum perpendicular to the wall.

$$\Delta p_\perp = mv_{\perp_{\text{final}}} - mv_{\perp_{\text{initial}}} = m\left(v\sin 45° - -v\sin 45°\right) = 2mv\sin 45°$$

$$= 2\left(6.0 \times 10^{-2}\,\text{kg}\right)\left(25\,\text{m/s}\right)\sin 45° = \boxed{2.1\,\text{kg·m/s, to the left}}$$

31. (a) Since the velocity changes direction, the momentum changes. Take the final velocity to be in the positive direction. Then the initial velocity is in the negative direction. The average force is the change in momentum divided by the time.

$$F_{\text{avg}} = \frac{\Delta p}{\Delta t} = \frac{\left(mv - -mv\right)}{\Delta t} = \boxed{2\frac{mv}{\Delta t}}$$

(b) Now, instead of the actual time of interaction, use the time between collisions in order to get the average force over a long time.

$$F_{\text{avg}} = \frac{\Delta p}{t} = \frac{\left(mv - -mv\right)}{t} = \boxed{2\frac{mv}{t}}$$

37. Let A represent the moving ball, and let B represent the ball initially at rest. The initial direction of the ball is the positive direction. We have:

$$v_A = 7.5\,\text{m/s},\ v_B = 0,\ \text{and}\ v_A' = -3.8\,\text{m/s}.$$

(a) Use Eq. 9-8 to obtain a relationship between the velocities.

$$v_A - v_B = -\left(v_A' - v_B'\right) \quad \rightarrow$$

$$v_B' = v_A - v_B + v_A' = 7.5\,\text{m/s} - 0 - 3.8\,\text{m/s} = \boxed{3.7\,\text{m/s}}$$

(b) Use momentum conservation to solve for the mass of the target ball.

$$m_A v_A + m_B v_B = m_A v_A' + m_B v_B' \quad \rightarrow$$

$$m_B = m_A \frac{\left(v_A - v_A'\right)}{\left(v_B' - v_B\right)} = \left(0.220\ \text{kg}\right)\frac{\left(7.5\,\text{m/s} - -3.8\,\text{m/s}\right)}{3.7\,\text{m/s}} = \boxed{0.67\ \text{kg}}$$

43. (a) In Example 9-11, $K_i = \frac{1}{2}mv^2$ and $K_f = \frac{1}{2}(m+M)v'^2$. The speeds are related by

$$v' = \frac{m}{m+M}v$$

$$\frac{\Delta K}{K_i} = \frac{K_f - K_i}{K_i} = \frac{\frac{1}{2}(m+M)v'^2 - \frac{1}{2}mv^2}{\frac{1}{2}mv^2} = \frac{(m+M)\left(\frac{m}{m+M}v\right)^2 - mv^2}{mv^2}$$

$$= \frac{\frac{m^2v^2}{m+M} - mv^2}{mv^2} = \frac{m}{m+M} - 1 = \boxed{\frac{-M}{m+M}}$$

(b) For the given values:

$$\frac{-M}{m+M} = \frac{-380\text{ g}}{396\text{ g}} = \boxed{-0.96}$$

Thus 96% of the energy is lost.

49. (a) For a perfectly elastic collision, Eq. 9-8 says $v_A - v_B = -\left(v'_A - v'_B\right)$. Substitute that into the coefficient of restitution definition.

$$e = \frac{v'_A - v'_B}{v_B - v_A} = -\frac{\left(v_A - v_B\right)}{v_B - v_A} = 1$$

For a completely inelastic collision, $v'_A = v'_B$. Substitute that into the coefficient of restitution definition.

$$e = \frac{v'_A - v'_B}{v_B - v_A} = 0$$

(b) Let A represent the falling object and B represent the heavy steel plate. The speeds of the steel plate are $v_B = 0$ and $v'_B = 0$. Thus $e = -v'_A/v_A$. Consider energy conservation during the falling or rising path. The potential energy of body A at height h is transformed into kinetic energy just before it collides with the plate. Choose down to be the positive direction.

$$mgh = \frac{1}{2}mv_A^2 \rightarrow$$
$$v_A = \sqrt{2gh}$$

The kinetic energy of body A immediately after the collision is transformed into potential energy as it rises. Also, since it is moving upwards, it has a negative velocity.

$$mgh' = \frac{1}{2}mv_A'^2 \rightarrow$$
$$v'_A = -\sqrt{2gh'}$$

Substitute the expressions for the velocities into the definition of the coefficient of restitution.

$$e = -v'_A/v_A = -\frac{-\sqrt{2gh'}}{\sqrt{2gh}} \rightarrow$$

$$\boxed{e = \sqrt{h'/h}}$$

55. Use this diagram for the momenta after the decay. Since there was no momentum before the decay, the three momenta shown must add to 0 in both the x and y directions.

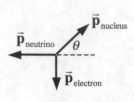

$$\left(p_{\text{nucleus}}\right)_x = p_{\text{neutrino}} \qquad \left(p_{\text{nucleus}}\right)_y = p_{\text{electron}}$$

$$p_{\text{nucleus}} = \sqrt{\left(p_{\text{nucleus}}\right)_x^2 + \left(p_{\text{nucleus}}\right)_y^2} = \sqrt{\left(p_{\text{neutrino}}\right)^2 + \left(p_{\text{electron}}\right)^2}$$

$$= \sqrt{\left(6.2\times10^{-23}\text{ kg}\cdot\text{m/s}\right)^2 + \left(9.6\times10^{-23}\text{ kg}\cdot\text{m/s}\right)^2} = \boxed{1.14\times10^{-22}\text{ kg}\cdot\text{m/s}}$$

$$\theta = \tan^{-1}\frac{\left(p_{\text{nucleus}}\right)_y}{\left(p_{\text{nucleus}}\right)_x} = \tan^{-1}\frac{\left(p_{\text{electron}}\right)}{\left(p_{\text{neutrino}}\right)} = \tan^{-1}\frac{\left(9.6\times10^{-23}\text{ kg}\cdot\text{m/s}\right)}{\left(6.2\times10^{-23}\text{ kg}\cdot\text{m/s}\right)} = 57°$$

The second nucleus' momentum is $\boxed{147° \text{ from the electron's momentum}}$, and is $\boxed{123° \text{ from the neutrino's momentum.}}$

61. To do this problem with only algebraic manipulations is complicated. We use a geometric approach instead. See the diagram of the geometry.

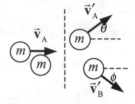

Momentum conservation: $m\vec{\mathbf{v}}_A = m\vec{\mathbf{v}}_A' + m\vec{\mathbf{v}}_B' \rightarrow \vec{\mathbf{v}}_A = \vec{\mathbf{v}}_A' + \vec{\mathbf{v}}_B'$

Kinetic energy conservation: $\frac{1}{2}mv_A^2 = \frac{1}{2}mv_A'^2 + \frac{1}{2}mv_B'^2 \rightarrow v_A^2 = v_A'^2 + v_B'^2$

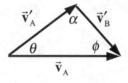

The momentum equation can be illustrated as a vector summation diagram, and the kinetic energy equation relates the magnitudes of the vectors in that summation diagram. Examination of the energy equation shows that it is identical to the Pythagorean theorem. The only way that the Pythagorean theorem can hold true is if the angle α in the diagram is a right angle. If α is a right angle, then $\theta + \phi = 90°$, and so the angle between the final velocity vectors must be 90°.

67. From the symmetry of the wire, we know that $x_{\text{CM}} = 0$. Consider an infinitesimal piece of the wire, with mass dm, and coordinates $(x, y) = (r\cos\theta, r\sin\theta)$. If the length of that piece of wire is dl, then since the wire is uniform, we have $dm = M\, dl/(\pi r)$. And from the diagram and the definition of radian angle measure, we have $dl = r\, d\theta$. Thus $dm = M r\, d\theta/(\pi r) = M\, d\theta/\pi$. Now, apply Eq. 9-13.

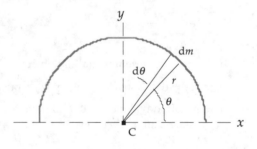

$$y_{\text{CM}} = \frac{1}{M}\int y\, dm = \frac{1}{M}\int_0^\pi r\sin\theta\,\frac{M}{\pi}\,d\theta = \frac{r}{\pi}\int_0^\pi \sin\theta\, d\theta = \frac{2r}{\pi}$$

Thus the coordinates of the center of mass are $\left(x_{\text{CM}}, y_{\text{CM}}\right) = \boxed{\left(0, \dfrac{2r}{\pi}\right)}$.

73. (*a*) Find the CM relative to the center of the Earth.

$$x_{CM} = \frac{m_E x_E + m_M x_M}{m_E + m_M} = \frac{\left(5.98\times10^{24}\text{ kg}\right)\left(0\right) + \left(7.35\times10^{22}\text{ kg}\right)\left(3.84\times10^8\text{ m}\right)}{5.98\times10^{24}\text{ kg} + 7.35\times10^{22}\text{ kg}}$$

$$= \boxed{4.66\times10^6\text{ m from the center of the Earth}}$$

This is actually inside the volume of the Earth, since $R_E = 6.38 \times 10^6$ m.

(*b*) It is this Earth–Moon CM location that actually traces out the orbit as discussed in an earlier chapter. The Earth and Moon will orbit about this orbit path in (approximately) circular orbits. The motion of the Moon, for example, around the Sun would then be a sum of two motions: i) the motion of the Moon about the Earth–Moon CM; and ii) the motion of the Earth–Moon CM about the Sun. To an external observer, the Moon's motion would appear to be a small radius, higher frequency circular motion (motion about the Earth–Moon CM) combined with a large radius, lower frequency circular motion (motion about the Sun). The Earth's motion would be similar, but since the center of mass of that Earth–Moon motion is inside the Earth, the Earth would be observed to "wobble" about that CM.

79. Call the origin of coordinates the CM of the balloon, gondola, and person at rest. Since the CM is at rest, the total momentum of the system relative to the ground is 0. The man climbing the rope cannot change the total momentum of the system, and so the CM must stay at rest. Call the upward direction positive. Then the velocity of the man with respect to the balloon is -v. Call the velocity of the balloon with respect to the ground v_{BG}. Then the velocity of the man with respect to the ground is $v_{MG} = -v + v_{BG}$. Apply conservation of linear momentum in one dimension.

$$0 = mv_{MG} + Mv_{BG} = m\left(-v + v_{BG}\right) + Mv_{BG} \quad \rightarrow$$

$$\boxed{v_{BG} = v\frac{m}{m+M} \text{ , upward}}$$

If the passenger stops, $\boxed{\text{the balloon also stops}}$, and the CM of the system remains at rest.

85. It is proven in the solution to problem 61 that in an elastic collision between two objects of equal mass, with the target object initially stationary, the angle between the final velocities of the objects is 90°.

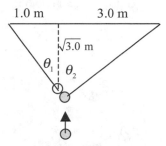

For this specific circumstance, see the diagram. We assume that the target ball is hit "correctly" so that it goes in the pocket. Find θ_1 from the geometry of the "left" triangle:

$$\theta_1 = \tan^{-1}\frac{1.0}{\sqrt{3.0}} = 30°$$

Find θ_2 from the geometry of the "right" triangle:

$$\theta_2 = \tan^{-1}\frac{3.0}{\sqrt{3.0}} = 60°$$

Since the balls will separate at a 90° angle, if the target ball goes in the pocket, this does appear to be a $\boxed{\text{good possibility of a scratch shot}}$.

91. The fraction of energy transformed is $\dfrac{K_{initial} - K_{final}}{K_{initial}}$.

$$\frac{K_{initial} - K_{final}}{K_{initial}} = \frac{\frac{1}{2}m_A v_A^2 - \frac{1}{2}\left(m_A + m_B\right)v'^2}{\frac{1}{2}m_A v_A^2} = \frac{m_A v_A^2 - \left(m_A + m_B\right)\left(\frac{m_A}{m_A + m_B}\right)^2 v_A^2}{m_A v_A^2}$$

$$= 1 - \frac{m_A}{m_A + m_B} = \frac{m_B}{m_A + m_B} = \boxed{\frac{1}{2}}$$

97. Since the only forces on the astronauts are internal to the 2-astronaut system, their CM will not change. Call the CM location the origin of coordinates. That is also the original location of the two astronauts.

$$x_{CM} = \frac{m_A x_A + m_B x_B}{m_A + m_B} \quad \rightarrow$$

$$0 = \frac{\left(60 \text{ kg}\right)\left(12 \text{ m}\right) + \left(80 \text{ kg}\right)x_B}{140 \text{ kg}} \quad \rightarrow$$

$$x = -9 \text{ m}$$

Their distance apart is $x_A - x_B = 12$ m - (-9 m) = $\boxed{21 \text{ m}}$.

103. The original horizontal distance can be found from the range formula from Example 3-10.

$$R = v_0^2 \sin 2\theta_0 / g = (25 \, \text{m/s})^2 (\sin 56°) / (9.8 \, \text{m/s}^2) = 52.87 \, \text{m}$$

The height at which the objects collide can be found from Eq. 2-12c for the vertical motion, with $v_y = 0$ at the top of the path. Take up to be positive.

$$v_y^2 = v_{y0}^2 + 2a(y - y_0) \;\rightarrow\; (y - y_0) = \frac{v_y^2 - v_{y0}^2}{2a} = \frac{0 - \left[(25 \, \text{m/s}) \sin 28°\right]^2}{2(-9.80 \, \text{m/s}^2)} = 7.028 \, \text{m}$$

Let m represent the bullet and M the skeet. When the objects collide, the skeet is moving horizontally at $v_0 \cos\theta =$ (25 m/s) cos(28°) = 22.07 m/s = v_x, and the bullet is moving vertically at v_y = 230 m/s. Write momentum conservation in both directions to find the velocities after the totally inelastic collision.

$$p_x: \; Mv_x = (M + m)v_x' \;\rightarrow\; v_x' = \frac{Mv_x}{M + m} = \frac{(0.25 \, \text{kg})(22.07 \, \text{m/s})}{(0.25 + 0.015) \, \text{kg}} = 20.82 \, \text{m/s}$$

$$p_y: \; mv_y = (M + m)v_y' \;\rightarrow\; v_y' = \frac{mv_y}{M + m} = \frac{(0.015 \, \text{kg})(230 \, \text{m/s})}{(0.25 + 0.015) \, \text{kg}} = 13.02 \, \text{m/s}$$

(*a*) The speed v'$_y$ can be used as the starting vertical speed in Eq. 2-12c to find the height that the skeet–bullet combination rises above the point of collision.

$$v_y^2 = v_{y0}^2 + 2a(y - y_0)_{\text{extra}} \;\rightarrow\;$$

$$(y - y_0)_{\text{extra}} = \frac{v_y^2 - v_{y0}^2}{2a} = \frac{0 - (13.02 \, \text{m/s})^2}{2(-9.80 \, \text{m/s}^2)} = 8.649 \, \text{m} \approx \boxed{8.6 \, \text{m}}$$

(*b*) From Eq. 2-12b applied to the vertical motion after the collision, we can find the time for the skeet–bullet combination to reach the ground.

$$y = y_0 + v_y' t + \tfrac{1}{2} a t^2 \;\rightarrow\; 0 = 8.649 \, \text{m} + (13.02 \, \text{m/s})t + \tfrac{1}{2}(-9.80 \, \text{m/s}^2)t^2 \;\rightarrow\;$$

$$4.9t^2 - 13.02t - 8.649 = 0 \;\rightarrow\; t = 3.207 \, \text{s}, \; -0.550 \, \text{s}$$

The positive time root is used to find the horizontal distance traveled by the combination after the collision.

$$x_{\text{after}} = v_x' t = (20.82 \, \text{m/s})(3.207 \, \text{s}) = 66.77 \, \text{m}$$

If the collision would not have happened, the skeet would have gone $R/2$ horizontally from this point.

$$\Delta x = x_{\text{after}} - \tfrac{1}{2} R = 66.77 \, \text{m} - \tfrac{1}{2}(52.87 \, \text{m}) = 40.33 \, \text{m} \approx \boxed{40 \, \text{m}}$$

Note that the answer is correct to 2 significant figures.

109.(*a*) The average force is the momentum change divided by the elapsed time.

$$F_{\text{avg}} = \frac{\Delta p}{\Delta t} = \frac{m \Delta v}{\Delta t} = \frac{(1500 \, \text{kg})(0 - 45 \, \text{km/h})\left(\dfrac{1 \, \text{m/s}}{3.6 \, \text{km/h}}\right)}{0.15 \, \text{s}} = -1.25 \times 10^5 \, \text{N} \approx \boxed{-1.3 \times 10^5 \, \text{N}}$$

The negative sign indicates direction - that the force is in the opposite direction to the original direction of motion.
(*b*) Use Newton's second law.

$$F_{\text{avg}} = m a_{\text{avg}} \;\rightarrow\; a_{\text{avg}} = \frac{F_{\text{avg}}}{m} = \frac{-1.25 \times 10^5 \, \text{N}}{1500 \, \text{kg}} = -83.33 \, \text{m/s}^2 \approx \boxed{-83 \, \text{m/s}^2}$$

Chapter 10: Rotational Motion

Chapter Overview and Objectives

In this chapter, the kinematics and dynamics of rotational motion about a fixed axis is described. The terms angular velocity, angular acceleration, torque, moment of inertia, and angular momentum are defined and Newton's laws of motion, the law of conservation of momentum, and the work-energy theorem are applied to rotational motion about a fixed axis.

After completing this chapter you should:
- Know the definition of the radian.
- Know the definitions of angular position, angular velocity, and angular acceleration.
- Know how to solve problems with constant angular acceleration.
- Know how translational kinematic quantities of a point on a rotating object are related to the rotational kinematical quantities of the object.
- Know the condition for rolling without slipping.
- Know the definition of torque and angular momentum.
- Know how to apply Newton's second law to problems involving rotational motion.
- Know the definition of angular momentum.
- Know how to apply the law of conservation of angular momentum to problems involving rotational motion.
- Know the expressions for work done by torque and rotational kinetic energy.
- Know how to apply the work-energy theorem to problems with rotational and translational motion.
- Know the dimensions and SI units of the rotational quantities introduced in this chapter.

Summary of Equations

Definition of angular position in terms of radians: $\theta = \dfrac{l}{R}$ (Section 10-1)

Definition of average angular velocity: $\bar{\omega} = \dfrac{\Delta \theta}{\Delta t}$ (Section 10-1)

Definition of instantaneous angular velocity: $\omega = \dfrac{d\theta}{dt}$ (Section 10-1)

Definition of average angular acceleration: $\bar{\alpha} = \dfrac{\Delta \omega}{\Delta t}$ (Section 10-1)

Definition of instantaneous angular acceleration: $\alpha = \dfrac{d\omega}{dt}$ (Section 10-1)

Relationship between angular velocity and velocity of a point on a rotating object:

$$v = R\omega$$ (Section 10-1)

Relationship between angular acceleration and tangential acceleration of a point on a rotating object:

$$a_{tan} = \frac{dv}{dt} = R\frac{d\omega}{dt} = R\alpha$$ (Section 10-1)

Linear acceleration of a point on a rotating object: $\vec{\mathbf{a}} = -\dfrac{v^2}{r}\hat{\mathbf{r}} + R\alpha\,\hat{\mathbf{r}}_\perp$ (Section 10-1)

Relationship between frequency and angular frequency:

$$f = \frac{\omega}{2\pi} \quad or \quad \omega = 2\pi f \qquad \text{(Section 10-1)}$$

Relationship between period and frequency: $T = \dfrac{1}{f}$ (Section 10-1)

Relationships between angular kinematical quantities for constant angular acceleration:

$$\omega = \omega_0 + \alpha t \qquad\qquad \theta = \omega_0 t + \alpha t^2 \qquad \text{(Section 10-3)}$$

$$\omega^2 = \omega_0^2 + 2\alpha\theta \qquad\qquad \bar{\omega} = \frac{\omega + \omega_0}{2} \qquad \text{(Section 10-3)}$$

Definition of torque: $\tau = R_\perp F = RF_\perp = RF\sin\theta$ (Section 10-4)

Definition of moment of inertia: $I = \sum m_i R_i^2$ (Section 10-5)

Newton's second law for rotation: $\sum \tau = I\alpha$ (Section 10-5)

Moment of inertia of a continuous mass distribution: $I = \int R^2\,dm = \int R^2 \rho\,dV$ (Section 10-5)

Parallel-axis theorem: $I = I_{CM} + Mh^2$ (Section 10-7)

Perpendicular-axis theorem: $I_z = I_x + I_y$ (Section 10-7)

Rotational kinetic energy: $K = \frac{1}{2}I\omega^2$ (Section 10-8)

Work done by torque: $W = \int \tau\,d\theta$ (Section 10-8)

Work-energy theorem for rotational motion: $W = \int \tau\,d\theta = \frac{1}{2}I\omega_2^{\,2} - \frac{1}{2}I\omega_1^{\,2}$ (Section 10-8)

Condition for rolling without slipping: $v_{axis} = \omega R_{contact}$ (Section 10-9)

Total kinetic energy of a rigid body: $K = \frac{1}{2}mv_{CM}^2 + \frac{1}{2}I_{CM}\omega^2$ (Section 10-9)

Chapter Summary

Section 10-1. Angular Quantities

When we describe the rotational motion we will always express the angles in terms of **radians**. One radian is defined as the angle subtended by an arc whose length is equal to its radius. An angle of 360° corresponds to 2π radians. Consider an arc length l along a circle of radius R. This arc length covers $l/(2\pi R)$ of the circumference of the circle. The angle subtended by this arc is thus

$$\theta = \left(\frac{l}{2\pi R}\right)2\pi = \frac{l}{R}$$

The angle θ is positive when it corresponds to a counterclockwise rotation and negative when it corresponds to a clockwise rotation.

The quantities that are used to describe rotational motion are angular position, angular velocity, and angular acceleration. These quantities are defined in analogous ways to how linear position, linear velocity, and linear acceleration are defined to describe linear motion.

The rate of change of the angular position θ is called the **angular velocity** ω. The most frequently used unit for angular velocity is the rad/s. The average angular velocity $\bar{\omega}$ is defined as

$$\bar{\omega} = \frac{\Delta\theta}{\Delta t}$$

where $\Delta\theta$ is the change in the angular position during a time interval Δt. The instantaneous angular velocity, ω, is the limit of the average angular velocity $\bar{\omega}$ as Δt goes to zero:

$$\omega = \lim_{\Delta t \to 0}\frac{\Delta\theta}{\Delta t} = \frac{d\theta}{dt}$$

The sign of angular velocity includes information about the rotational direction. An increasing angular position with increasing time results in a positive angular velocity and a decreasing angular position with increasing time results in a negative angular velocity. If the rotation is counterclockwise the angular velocity is positive and if the rotation is clockwise the angular velocity is negative.

The rate of change of the angular velocity ω is called the **angular acceleration** α. The most frequently used unit for angular acceleration is the rad/s^2. The average angular acceleration $\bar{\alpha}$ is defined as

$$\bar{\alpha} = \frac{\Delta\omega}{\Delta t}$$

where $\Delta\omega$ is the change in the angular velocity during a time interval Δt. The instantaneous angular acceleration is the limit of the average angular acceleration as Δt goes to zero:

$$\alpha = \lim_{\Delta t \to 0}\frac{\Delta\omega}{\Delta t} = \frac{d\omega}{dt}$$

An increasing angular velocity with increasing time results in a positive angular acceleration and a decreasing angular velocity with increasing time results in a negative acceleration.

Rotational variables are most useful in our description of rotating rigid objects. If the rotation axis is fixed, each part of the body will have the same angular velocity and angular acceleration. However, different parts will have different linear velocities and linear accelerations. The path followed by a point on a rigid object is a circle, which is centered on the rotation axis. If the radius of the circle is R then the linear displacement l of this point is related to the angular displacement θ in the following manner

$$l = R\theta$$

The linear velocity can be obtained by differentiating the linear displacement with respect to time:

$$v = \frac{dl}{dt} = R\frac{d\theta}{dt} = R\omega$$

Note that in this relationship, the units of the left-hand side of the equation are [L]/[T]. The units of the right-hand side of this equation are [L][θ]/[T]. Do we conclude that angles are dimensionless quantities? This indeed turns out to be correct, and we could have realized this earlier if we had considered the units associated with our definition of the angle θ. The expression l/R is dimensionless and so is θ. We need to keep this in mind when we carry out dimensional analyses of our equations.

If the angular velocity changes as a function of time, the linear velocity will also change and there will be a non-zero linear acceleration. This acceleration is directed parallel or anti-parallel to the direction of motion and is thus tangential to the circular orbit of the part of the object we are looking at:

$$a_t = \frac{dv}{dt} = R\frac{d\omega}{dt} = R\alpha$$

However, even when the angular velocity is constant, the acceleration of the part of the object we are focusing on will be non-zero since we observe that it carries out circular motion. There must thus be a component of the acceleration that points towards the center of the circle and has a magnitude equal to

$$a_r = \frac{v^2}{R}$$

The total acceleration of a point on the rotating rigid object is the vector sum of these two components:

$$\vec{a} = -\frac{v^2}{r}\hat{r} + R\alpha\,\hat{r}_\perp$$

where $\hat{r}$ is a unit vector pointing from the axis of rotation to the point on the object and $\hat{r}_\perp$ is a unit vector pointing tangent to the circle described by the point on the rotating rigid object in the direction of positive angle measure.

The quantity **frequency** of rotation f is a measure of the number of revolutions of the object per unit time. If the rotational angular velocity of an object is ω, the object makes $\omega/2\pi$ revolutions per second. The frequency of rotation is thus

$$f = \frac{\omega}{2\pi}$$

The unit of frequency is s⁻¹ or the Hertz (Hz): 1 Hz = 1 s⁻¹. The time required for one revolution of the object is called the **period** of revolution T. If the object makes f revolutions per second, the time required for one complete revolution is thus $1/f$.

Example 10-1-A. Converting degrees to radians. How many radians correspond to one minute of arc? How many radians correspond to one second of arc?

Approach: When we convert from degrees to radians we use the conversion factor based on the following relation: $360° = 2\pi$ radians.

Solution: One minute of an arc is 1/60 of a degree. Thus

$$1' = \left(\frac{1}{60}°\right) = \left(\frac{1}{60}°\right) \times \left(\frac{2\pi\,\text{rad}}{360°}\right) = 2.91 \times 10^{-4}\ \text{rad}$$

One second of an arc is 1/3600 of a degree. Thus

$$1'' = \left(\frac{1}{3600}°\right) = \left(\frac{1}{3600}°\right) \times \left(\frac{2\pi\,\text{rad}}{360°}\right) = 4.85 \times 10^{-6}\ \text{rad}$$

Example 10-1-B: Angular velocity and angular acceleration of a bicycle wheel. A bicyclist accelerates from rest to a speed of $v_f = 5.38$ m/s in a time of $t = 8.93$ s. The diameter of his bicycle's tires is $d = 27$ inches. What is the angular velocity of the bicycle wheels when he is moving at its final speed? What was the average angular acceleration of the bicycle wheels?

Approach: In this problem we are not dealing with the rotation of a rigid object about a fixed rotation axis. Instead, the rotation axes are moving with the velocity of the bicycle. We will assume that the wheels do not slip. For every rotation of the bicycle wheels the bicycle will move a distance πd and we can thus relate the angular velocity of the wheels to the linear velocity of the bicycle.

Solution: Consider one complete revolution of the bicycle wheel. If this revolution takes place during a time interval Δt we can determine the linear velocity of the bicycle:

$$v = \frac{\pi d}{\Delta t}$$

Since the wheel makes one complete revolution, its angular velocity can now be determined:

$$\omega = \frac{2\pi}{\Delta t}$$

The angular velocity can be related to the linear velocity in the following manner:

$$\omega = \frac{2\pi}{\Delta t} = \frac{2\pi}{\left(\dfrac{\pi d}{v}\right)} = \frac{2v}{d} = \frac{2 \times 5.38}{27 \times 0.0254} = 15.7 \text{ rad/s}$$

Since the bicycle wheel starts from rest, its initial angular velocity is equal to 0 rad/s. To determine the average angular acceleration, we divide the change in angular velocity by the time interval

$$\bar{\alpha} = \frac{\omega - \omega_0}{t} = \frac{15.7 - 0}{8.93} = 1.76 \text{ rad/s}^2$$

Section 10-2. Vector Nature of Angular Quantities

It might seem to you that there is a directional nature to rotation; the rotation axis has a direction and certainly a rotation of an object about a vertical axis is very different from a rotation about a horizontal axis. It is thus clear that in order to specify rotational motion we must specify not only the magnitude of the angular velocity/acceleration but also the direction of the rotation axis. The angular velocity vector is directed along the rotation axis. Its direction can be found by using the **right-hand rule**. To determine the direction of the angular velocity vector (1) curl up the fingers of your right hand with the thumb sticking out away from your hand, (2) orient the curling fingers in the direction of rotation, and (3) your thumb will point in the direction of the angular velocity vector. The magnitude of the angular velocity can be found by using the procedures outlined in Section 10.1.

Using the definition of angular acceleration in terms of the change in the angular velocity we immediately conclude that since the angular velocity vector is directed along the rotation axis, the angular acceleration velocity will also be directed along the rotation axis. The angular acceleration vector is thus either parallel or anti-parallel to the angular velocity vector.

Mathematically, the angular velocity vector defined in this manner is not a vector. Vectors are defined by how they transform under changes in the coordinate system. One of the properties that vectors have is that if all the coordinate directions are reversed, the vector is unchanged. If all coordinate directions are reversed, then right- and left-handedness are reversed. This would result in the angular velocity vector being defined in the opposite direction and its definition is thus not invariant under coordinate reversal. It is not a vector. It is what is called an **axial vector** or **pseudo-vector**. However, for our purposes, we can treat the angular velocity and angular acceleration vectors in the same way as we treat normal vectors.

Section 10-3. Constant Angular Acceleration

Because the mathematical relationships between angular position, angular velocity, and angular acceleration are identical to the relationships between linear position, linear velocity, and linear acceleration, the kinematic equations relating the rotational parameters have forms similar to those relating the linear parameters. For the case of constant angular acceleration (α = constant) or constant linear acceleration (a = constant) we obtain the following kinematic equations (see also Section 2.5):

Rotation: Translation:

$$\omega = \omega_0 + \alpha t$$ $$v = v_0 + at$$

$$\theta = \omega_0 t + \frac{1}{2}\alpha t^2$$ $$x = v_0 t + \frac{1}{2}at^2$$

$$\omega^2 = \omega_0^2 + 2\alpha\theta$$ $$v^2 = v_0^2 + 2ax$$

$$\overline{\omega} = \frac{\omega + \omega_0}{2}$$ $$\overline{v} = \frac{v + v_0}{2}$$

Example 10-3-A. Constant angular acceleration. A winch is used to wind up cable. It is desired to wind up an additional length $l = 13.4$ m of cable onto the drum of diameter $D = 0.364$ m. The current angular velocity of the drum is $\omega = 10.4$ rev/min. What must the constant angular acceleration of the drum be so that the drum comes to rest when the additional cable is wound on the drum? Assume the additional cable on the drum does not change its diameter.

Approach: Since we know the diameter of the drum and the length of the cable to be wound on the drum we know how many degrees the drum has to turn. Since we know the initial and the final angular velocity and the angular displacement we can calculate the required angular acceleration.

Solution: The angle that the drum must turn through is equal to

$$\Delta\theta = \frac{l}{R} = \frac{l}{(D/2)} = \frac{13.4}{(0.364/2)} = 73.6 \, \text{rad}$$

Now consider the kinematic equations for constant angular acceleration α:

$$\theta(t) = \omega_0 t + \frac{1}{2}\alpha t^2$$

$$\omega(t) = \omega_0 + \alpha t$$

where ω_0 is the angular velocity at time $t = 0$ s and we have assumed that the angular position at time $t = 0$ s is 0 rad. Using the last equation we can determine the time t at which the angular velocity will be 0 rad/s:

$$t = -\frac{\omega_0}{\alpha}$$

The angular position at this time is equal to

$$\theta\left(t = -\frac{\omega_0}{\alpha}\right) = \omega_0\left(-\frac{\omega_0}{\alpha}\right) + \frac{1}{2}\alpha\left(-\frac{\omega_0}{\alpha}\right)^2 = -\frac{\omega_0^2}{\alpha} + \frac{1}{2}\frac{\omega_0^2}{\alpha} = -\frac{1}{2}\frac{\omega_0^2}{\alpha}$$

The angular displacement $\Delta\theta$ is equal to

$$\Delta\theta = \theta\left(t = -\frac{\omega_0}{\alpha}\right) - \theta(0) = -\frac{1}{2}\frac{\omega_0^2}{\alpha}$$

This equation can be used to determine the angular acceleration α:

$$\alpha = -\frac{1}{2}\frac{\omega_0^2}{\Delta\theta}$$

In order to use this equation we must convert the units of the initial angular velocity from rev/min to rad/s:

$$10.4 \text{ rev/min} = \left(10.4 \frac{\text{rev}}{\text{min}}\right) \times \left(\frac{2\pi \text{ rad}}{1 \text{ rev}}\right) \times \left(\frac{1 \text{ min}}{60 \text{ s}}\right) = 1.09 \text{ rad/s}$$

The angular acceleration is thus equal to

$$\alpha = -\frac{1}{2}\frac{\omega_0^2}{\Delta\theta} = -\frac{1}{2}\frac{1.09^2}{73.6} = -8.05\times10^{-3} \text{ rad/s}^2$$

Section 10-4. Torque

The rotational motion of a body can be changed if we apply a force. However, when we study rotational motion the point of application of the force and its direction are both important in determining how the force changes the rotational motion. Consider the two situations shown in the Figure to the right, which shows a disk that can rotate around an axis through its center. Assume the disk is initially at rest. When an external force is applied in a direction tangential to the edge of the disk, the disk will start to rotate in the counterclockwise direction (top Figure). When the external force is applied in a direction perpendicular to the edge of the disk, the disk will remain at rest (bottom Figure). The perpendicular distance of the rotation axis to the line the force acts along is called the **moment arm** $R_\perp$. In the top Figure the moment arm is equal to R, the radius of the disk, while in the bottom Figure the moment arm is equal to 0 m. The magnitude of the **torque** τ associated with an applied force F is defined to be the product of the moment arm $R_\perp$ and magnitude of the force F:

$$\tau = R_\perp F$$

An equivalent way of calculating the torque is to write it as

$$\tau = RF_\perp$$

where R is the distance from the axis of rotation to the point of application of the force on the object and $F_\perp$ is the component of the force perpendicular to the line from the axis of rotation to the point of application of the force. Another way to calculate the torque is

$$\tau = RF\sin\theta$$

where θ is the angle between the line joining the axis of rotation and the point of application of the force with the direction of the force.

Example 10-4-A. The total torque on a bolt. A wrench is used to loosen a bolt as shown in the diagram. The total frictional torque is equivalent to a force $f = 1{,}890$ N acting with a moment arm equal to the radius of the bolt $R_\perp = 0.00432$ m. The person loosening the bolt is pushing with a force $F = 132$ N at an angle $\theta = 59°$, as shown in the diagram. The bolt is rotating with a constant angular velocity; this implies that the total torque on the bolt is zero. What is the distance from the axis of rotation to the point at which the person is applying the force to the wrench?

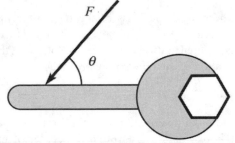

Approach: The total torque acting on the bolt is the sum of the frictional torque and the applied torque. Since the problems tells us that the total torque is 0 Nm, and since we can calculate the frictional torque easily, we can determine the applied torque. Using the magnitude and the direction of the applied force, we can calculate the distance of application.

Solution: The torque associated with the frictional force is equal to

$$\tau_{friction} = -R_\perp F = -\left(0.00432\times1890\right) = -8.16 \text{ Nm}$$

where the minus sign is a result from the fact that this torque would produce a rotation in the clockwise direction. The torque associated with the applied force must thus be $+8.16$ Nm. The torque due to the applied force on the wrench is given by

$$\tau_{wrench} = +dF_{\perp} = +dF\sin\theta$$

where the positive sign is a result from the fact that this torque would produce a rotation in the counterclockwise direction. Since we know the torque associated with applied force we can determine d:

$$d = \frac{\tau_{wrench}}{F\sin\theta} = \frac{8.16}{132\times\sin(59°)} = 0.0721 \text{ m}$$

Section 10-5. Rotational Dynamics; Torque and Rotational Inertia

Consider a force F acting on a particle of mass m that carries out circular motion with radius R (see Figure). At the moment indicated in the figure, we can apply Newton's second law to determine the tangential linear acceleration of the particle:

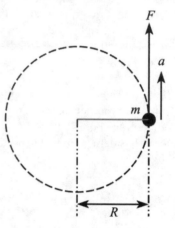

$$F = ma$$

The torque associated with the force F is FR. Rewriting Newton's second law in terms of the torque we find

$$\tau = RF = mRa$$

Since the particle is going to carry out circular motion, it may be more appropriate to use angular variables. By replacing the linear acceleration in the previous equation with the angular acceleration we obtain

$$\tau = RF = mR(R\alpha) = (mR^2)\alpha$$

This relation looks very similar to Newton's second law for linear motion. We see that the force is replaced by the torque associated with the force, the linear acceleration is replaced by the angular acceleration, and the mass is replaced by what is called the **moment of inertia** $I = mR^2$ about the axis of rotation. Since the moment of inertia depends on the distance to the rotation axis, it is not an intrinsic property of an object; it will differ for different rotation axes.

A rigid object can be considered to consist of a large number of discrete masses, each carrying out circular motion around the rotation axis. The moment of inertia about the rotation axis for this collection of discrete masses is equal to

$$I = \sum_i m_i R_i^2$$

where m_i is the mass of the ith particle located a distance R_i from the axis of rotation. The sum is over each of the discrete masses that make up the object. If more than one force is acting on the object we need to calculate the torque associated with each force to calculate the total torque associated with these forces. Newton's second law for the rotation of the object now becomes

$$\sum \tau = I\alpha$$

Although this relation is valid for a rotation around a fixed axis, it can also be applied to a moving axis of rotation if the axis of rotation is through the center of mass of the body and the axis does not change direction in space during the motion. In that case

$$\left(\sum \tau\right)_{CM} = I_{CM}\alpha_{CM}$$

where the CM subscript refers to the center of mass of the object.

Section 10-6. Solving Problems in Rotational Dynamics

The procedure used to solve problems involving rotational dynamics is similar to the procedure followed to solve problems involving linear dynamics. In order to solve problems involving rotational dynamics the following steps should be followed:

1. As in any problem, draw a clear and complete diagram of the situation. Identify the known and unknown quantities in the problem.
2. Draw a free-body diagram for each object in the problem. In problems involving rotational dynamics it is important to place the point of action of the force on the body correctly so that the torque is correctly calculated.
3. Identify the axis of rotation of the problem. Remember that we have adopted the convention that a counterclockwise rotation corresponds to a positive angular velocity and clockwise rotation corresponds to a negative angular acceleration. Similarly, any torque that tends to rotate the object counterclockwise is positive and any torque that tends to rotate the object clockwise is negative.
4. Apply Newton's second law of motion for rotation, $\Sigma\tau = I\alpha$. Remember to express all quantities in a consistent set of units.
5. Apply Newton's second law for translational motion also, if necessary.
6. Solve the equations of motion for the unknown quantity.
7. Check your answer by making order of magnitude estimates and asking yourself if the answer is sensible or not.

Example 10-6-A. Motion down an inclined plane. A rope is wound around a spool with a moment of inertia $I = 3.45$ kg·m^2 and radius $R = 0.187$ m. One end of the rope is attached to a block of mass $m = 6.45$ kg that is free to slide down a frictionless plane with an angle of inclination $\theta = 32°$, as shown in the diagram. What is the acceleration of the block down the plane?

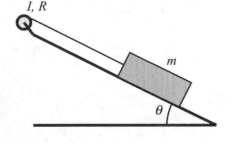

Approach: In order to solve this problem we need to follow the steps outlined in Section 10.6. We will assume that the rope does not slip on the spool and the angular acceleration of the spool is thus directly related to the linear acceleration of the block.

Solution: Let's follow the steps outlined in Section 10.6.
1. *As in any problem, draw a clear and complete diagram of the situation. Identify the known and unknown quantities in the problem.* The diagram shown on the right shows the situation and defines the direction of the linear and angular acceleration that are to be determined.
2. *Draw a free-body diagram for each object in the problem. In problems involving rotational dynamics it is important to place the point of action of the force on the body correctly so that the torque is correctly calculated.*
   In this problem we have two objects to consider: the spool and the block. There are three forces acting on the spool: the force T exerted by the rope on the spool, the gravitational force mg due to the mass of the spool, and the normal force P exerted by the axis on the spool. There are also three forces acting on the block: the gravitational force Mg due to the mass of the block, the normal force N exerted by the inclined plane on the block, and the force T exerted by the rope on the block. The free-body diagrams for the two objects are shown in the Figure on the right.
3. *Identify the axis of rotation of the problem. Remember that we have adopted the convention that a counterclockwise rotation corresponds to a positive angular velocity and clockwise rotation corresponds to a negative angular acceleration. Similarly, any torque that tends to rotate the object counterclockwise is positive and any torque that tends to rotate the object clockwise is negative.* The tension T acts on the spool and will produce a clockwise rotation. Since both the gravitational force mg and the normal force P are acting at the location of the rotation axis, the torque associated with these two forces is equal to 0 Nm. The total torque acting on the spool is thus the torque due to the tension T; the torque is equal to $-TR$.
4. *Apply Newton's second law of motion for rotation, $\Sigma\tau = I\alpha$. Remember to express all quantities in a consistent set of units.* Using the torque calculated in step 3, we can now determine the angular acceleration:

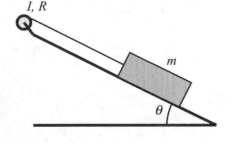
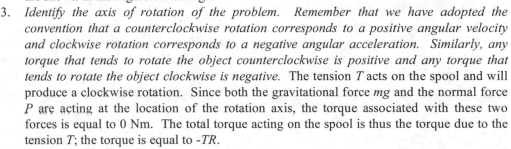
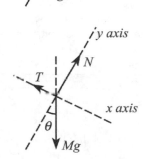

$$\alpha = \frac{\Sigma\tau}{I} = -\frac{RT}{I}$$

5. *Apply Newton's second law for translational motion also, if necessary.* The block will move down the incline with a linear acceleration a along the x axis. The component of the net force along the x axis is equal to

$$F_x = Mg\sin\theta - T$$

The linear acceleration of the block is thus equal to

$$a = \frac{F_x}{M} = \frac{Mg\sin\theta - T}{M} = g\sin\theta - \frac{T}{M}$$

6. *Solve the equations of motion for the unknown quantity.* If we look at the two expressions for the linear and the angular acceleration we see two equations with three unknowns (α, a, T). We cannot solve these equations unless we have a third equation. The third equation is the equation that relates the linear and the angular acceleration. Since the rope does not slip, we know that the linear acceleration of the rim of the spool must be equal to the linear acceleration of the block: $a = -\alpha R$. The minus sign is used in this equation to ensure consistency of the signs of the accelerations (a positive a will result in a negative α.) Using our expressions for the linear and angular acceleration we find

$$a = g\sin\theta - \frac{T}{M} = -\alpha R = \frac{R^2 T}{I}$$

or

$$g\sin\theta = \frac{R^2 T}{I} + \frac{T}{M} = T\left\{\frac{MR^2 + I}{IM}\right\}$$

The tension T is equal to

$$T = \frac{g\sin\theta}{\left\{\dfrac{MR^2 + I}{IM}\right\}} = \left\{\frac{IM}{MR^2 + I}\right\}g\sin\theta$$

Using this tension we can now determine the acceleration of the block

$$a = g\sin\theta - \frac{T}{M} = g\sin\theta - \left\{\frac{I}{MR^2 + I}\right\}g\sin\theta = \left\{\frac{MR^2}{MR^2 + I}\right\}g\sin\theta = 0.319\text{ m/s}^2$$

7. *Check your answer by making order of magnitude estimates and asking yourself if the answer is sensible or not.* We can check our answer by comparing it with what we know about the acceleration of a free box on an inclined plane. In that case, the acceleration is $g\sin\theta$. We see that our answer in part 6 reduces to this case when $I = 0$ kg m^2.

Section 10-7. Determining Moments of Inertia

The moment of inertia of a collection of point masses was discussed in Section 10-5. To determine the moment of inertia about a given axis for an extended body, calculus must be used in a way similar to determining the center of mass of an extended body. We think of the extended body as a collection of infinitesimal masses of size dm. To determine the moment of inertia of the extended body, we add the moment of inertia of each piece, $R^2 dm$, by integrating over the body:

$$I = \int R^2\, dm$$

Here R is the distance from the given moment of inertia to the location of the mass dm. As in the case of determining the center of mass of an extended body, we usually write the dm as a density times an infinitesimal volume element, $\rho\, dV$. In this form the integral appears as

$$I = \int R^2 \rho\, dV$$

There are two important theorems that can sometimes be used to assist in determining the moment of inertia of objects about particular axes. The **parallel-axis theorem** states that the moment of inertia about an axis that is displaced a distance h from a parallel axis through the center of mass is the moment of inertia about the axis through the center of mass plus the mass of the body, M, multiplied by h^2:

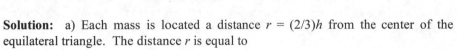

$$I = I_{CM} + Mh^2$$

The second theorem is the **perpendicular axis theorem**. It applies only to effectively two-dimensional objects, objects whose thickness is very small compared to their dimensions in the plane perpendicular to their thickness. The theorem states that the moment of inertia of the body about an axis perpendicular to the body is equal to the sum of the moments of inertia about any two perpendicular axes in the plane of the body that intersect with the first axis in the plane of the body. We write this as

$$I_z = I_x + I_y$$

where I_z is the moment of inertia about the axis perpendicular to the plane of the body, and I_x and I_y are moments of inertia of the body about two perpendicular axes that lie in the plane of the body and intersect the z axis in the plane of the body.

Example 10-7-A. Calculating the moment of inertia of a point-mass distribution. Three equal masses, m, lie on the corners of an equilateral triangle with sides of length a. Determine the moment of inertia about axes perpendicular to the plane of the triangle a) through the center of the triangle, b) through a mid-point of one edge of the triangle, and c) through a vertex of the triangle.

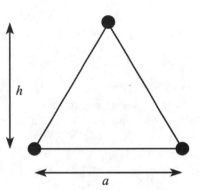

Approach: For each mass we have to calculate the distance to the rotation axis and use it to calculate its contribution to the moment of inertia. The total moment of inertia is the sum of the contributions from each mass.

Solution: a) Each mass is located a distance $r = (2/3)h$ from the center of the equilateral triangle. The distance r is equal to

$$r = \frac{2}{3} a \sin 60° = \sqrt{\frac{1}{3}}\, a$$

The moment of inertia of each mass with respect to the rotation axis is equal to

$$dI = mr^2 = \frac{1}{3} m a^2$$

Adding up the moments of inertia for each mass can determine the total moment of inertia with respect to this rotation axis

$$I = \sum dI = ma^2$$

b) Consider a rotation axis through a mid-point of one edge of the triangle. Two masses are located a distance $a/2$ from the rotation axis, and one mass is located a distance h from the rotation axis. The moment of inertia of the system around this rotation axis is equal to

$$I = \sum mr^2 = m\left(\frac{a}{2}\right)^2 + m\left(\frac{a}{2}\right)^2 + mh^2 = \frac{1}{2} m a^2 + m\left(a \sin 60°\right)^2 = \frac{5}{4} m a^2$$

c) Consider an axis located through one of the vertices of the triangle. One of the masses is on the rotation axis and two are a distance a from the rotation axis. The moment of inertia of the system around this rotation axis is equal to

$$I = \sum mr^2 = ma^2 + ma^2 + 0 = 2ma^2$$

Example 10-7-B. Applying the parallel-axis theorem. Obtain the answers for parts b) and c) for Example 10-7-A using the answer to part a) and the parallel-axis theorem.

Approach: Using the answer to part a) of Example 10-7-A we can calculate the moment of inertia with respect to any parallel axis.

Solution: The moment of inertia about the rotation axis through the center of mass is determined in part a):

$$I_{CM} = \sum dI = ma^2$$

Now consider part b). The distance between a rotation axis through a mid-point of one edge of the triangle and the rotation axis through the center of mass is $(1/3)h$. Applying the parallel-axis theorem we can calculate the moment of inertia with respect to this axis:

$$I = I_{CM} + 3m\left(\frac{1}{3}h\right)^2 = ma^2 + \frac{1}{3}m\left(a\sin 60°\right)^2 = ma^2 + \frac{1}{4}ma^2 = \frac{5}{4}ma^2$$

Now consider part c). The distance between a rotation axis through one of the masses of the triangle and the rotation axis through the center of mass is $(2/3)h$. Applying the parallel-axis theorem we can calculate the moment of inertia with respect to this axis:

$$I = I_{CM} + 3m\left(\frac{2}{3}h\right)^2 = ma^2 + \frac{4}{3}m\left(a\sin 60°\right)^2 = ma^2 + ma^2 = 2ma^2$$

Example 10-7-C. Calculating the moment of inertia. A cylinder of radius $R = 0.127$ m and height $h = 0.243$ m has a density that varies with radius from the axis of the cylinder as $\rho(r) = (a - br)$ where $a = 3200$ kg/m^3 and $b = 1000$ kg/m^4. What is the moment of inertia of this cylinder about its axis?

Approach: In order to solve this problem we consider the cylinder as a collection of cylindrical shells of thickness dr. The moment of inertia of each shell is easy to calculate since each element of the shell is located the same distance from the rotation axis. The total moment of inertia can be obtained by adding (integrating) the contribution of each cylindrical shell.

Solution: The volume of a cylindrical shell of inner radius r, outer radius $r + dr$, and height h is equal to dV where

$$dV = \left\{\pi\left(r + dr\right)^2 - \pi r^2\right\}h \approx 2\pi rhdr$$

The mass of this cylindrical shell is equal to

$$dM = \rho(r)dV = 2\pi rh\rho(r)dr$$

The moment of inertia of this cylinder is its mass times its radius squared (since each mass element is located the same distance from the rotation axis). Thus

$$dI = r^2 dM = 2\pi r^3 h\rho(r)dr$$

The moment of inertia of the cylinder can be found by summing the contributions from each cylindrical shell:

$$I = \int dI = \int_0^R 2\pi r^3 h\rho(r)dr = \int_0^R 2\pi r^3 h\left(a - br\right)dr = 2\pi h\left\{\frac{1}{4}ar^4 - \frac{1}{5}br^5\right\}\Big|_0^R = 2\pi h\left\{\frac{1}{4}aR^4 - \frac{1}{5}bR^5\right\} = 0.308 \text{ kg m}^2$$

Example 10-7-D. Calculating the moment of inertia using the parallel-axis theorem. Determine the moment of inertia of a uniform solid sphere of radius R and mass M about an axis that is tangent to the surface of the sphere.

Approach: The moment of inertia of a uniform solid sphere of radius R and mass M around a rotation axis through its center can be calculated using a procedure similar to the procedure followed in Example 10-7-C. The result of this calculation can be found in Figure 10-20 of the textbook. By applying the parallel-axis theorem we can determine the moment of inertia about an axis that is tangent to the surface of the sphere.

Solution: The moment of inertia of the sphere about an axis through the center of the sphere is equal to

$$I_{CM} = \frac{2}{5} MR^2$$

We can always find an axis through the center of mass that is parallel to the rotation axis that is tangent to the surface of the sphere. The distance between these two axes is equal to the radius of the sphere. Applying the parallel-axis theorem we find that the moment of inertia about the axis tangent to the surface of the sphere is equal to

$$I_{tangent} = I_{CM} + MR^2 = \frac{2}{5} MR^2 + MR^2 = \frac{7}{5} MR^2$$

Example 10-7-E. Calculating the moment of inertia using the perpendicular-axis theorem. Use the perpendicular-axis theorem to determine the moment of inertia of a flat metal washer of mass M with an inner radius of R_1 and outer radius R_2 about an axis in the plane of the washer, passing through a diameter of the washer.

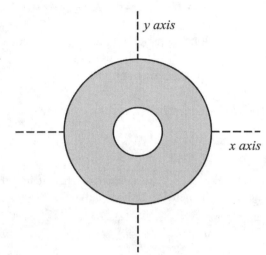

Approach: Consider the washer shown in the Figure. We are asked to calculate the moment of inertia with respect to the x axis. It will be difficult to calculate the moment of inertia with respect to this axis directly. We can easily determine the moment of inertia with respect to an axis through the center of mass, perpendicular to the xy plane - the z axis (see Figure 10.20 in the textbook).

Solution: The moment of inertia with respect to the z axis is equal to

$$I_z = \frac{1}{2} M \left\{ R_1^2 + R_2^2 \right\}$$

Due to the symmetry of the system, the moment of inertia with respect to the x axis is the same as the moment of inertia with respect to the y axis. Using the perpendicular-axis theorem we can express the moment of inertia with respect to the z axis in terms of the moment of inertia with respect to the x and y axes:

$$I_z = I_x + I_y = 2I_x$$

or

$$I_x = \frac{1}{2} I_z = \frac{1}{2} \left[\frac{1}{2} M \left(R_1^2 + R_2^2 \right) \right] = \frac{1}{4} M \left(R_1^2 + R_2^2 \right)$$

Section 10-8. Rotational Kinetic Energy

The work-energy theorem can be written in terms of rotational quantities. The rotational kinetic energy K of a body is equal to

$$K = \frac{1}{2} I \omega^2$$

The work W done by a torque on a body is given by

$$W = \int \tau \, d\theta$$

The work-energy principle can be written in terms of these quantities as

$$W = \int \tau \, d\theta = \frac{1}{2} I \omega_2^2 - \frac{1}{2} I \omega_1^2 = \Delta K$$

Example 10-8-A. Applying the work-energy theorem for rotational motion. Stand a pencil on its point and let it go. It will fall over and strike the table. What is its angular velocity as it strikes the table? Assume that it rotates about the point in contact with the table and its moment of inertia is that of a rod rotating about an axis through its end.

Approach: During the motion of the pencil, the gravitational force does work on the pencil. The work can be determined by determining the change in the potential energy of the pencil. Since the pencil is initially at rest, its initial angular velocity is equal to 0 rad/s. The work-energy theorem can now be used to determine the final angular velocity.

Solution: Let us assume that the length of the pencil is L and its mass is M. If the pencil is uniform, its center of mass is located at its center. The change in the potential energy of the pencil is

$$\Delta U = U_f - U_i = 0 - Mg\frac{L}{2} = -\frac{1}{2}MgL$$

The work done by the gravitational force is equal to

$$W = -\Delta U = \frac{1}{2}MgL$$

The work-energy theorem tells us that the change in the kinetic energy is

$$\Delta K = W = \frac{1}{2}MgL$$

The kinetic energy of the pencil when it hits the table is in the form of rotational kinetic energy. The initial kinetic energy of the pencil will be equal to 0 J. The change in the kinetic energy is thus equal to

$$\Delta K = K_f - K_i = \frac{1}{2}I\omega_f^2 = \frac{1}{2}\left(\frac{1}{3}ML^2\right)\omega_f^2 = \frac{1}{6}ML^2\omega_f^2$$

Using this expression and the work-energy theorem we can determine the final angular velocity:

$$\omega_f = \sqrt{\frac{6\Delta K}{ML^2}} = \sqrt{\frac{6\left(\frac{1}{2}MgL\right)}{ML^2}} = \sqrt{\frac{3g}{L}}$$

Section 10-9. Rotational Plus Translational Motion; Rolling

When we consider rolling motion, we will in general assume that the rolling object is not slipping. This requires that the linear velocity of the center of mass v_{CM} is related to the angular velocity ω around the center of mass in the following way

$$v_{CM} = R\omega$$

where R is the radius of the rolling object. Rolling motion without slipping is only possible if there is sufficient friction between the wheel and the surface.

An object can have non-zero center-of-mass translational and rotational motion. An object carrying out rolling motion is an example of an object that has both translational and rotational motion. If the axis of rotation of the object maintains a fixed direction in space, the total kinetic energy of the object can be written as the sum of translational and rotational kinetic energy

$$K = \frac{1}{2}mv_{CM}^2 + \frac{1}{2}I_{CM}\omega^2$$

where v_{CM} is the velocity of the center of mass of the object, I_{CM} is the moment inertia about an axis through the center of mass parallel to the axis of rotation of the object, m is the mass of the object, and ω is the angular velocity of the object. Be careful to make sure that you only use $\Sigma\tau = I\alpha$ if (1) the axis of rotation is fixed in an inertial reference frame or (2) the axis is fixed in direction and passes through the center of mass of the object.

Section 10-10. Why Does a Rolling Sphere Slow Down?

If the only force on a rolling sphere acts at the point of contact with the surface it is rolling on, no work would be done on the sphere because there is no displacement at the point of contact if the sphere rolls without slipping. However, every real sphere and surface distort slightly on contact between the sphere and the surface. If the object is in motion, slippage has to occur at some points along the area of contact. The frictional force at the surface will do negative work on the object and slow it down.

Practice Quiz

1. What are the x and y components of the linear velocity of a point a distance R from the axis of a rotating object that is rotating with a constant angular velocity ω if $\theta = 0$ rad at time $t = 0$ s?

2. On a rotating object with a fixed axis, is it possible for the tangential and centripetal acceleration to add to each other so that a point a distance R from the axis has zero total acceleration?

3. On a rotating object with a fixed axis, is it possible for the acceleration of the center of mass to add to the centripetal acceleration so that a point a distance R from the axis of rotation has zero total acceleration?

4. What is the direction of the angular velocity of the second hand of a clock?

5. The velocity of the contact point of a wheel that rolls without slipping and the ground is 0 m/s, relative to the ground. What is the velocity of the point at the top of the wheel, relative to the ground?

6. What is incorrect about the statement "The moment of inertia of the object is 24 kg·m²"?

7. Sketch the path of three different points on a wheel as it rolls without slipping: the center of the wheel, a point on the outside circumference of the wheel, and a point halfway from the center to the outside edge.

8. Find the fallacy in the following. A perpetual motion apparatus can be constructed by hanging two masses on springs that slide on arms as shown in the diagram. If the apparatus is initially spun to some angular velocity, it will spin forever because if it starts to slow down, the masses will move inward due to reduced centripetal force. If the masses move inward, the angular momentum is reduced and the angular velocity must increase to conserve angular momentum. Thus the angular velocity can never decrease so the system will spin forever at constant speed.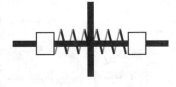

9. For a given direction axis of rotation, through which point does the axis pass that minimizes the moment of inertia?

10. Some fancy doors have the doorknob in the middle of the door rather than near the edge of the door opposite the hinges. Why does the knob in this position make it more difficult to pull the door open quickly?

11. A car accelerates from rest to 3.47 m/s in a time of 2.17 s while in first gear. If the car moves 43.7 cm per revolution of the car engine, what was the constant angular acceleration of the car engine during the car's acceleration?

12. A cylindrical wheel with a diameter of 28.2 cm and a mass of 12.7 kg has a force acting on it as shown in the diagram. The point of application of the force is 19.5 cm from the axis of the wheel. What is the angular acceleration of the wheel?

13. Three 2.4 kg masses are located in a plane as shown in the diagram. Determine the location of the center of mass and the moment of inertia about an axis perpendicular to the page through the center of mass.

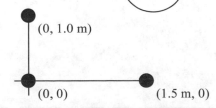

14. The purpose of tail rotors on helicopters is to prevent conservation of angular momentum from rotating the body of the helicopter as the lift rotor changes angular speed. Inexpensive toy helicopters usually have no such tail rotor. A toy helicopter has its rotor with moment of inertia of 0.032 kg·m² about its rotation axis through the center of mass of the helicopter. The rest of the helicopter has a moment of inertia of 0.445 kg·m² about the same axis. The helicopter is initially at rest and a motor increases the angular velocity of the rotor to 800 revolutions per minute relative to the body of the helicopter. If no external torques act on the helicopter, what will the angular velocity of the body of the helicopter be?

15. A yo-yo has a moment of inertia of 1.45×10^{-6} kg·m² and a mass 0.0674 kg. Its string is wrapped around an axle of diameter 0.832 cm. What is the angular velocity of the yo-yo when it has fallen 50.0 cm from its resting position?

Responses to Select End-of-Chapter Questions

1. The odometer will register a distance greater than the distance actually traveled. The odometer counts the number of revolutions and the calibration gives the distance traveled per revolution ($2\pi r$). The smaller tire will have a smaller radius, and a smaller actual distance traveled per revolution.

7. No. If two equal and opposite forces act on an object, the net force will be zero. If the forces are not co-linear, the two forces will produce a torque. No. If an unbalanced force acts through the axis of rotation, there will be a net force on the object, but no net torque.

13. The moment of inertia will be least about an axis parallel to the spine of the book, passing through the center of the book. For this choice, the mass distribution for the book will be closest to the axis.

Solutions to Select End-of-Chapter Problems

1. (a) $(45.0°)(2\pi \text{ rad}/360°) = \boxed{\pi/4 \text{ rad}} = \boxed{0.785 \text{ rad}}$

   (b) $(60.0°)(2\pi \text{ rad}/360°) = \boxed{\pi/3 \text{ rad}} = \boxed{1.05 \text{ rad}}$

   (c) $(90.0°)(2\pi \text{ rad}/360°) = \boxed{\pi/2 \text{ rad}} = \boxed{1.57 \text{ rad}}$

   (d) $(360.0°)(2\pi \text{ rad}/360°) = \boxed{2\pi \text{ rad}} = \boxed{6.283 \text{ rad}}$

   (e) $(445°)(2\pi \text{ rad}/360°) = \boxed{89\pi/36 \text{ rad}} = \boxed{7.77 \text{ rad}}$

7. The angular velocity is expressed in radians per second. The second hand makes 1 revolution every 60 seconds, the minute hand makes 1 revolution every 60 minutes, and the hour hand makes 1 revolution every 12 hours.

   (a) Second hand: $\omega = \left(\dfrac{1 \text{ rev}}{60 \text{ sec}}\right)\left(\dfrac{2\pi \text{ rad}}{1 \text{ rev}}\right) = \boxed{\dfrac{\pi}{30} \text{ rad/sec}} \approx \boxed{1.05 \times 10^{-1} \dfrac{\text{rad}}{\text{sec}}}$

   (b) Minute hand: $\omega = \left(\dfrac{1 \text{ rev}}{60 \text{ min}}\right)\left(\dfrac{2\pi \text{ rad}}{1 \text{ rev}}\right)\left(\dfrac{1 \text{ min}}{60 \text{ s}}\right) = \boxed{\dfrac{\pi}{1800} \dfrac{\text{rad}}{\text{sec}}} \approx \boxed{1.75 \times 10^{-3} \dfrac{\text{rad}}{\text{sec}}}$

   (c) Hour hand: $\omega = \left(\dfrac{1 \text{ rev}}{12 \text{ h}}\right)\left(\dfrac{2\pi \text{ rad}}{1 \text{ rev}}\right)\left(\dfrac{1 \text{ h}}{3600 \text{ s}}\right) = \boxed{\dfrac{\pi}{21,600} \dfrac{\text{rad}}{\text{sec}}} \approx \boxed{1.45 \times 10^{-4} \dfrac{\text{rad}}{\text{sec}}}$

   (d) The angular acceleration in each case is $\boxed{0}$, since the angular velocity is constant.

13. (*a*) The angular rotation can be found from Eq. 10-3a. The initial angular frequency is 0 and the final frequency is 1 rpm.

$$\alpha = \frac{\omega - \omega_0}{t} = \frac{\left(1.0\dfrac{\text{rev}}{\text{min}}\right)\left(\dfrac{2\pi\,\text{rad}}{1\,\text{rev}}\right)\left(\dfrac{1.0\,\text{min}}{60\,\text{s}}\right) - 0\,\text{rad/s}}{720\,\text{s}} = 1.454 \times 10^{-4}\,\text{rad/s}^2 \approx \boxed{1.5 \times 10^{-4}\,\text{rad/s}^2}$$

(*b*) After 7.0 min (420 s), the angular speed is as follows.

$$\omega = \omega_0 + \alpha t = 0 + \left(1.454 \times 10^{-4}\,\text{rad/s}^2\right)(420\,\text{s}) = 6.107 \times 10^{-2}\,\text{rad/s}$$

Find the components of the acceleration of a point on the outer skin from the angular speed and the radius.

$$a_{\text{tan}} = \alpha R = \left(1.454 \times 10^{-4}\,\text{rad/s}^2\right)(4.25\,\text{m}) = \boxed{6.2 \times 10^{-4}\,\text{m/s}^2}$$

$$a_{\text{rad}} = \omega^2 R = \left(6.107 \times 10^{-2}\,\text{rad/s}\right)^2 (4.25\,\text{m}) = \boxed{1.6 \times 10^{-2}\,\text{m/s}^2}$$

19. (*a*) The angular acceleration can be found from Eq. 10-9c.

$$\alpha = \frac{\omega^2 - \omega_o^2}{2\theta} = \frac{0 - (850\,\text{rev/min})^2}{2(1350\,\text{rev})} = \left(-267.6\,\frac{\text{rev}}{\text{min}^2}\right)\left(\frac{2\pi\,\text{rad}}{1\,\text{rev}}\right)\left(\frac{1\,\text{min}}{60\,\text{s}}\right)^2 = \boxed{-0.47\,\frac{\text{rad}}{\text{s}^2}}$$

(*b*) The time to come to a stop can be found from $\theta = \frac{1}{2}\left(\omega_o + \omega\right)t$.

$$t = \frac{2\theta}{\omega_o + \omega} = \frac{2(1350\,\text{rev})}{850\,\text{rev/min}}\left(\frac{60\,\text{s}}{1\,\text{min}}\right) = \boxed{190\,\text{s}}$$

25. Each force is oriented so that it is perpendicular to its lever arm. Call counterclockwise torques positive. The torque due to the three applied forces is given by the following.

$$\tau_{\substack{\text{applied} \\ \text{forces}}} = (28\,\text{N})(0.24\,\text{m}) - (18\,\text{N})(0.24\,\text{m}) - (35\,\text{N})(0.12\,\text{m}) = -1.8\,\text{m}\cdot\text{N}$$

Since this torque is clockwise, we assume the wheel is rotating clockwise, and so the frictional torque is counterclockwise. Thus the net torque is as follows.

$$\tau_{\text{net}} = (28\,\text{N})(0.24\,\text{m}) - (18\,\text{N})(0.24\,\text{m}) - (35\,\text{N})(0.12\,\text{m}) + 0.40\,\text{m}\cdot\text{N} = -1.4\,\text{m}\cdot\text{N}$$

$$= \boxed{1.4\,\text{m}\cdot\text{N},\ \text{clockwise}}$$

31. For a sphere rotating about an axis through its center, the moment of inertia is as follows.

$$I = \tfrac{2}{5}MR^2 = \tfrac{2}{5}(10.8\,\text{kg})(0.648\,\text{m})^2 = \boxed{1.81\,\text{kg}\cdot\text{m}^2}$$

37. (*a*) The small ball can be treated as a particle for calculating its moment of inertia.

$$I = MR^2 = (0.650\,\text{kg})(1.2\,\text{m})^2 = \boxed{0.94\,\text{kg}\cdot\text{m}^2}$$

(*b*) To keep a constant angular velocity, the net torque must be zero, and so the torque needed is the same magnitude as the torque caused by friction.

$$\sum \tau = \tau_{\text{applied}} - \tau_{\text{fr}} = 0 \quad \rightarrow \quad \tau_{\text{applied}} = \tau_{\text{fr}} = F_{\text{fr}}r = (0.020\,\text{N})(1.2\,\text{m}) = \boxed{2.4 \times 10^{-2}\,\text{m}\cdot\text{N}}$$

43. The applied force causes torque, which gives the pulley an angular acceleration. Since the applied force varies with time, so will the angular acceleration. The variable acceleration will be integrated to find the angular velocity. Finally, the speed of a point on the rim is the tangential velocity of the rim of the wheel.

$$\sum \tau = R_0 F_T = I\alpha \ \rightarrow \ \alpha = \frac{R_0 F_T}{I} = \frac{d\omega}{dt} \ \rightarrow \ d\omega = \frac{R_0 F_T}{I} dt \ \rightarrow \ \int_{\omega_0}^{\omega} d\omega = \int_0^t \frac{R_0 F_T}{I} dt \ \rightarrow$$

$$\omega = \frac{v}{R_0} = \omega_0 + \frac{R_0}{I} \int_0^t F_T dt = \frac{R_0}{I} \int_0^t F_T dt \ \rightarrow$$

$$v_T = \omega R_0 = \frac{R_0^2}{I} \int_0^t F_T dt = \frac{R_0^2}{I} \int_0^t \left(3.00t - 0.20t^2\right) dt \ = \frac{R_0^2}{I}\left[\left(\tfrac{3}{2}t^2 - \tfrac{0.20}{3}t^3\right)\text{N}\bullet\text{s}\right]$$

$$v\left(t = 8.0\,\text{s}\right) = \frac{\left(0.330\,\text{m}\right)^2}{\left(0.385\,\text{kg}\bullet\text{m}^2\right)}\left[\left(\tfrac{3}{2}\left(8.0\,\text{s}\right)^2 - \tfrac{0.20}{3}\left(8.0\,\text{s}\right)^3\right)\text{N}\bullet\text{s}\right] = 17.499\,\text{m/s} \approx \boxed{17\,\text{m/s}}$$

49. (a) Thin hoop, radius R_0

$\qquad\qquad\qquad\qquad\qquad\qquad$ $I = Mk^2 = MR_0^2 \ \rightarrow \ k = \boxed{R_0}$

(b) Thin hoop, radius R_0, width w

$\qquad\qquad\qquad\qquad\qquad\qquad$ $I = Mk^2 = \tfrac{1}{2} MR_0^2 + \tfrac{1}{12} Mw^2 \ \rightarrow \ k = \boxed{\sqrt{\tfrac{1}{2} R_0^2 + \tfrac{1}{12} w^2}}$

(c) Solid cylinder

$\qquad\qquad\qquad\qquad\qquad\qquad$ $I = Mk^2 = \tfrac{1}{2} MR_0^2 \ \rightarrow \ k = \boxed{\sqrt{\tfrac{1}{2}}\,R_0}$

(d) Hollow cylinder

$\qquad\qquad\qquad\qquad\qquad\qquad$ $I = Mk^2 = \tfrac{1}{2} M\left(R_1^2 + R_2^2\right) \ \rightarrow \ k = \boxed{\sqrt{\tfrac{1}{2}\left(R_1^2 + R_2^2\right)}}$

(e) Uniform sphere

$\qquad\qquad\qquad\qquad\qquad\qquad$ $I = Mk^2 = \tfrac{2}{5} Mr_0^2 \ \rightarrow \ k = \boxed{\sqrt{\tfrac{2}{5}}\,r_0}$

(f) Long rod, through center

$\qquad\qquad\qquad\qquad\qquad\qquad$ $I = Mk^2 = \tfrac{1}{12} Ml^2 \ \rightarrow \ k = \boxed{\sqrt{\tfrac{1}{12}}\,l}$

(g) Long rod, through end

$\qquad\qquad\qquad\qquad\qquad\qquad$ $I = Mk^2 = \tfrac{1}{3} Ml^2 \ \rightarrow \ k = \boxed{\sqrt{\tfrac{1}{3}}\,l}$

(h) Rectangular thin plate

$\qquad\qquad\qquad\qquad\qquad\qquad$ $I = Mk^2 = \tfrac{1}{12} M\left(l^2 + w^2\right) \ \rightarrow \ k = \boxed{\sqrt{\tfrac{1}{12}\left(l^2 + w^2\right)}}$

55. The parallel-axis theorem is given in Eq. 10-17. The distance from the center of mass of the rod to the end of the rod is $h = l/2$.

$$I = I_{CM} + Mh^2 = \tfrac{1}{12} Ml^2 + M\left(\tfrac{1}{2}l\right)^2 = \left(\tfrac{1}{12} + \tfrac{1}{4}\right) Ml^2 = \boxed{\tfrac{1}{3} Ml^2}$$

61. (*a*) We choose coordinates so that the center of the plate is at the origin. Divide the plate up into differential rectangular elements, each with an area of $dA = dxdy$.

The mass of an element is $dm = \left(\dfrac{M}{lw}\right) dxdy$. The distance of that element from the axis of rotation is $R = \sqrt{x^2 + y^2}$. Use Eq. 10-16 to calculate the moment of inertia.

$$I_{center} = \int R^2 dM = \int_{-w/2}^{w/2} \int_{-l/2}^{l/2} \left(x^2 + y^2\right)\frac{M}{lw} dxdy = \frac{4M}{lw} \int_0^{w/2} \int_0^{l/2} \left(x^2 + y^2\right) dxdy$$

$$= \frac{4M}{lw} \int_0^{w/2} \left[\tfrac{1}{3}\left(\tfrac{1}{2}l\right)^3 + \left(\tfrac{1}{2}l\right)y^2\right] dy = \frac{2M}{w} \int_0^{w/2} \left[\tfrac{1}{12}l^2 + y^2\right] dy$$

$$= \frac{2M}{w}\left[\tfrac{1}{12}l^2\left(\tfrac{1}{2}w\right) + \tfrac{1}{3}\left(\tfrac{1}{2}w\right)^3\right] = \boxed{\tfrac{1}{12} M\left(l^2 + w^2\right)}$$

(b) For the axis of rotation parallel to the w dimension (so the rotation axis is in the y direction), we can consider the plate to be made of a large number of thin rods, each of length l, rotating about an axis through their center. The moment of inertia of one of these rods is $\frac{1}{12} m_i l^2$, where m_i is the mass of a single rod. For a collection of identical rods, then, the moment of inertia would be $I_y = \sum_i \frac{1}{12} m_i l^2 = \boxed{\frac{1}{12} M l^2}$. A similar argument would give

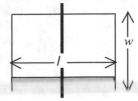

$I_x = \boxed{\frac{1}{12} M w^2}$. This illustrates the perpendicular-axis theorem, Eq. 10-18, $I_z = I_x + I_y$.

67. The only force doing work in this system is gravity, so mechanical energy is conserved. The initial state of the system is the configuration with m_A on the ground and all objects at rest. The final state of the system has m_B just reaching the ground, and all objects in motion. Call the zero level of gravitational potential energy to be the ground level. Both masses will have the same speed since they are connected by the rope. Assuming that the rope does not slip on the pulley, the angular speed of the pulley is related to the speed of the masses by $\omega = v/R$. All objects have an initial speed of 0.

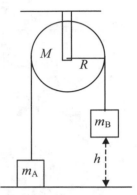

$$E_i = E_f \quad \rightarrow$$

$$\tfrac{1}{2} m_A v_i^2 + \tfrac{1}{2} m_B v_i^2 + \tfrac{1}{2} I \omega_i^2 + m_A g y_{1i} + m_B g y_{2i} =$$
$$\tfrac{1}{2} m_A v_f^2 + \tfrac{1}{2} m_B v_f^2 + \tfrac{1}{2} I \omega_f^2 + m_A g y_{1f} + m_B g y_{2f}$$

$$m_B g h = \tfrac{1}{2} m_A v_f^2 + \tfrac{1}{2} m_B v_f^2 + \tfrac{1}{2} \left(\tfrac{1}{2} M R^2 \right) \left(\frac{v_f^2}{R^2} \right) + m_A g h$$

$$v_f = \sqrt{ \frac{2 (m_B - m_A) g h}{(m_A + m_B + \tfrac{1}{2} M)} } = \sqrt{ \frac{2 (38.0\,\text{kg} - 35.0\,\text{kg})(9.80\,\text{m/s}^2)(2.5\,\text{m})}{(38.0\,\text{kg} + 35.0\,\text{kg} + (\tfrac{1}{2}) 3.1\,\text{kg})} } = \boxed{1.4\,\text{m/s}}$$

73. (a) Mechanical energy is conserved as the sphere rolls without slipping down the plane. Take the zero level of gravitational potential energy to the level of the center of mass of the sphere when it is on the level surface at the bottom of the plane. All of the energy is potential energy at the top, and all is kinetic energy (of both translation and rotation) at the bottom.

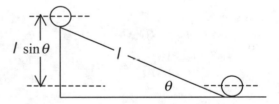

$$E_{\text{intial}} = E_{\text{final}} \quad \rightarrow \quad U_{\text{initial}} = K_{\text{final}} = K_{\text{CM}} + K_{\text{rot}} \quad \rightarrow$$

$$mgh = mgl \sin\theta = \tfrac{1}{2} m v_{\text{bottom}}^2 + \tfrac{1}{2} I \omega_{\text{bottom}}^2 = \tfrac{1}{2} m v_{\text{bottom}}^2 + \tfrac{1}{2} \left(\tfrac{2}{5} m r_0^2 \right) \left(\frac{v_{\text{bottom}}}{r_0} \right)^2 \quad \rightarrow$$

$$v_{\text{bottom}} = \sqrt{\tfrac{10}{7} g h} = \sqrt{\tfrac{10}{7} g l \sin\theta} = \sqrt{\tfrac{10}{7} (9.80\,\text{m/s}^2)(10.0\,\text{m}) \sin 30.0°} = 8.367\,\text{m/s}$$
$$\approx \boxed{8.37\,\text{m/s}}$$

$$\omega_{\text{bottom}} = \frac{v_{\text{bottom}}}{r_0} = \frac{8.367\,\text{m/s}}{0.254\,\text{m}} = \boxed{32.9\,\text{rad/s}}$$

(b) $$\frac{K_{\text{CM}}}{K_{\text{rot}}} = \frac{\tfrac{1}{2} m v_{\text{bottom}}^2}{\tfrac{1}{2} I \omega_{\text{bottom}}^2} = \frac{\tfrac{1}{2} m v_{\text{bottom}}^2}{\tfrac{1}{2} \left(\tfrac{2}{5} m r_0^2 \right) \left(\frac{v_{\text{bottom}}}{r_0} \right)^2} = \boxed{\frac{5}{2}}$$

(c) The translational speed at the bottom, and the ratio of kinetic energies, are both independent of the radius and the mass. The rotational speed at the bottom depends on the radius.

79. (*a*) The total kinetic energy includes the translational kinetic energy of the car's total mass, and the rotational kinetic energy of the car's wheels. The wheels can be treated as one cylinder. We assume the wheels are rolling without slipping, so that $v_{CM} = \omega / R_{wheels}$.

$$K_{tot} = K_{CM} + K_{rot} = \tfrac{1}{2} M_{tot} v_{CM}^2 + \tfrac{1}{2} I_{wheels} \omega^2 = \tfrac{1}{2} M_{tot} v_{CM}^2 + \tfrac{1}{2} \left(\tfrac{1}{2} M_{wheels} R_{wheels}^2 \right) \frac{v_{CM}^2}{R_{wheels}^2}$$

$$= \tfrac{1}{2} \left(M_{tot} + \tfrac{1}{2} M_{wheels} \right) v_{CM}^2 = \tfrac{1}{2} \left(1170\,\text{kg} \right) \left[\left(95\,\text{km/h} \right) \left(\frac{1\,\text{m/s}}{3.6\,\text{km/h}} \right) \right]^2 = 4.074 \times 10^5\,\text{J}$$

$$\approx \boxed{4.1 \times 10^5\,\text{J}}$$

(*b*) The fraction of kinetic energy in the tires and wheels is $\dfrac{K_{rot} + K_{trans\,wheels}}{K_{tot}}$.

$$\frac{K_{rot}}{K_{tot}} = \frac{\tfrac{1}{2} I_{wheels} \omega^2 + \tfrac{1}{2} M_{wheels} v_{CM}^2}{\tfrac{1}{2} M_{tot} v_{CM}^2 + \tfrac{1}{2} I_{wheels} \omega^2} = \frac{\tfrac{1}{2} \left(\tfrac{1}{2} M_{wheels} + M_{wheels} \right) v_{CM}^2}{\tfrac{1}{2} \left(M_{tot} + \tfrac{1}{2} M_{wheels} \right) v_{CM}^2} = \frac{\left(\tfrac{3}{2} M_{wheels} \right)}{\left(M_{tot} + \tfrac{1}{2} M_{wheels} \right)}$$

$$= \frac{210\,\text{kg}}{1170\,\text{kg}} = \boxed{0.18}$$

(*c*) A free-body diagram for the car is shown, with the frictional force of $\vec{F}_{fr}$ at each wheel to cause the wheels to roll. A separate diagram of one wheel is also shown. Write Newton's second law for the horizontal motion of the car as a whole, and the rotational motion of one wheel. Take clockwise torques as positive. Since the wheels are rolling without slipping, $a_{CM} = \alpha / R_{wheels}$.

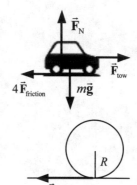

$$\sum \tau = 4 F_{fr} R = I_{wheels} \alpha = \tfrac{1}{2} M_{wheels} R_{wheels}^2 \frac{a_{CM}}{R_{wheels}} \quad \rightarrow$$

$$F_{fr} = \tfrac{1}{8} M_{wheels} a_{CM}$$

$$\sum F_x = F_{tow} - 4 F_{fr} = M_{tot} a_{CM} \quad \rightarrow$$

$$F_{tow} - 4 \left(\tfrac{1}{8} M_{wheels} a_{CM} \right) = M_{tot} a_{CM} \quad \rightarrow$$

$$a_{CM} = \frac{F_{tow}}{\left(M_{tot} + \tfrac{1}{2} M_{wheels} \right)} = \frac{1500\,\text{N}}{\left(1170\,\text{kg} \right)} = 1.282\,\text{m/s}^2 \approx \boxed{1.3\,\text{m/s}^2}$$

(*d*) If the rotational inertia were ignored, we would have the following.

$$\sum F_x = F_{tow} = M_{tot} a_{CM} \quad \rightarrow$$

$$a_{CM} = \frac{F_{tow}}{M_{tot}} = \frac{1500\,\text{N}}{1100\,\text{kg}} = 1.364\,\text{m/s}^2$$

$$\% \text{ error} = \frac{\Delta a_{CM}}{a_{CM}} \times 100 = \frac{1.364\,\text{m/s}^2 - 1.282\,\text{m/s}^2}{1.282\,\text{m/s}^2} \times 100 = \boxed{6\%}$$

85. (*a*) There are two forces on the yo-yo: gravity and string tension. If the top of the string is held fixed, then the tension does no work, and so mechanical energy is conserved. The initial gravitational potential energy is converted into rotational and translational kinetic energy. Since the yo-yo rolls without slipping at the point of contact of the string, the velocity of the CM is related to the angular velocity of the yo-yo by $v_{CM} = r\omega$, where r is the radius of the inner hub. Let m be the mass of the inner hub, and M and R be the mass and radius of each outer disk. Calculate the rotational inertia of the yo-yo about its CM, and then use conservation of energy to find the linear speed of the CM. We take the 0 of gravitational potential energy to be at the bottom of its fall.

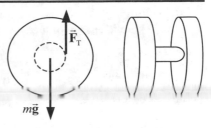

$$I_{CM} = \tfrac{1}{2} mr^2 + 2\left(\tfrac{1}{2} MR^2\right) = \tfrac{1}{2} mr^2 + MR^2$$

$$= \tfrac{1}{2}\left(5.0 \times 10^{-3}\,\text{kg}\right)\left(5.0 \times 10^{-3}\,\text{m}\right)^2 + \left(5.0 \times 10^{-2}\,\text{kg}\right)\left(3.75 \times 10^{-2}\,\text{m}\right)^2 = 7.038 \times 10^{-5}\,\text{kg•m}^2$$

$$m_{total} = m + 2M = 5.0 \times 10^{-3}\,\text{kg} + 2\left(5.0 \times 10^{-2}\,\text{kg}\right) = 0.105\,\text{kg}$$

$$U_{initial} = K_{final} \quad \rightarrow$$

$$m_{total}gh = \tfrac{1}{2} m_{total} v_{CM}^2 + \tfrac{1}{2} I_{CM}\omega^2 = \tfrac{1}{2} m_{total} v_{CM}^2 + \tfrac{1}{2}\frac{I_{CM}}{r^2} v_{CM}^2 = \left(\tfrac{1}{2} m_{total} + \tfrac{1}{2}\frac{I_{CM}}{r^2}\right) v_{CM}^2 \quad \rightarrow$$

$$v_{CM} = \sqrt{\frac{m_{total}gh}{\tfrac{1}{2}\left(m_{total} + \dfrac{I_{CM}}{r^2}\right)}} = \sqrt{\frac{\left(0.105\,\text{kg}\right)\left(9.80\,\text{m/s}^2\right)\left(1.0\,\text{m}\right)}{\tfrac{1}{2}\left[\left(0.105\,\text{kg}\right) + \dfrac{\left(7.038 \times 10^{-5}\,\text{kg•m}^2\right)}{\left(5.0 \times 10^{-3}\,\text{m}\right)^2}\right]}} = 0.8395 = \boxed{0.84\,\text{m/s}}$$

(*b*) Calculate the ratio K_{rot}/K_{tot}.

$$\frac{K_{rot}}{K_{tot}} = \frac{K_{rot}}{U_{initial}} = \frac{\tfrac{1}{2} I_{CM}\omega^2}{m_{total}gh} = \frac{\tfrac{1}{2}\dfrac{I_{CM}}{r^2} v_{CM}^2}{m_{total}gh} = \frac{I_{CM} v_{CM}^2}{2r^2 m_{total}gh}$$

$$= \frac{\left(7.038 \times 10^{-5}\,\text{kg•m}^2\right)\left(0.8395\,\text{m/s}\right)^2}{2\left(5.0 \times 10^{-3}\,\text{m}\right)^2 \left(0.105\,\text{kg}\right)\left(9.8\,\text{m/s}^2\right)\left(1.0\,\text{m}\right)} = 0.96 = \boxed{96\%}$$

91. (*a*) The initial energy of the flywheel is used for two purposes – to give the car translational kinetic energy 20 times, and to replace the energy lost due to friction, from air resistance and from braking. The statement of the problem leads us to ignore any gravitational potential energy changes.

$$W_{fr} = K_{final} - K_{initial} \quad \rightarrow \quad F_{fr}\Delta x \cos 180° = \tfrac{1}{2} M_{car} v_{car}^2 - K_{flywheel}$$

$$K_{flywheel} = F_{fr}\Delta x + \tfrac{1}{2} M_{car} v_{car}^2$$

$$= \left(450\,\text{N}\right)\left(3.5 \times 10^5\,\text{m}\right) + \left(20\right)\tfrac{1}{2}\left(1100\,\text{kg}\right)\left[\left(95\,\text{km/h}\right)\left(\frac{1\,\text{m/s}}{3.6\,\text{km/h}}\right)\right]^2$$

$$= 1.652 \times 10^8\,\text{J} \approx \boxed{1.7 \times 10^8\,\text{J}}$$

(*b*) $K_{flywheel} = \tfrac{1}{2} I\omega^2$

$$\omega = \sqrt{\frac{2\,KE}{I}} = \sqrt{\frac{2\,KE}{\tfrac{1}{2} M_{flywheel} R_{flywheel}^2}} = \sqrt{\frac{2\left(1.652 \times 10^8\,\text{J}\right)}{\tfrac{1}{2}\left(240\,\text{kg}\right)\left(0.75\,\text{m}\right)^2}} = \boxed{2200\,\text{rad/s}}$$

(*c*) To find the time, use the relationship that power = work/time, where the work done by the motor will be equal to the kinetic energy of the flywheel.

$$P = \frac{W}{t} \rightarrow t = \frac{W}{P} = \frac{\left(1.652 \times 10^8 \,\mathrm{J}\right)}{\left(150\,\mathrm{hp}\right)\left(746\ \mathrm{W/hp}\right)} = 1.476 \times 10^3 \,\mathrm{s} \approx \boxed{25\,\mathrm{min}}$$

97. Each wheel supports ¼ of the weight of the car. For rolling without slipping, there will be static friction between the wheel and the pavement. So for the wheel to be on the verge of slipping, there must be an applied torque that is equal to the torque supplied by the static frictional force. We take counterclockwise torques to the right in the diagram. The bottom wheel would be moving to the left relative to the pavement if it started to slip, so the frictional force is to the right. See the free-body diagram.

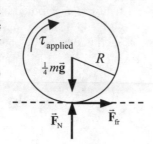

$$\tau_{\substack{\text{applied} \\ \text{min}}} = \tau_{\substack{\text{static} \\ \text{friction}}} = RF_{\text{fr}} = R\mu_s F_{\text{N}} = R\mu_s \tfrac{1}{4} mg$$

$$= \tfrac{1}{4}\left(0.33\,\mathrm{m}\right)\left(0.65\right)\left(950\,\mathrm{kg}\right)\left(9.80\,\mathrm{m/s}^2\right) = \boxed{5.0 \times 10^2 \,\mathrm{m\cdot N}}$$

103. Since there is no friction at the table, there are no horizontal forces on the rod, and so the center of mass will fall straight down. The moment of inertia of the rod about its center of mass is $\frac{1}{12} M l^2$. Since there are no dissipative forces, energy will be conserved during the fall. Take the zero level of gravitational potential energy to be at the tabletop. The angular velocity and the center of mass velocity are related by $\omega_{\text{CM}} = \dfrac{v_{\text{CM}}}{\left(\frac{1}{2} l\right)}$.

$$E_{\text{initial}} = E_{\text{final}} \rightarrow U_{\text{release}} = K_{\text{final}} \rightarrow Mg\left(\tfrac{1}{2} l\right) = \tfrac{1}{2} M v_{\text{CM}}^2 + \tfrac{1}{2} I \omega_{\text{CM}}^2 \rightarrow$$

$$Mg\left(\tfrac{1}{2} l\right) = \tfrac{1}{2} M v_{\text{CM}}^2 + \tfrac{1}{2}\left(\tfrac{1}{12} M l^2\right)\left[\frac{v_{\text{CM}}}{\left(\frac{1}{2} l\right)}\right]^2 \rightarrow gl = \tfrac{4}{3} v_{\text{CM}}^2 \rightarrow v_{\text{CM}} = \boxed{\sqrt{\tfrac{3}{4} gl}}$$

Chapter 11: Angular Momentum; General Rotation

Chapter Overview and Objectives

In this chapter, the vector cross product is introduced and its application to problems of general rotation are discussed. The rotational quantities of Chapter 10 are redefined as vector quantities in this chapter.

After completing this chapter you should:
- Know what a vector cross product is and be able to calculate vector cross products.
- Know the definitions of torque and angular momentum as vector quantities.
- Know how Newton's second law for general rotation can be applied to problems.
- Understand the concept of angular momentum and how it can be used to solve problems.
- Understand how Newton's laws do not apply to observers in noninertial reference frames and why pseudoforces are introduced by observers in these reference frames.

Summary of Equations

Definition of angular momentum (magnitude): $L = I\omega$ (Section 11-1)

Newton's second law written in terms of L: $\sum \tau = \dfrac{dL}{dt}$ (Section 11-1)

Definition of angular momentum (direction): $\vec{\mathbf{L}} = I\vec{\omega}$ (Section 11-1)

Magnitude of vector cross product: $\left| \vec{\mathbf{A}} \times \vec{\mathbf{B}} \right| = AB \sin\theta$ (Section 11-2)

Vector cross product in terms of Cartesian components:

$$\vec{\mathbf{A}} \times \vec{\mathbf{B}} = \begin{vmatrix} \hat{\mathbf{i}} & \hat{\mathbf{j}} & \hat{\mathbf{k}} \\ A_x & A_y & A_z \\ B_x & B_y & B_z \end{vmatrix} = \left(A_y B_z - A_z B_y \right) \hat{\mathbf{i}} + \left(A_z B_x - A_x B_z \right) \hat{\mathbf{j}} + \left(A_x B_y - A_y B_x \right) \hat{\mathbf{k}} \quad \text{(Section 11-2)}$$

Properties of vector cross product: $\vec{\mathbf{A}} \times \vec{\mathbf{A}} = 0$ (Section 11-2)

$\vec{\mathbf{A}} \times \vec{\mathbf{B}} = -\vec{\mathbf{B}} \times \vec{\mathbf{A}}$ (Section 11-2)

$\vec{\mathbf{A}} \times \left(\vec{\mathbf{B}} + \vec{\mathbf{C}} \right) = \vec{\mathbf{A}} \times \vec{\mathbf{B}} + \vec{\mathbf{A}} \times \vec{\mathbf{C}}$ (Section 11-2)

$\dfrac{d}{dt}\left(\vec{\mathbf{A}} \times \vec{\mathbf{B}} \right) = \dfrac{d\vec{\mathbf{A}}}{dt} \times \vec{\mathbf{B}} + \vec{\mathbf{A}} \times \dfrac{d\vec{\mathbf{B}}}{dt}$ (Section 11-2)

Definition of angular momentum of a particle: $\vec{\mathbf{l}} = \vec{\mathbf{r}} \times \vec{\mathbf{p}}$ (Section 11-3)

Newton's second law for a system of particles: $\sum \vec{\tau}_{CM} = \dfrac{d\vec{\mathbf{L}}_{CM}}{dt}$ (Section 11-4)

Newton's second law for a rotating rigid object with a fixed axis:

$$\sum \tau_{axis} = \frac{dL_\omega}{dt} = I\alpha$$

(Section 11-5)

Precession rate of a spinning top:

$$\Omega = \frac{Mgr}{L}$$

(Section 11-7)

Magnitude of the Coriolis acceleration:

$$a_{Coriolis} = 2\omega v_\perp$$

(Section 11-9)

Chapter Summary

Section 11-1. Angular Momentum - Objects Rotating About a Fixed Axis

The **angular momentum L** of a rigid object rotating about a fixed axis with angular velocity ω is defined as

$$L = I\omega$$

where I is the moment of inertia of the object with respect to the rotation axis. The units for angular momentum are kg m^2/s.

The rate of change of the angular momentum is related to the torque acting on the object:

$$\sum \tau = \frac{dL}{dt}$$

This equation is sometimes called **Newton's second law for rotation**. If the net torque on the object is 0, $dL/dt = 0$ and L = constant. We thus conclude that

> **The total angular momentum of a rotating object remains constant if the net external torque acting on it is zero.**

This is the **law of conservation of angular momentum**. Since the internal forces of the object come in pairs (Newton's third law), the torque associated with them is 0 Nm.

Section 11-2. Vector Cross Product; Torque as a Vector

The **vector product** or **cross product** of two vectors is a vector. The cross product of vectors $\vec{\mathbf{A}}$ and $\vec{\mathbf{B}}$ is written as $\vec{\mathbf{A}} \times \vec{\mathbf{B}}$. The cross product is a vector with magnitude $AB\sin\theta$, where θ is the angle between the directions of vectors $\vec{\mathbf{A}}$ and $\vec{\mathbf{B}}$ in the plane of the two vectors

$$\left| \vec{\mathbf{A}} \times \vec{\mathbf{B}} \right| = AB\sin\theta$$

The direction of $\vec{\mathbf{A}} \times \vec{\mathbf{B}}$ is perpendicular to the plane of the two vectors and points in the direction given by the right-hand rule. To apply the right-hand rule to find the direction of the cross product $\vec{\mathbf{A}} \times \vec{\mathbf{B}}$, place your fingers of your right hand in the direction of the first factor, $\vec{\mathbf{A}}$, so that they can curl toward the direction of the second factor, $\vec{\mathbf{B}}$. Extend your thumb and it will point in the direction of the cross product.

If the vectors $\vec{\mathbf{A}}$ and $\vec{\mathbf{B}}$ are written as a linear combination of the Cartesian axes unit vectors,

$$\vec{\mathbf{A}} = A_x\hat{\mathbf{i}} + A_y\hat{\mathbf{j}} + A_z\hat{\mathbf{k}} \quad \text{and} \quad \vec{\mathbf{B}} = B_x\hat{\mathbf{i}} + B_y\hat{\mathbf{j}} + B_z\hat{\mathbf{k}},$$

then the cross product of the two vectors can be written as

$$\vec{\mathbf{A}} \times \vec{\mathbf{B}} = \begin{vmatrix} \hat{\mathbf{i}} & \hat{\mathbf{j}} & \hat{\mathbf{k}} \\ A_x & A_y & A_z \\ B_x & B_y & B_z \end{vmatrix} = \left(A_y B_z - A_z B_y \right)\hat{\mathbf{i}} + \left(A_z B_x - A_x B_z \right)\hat{\mathbf{j}} + \left(A_x B_y - A_y B_x \right)\hat{\mathbf{k}}$$

Some of the properties of cross product are

$$\vec{A} \times \vec{A} = 0$$

$$\vec{A} \times \vec{B} = -\vec{B} \times \vec{A}$$

$$\vec{A} \times \left(\vec{B} + \vec{C}\right) = \vec{A} \times \vec{B} + \vec{A} \times \vec{C}$$

$$\frac{d}{dt}\left(\vec{A} \times \vec{B}\right) = \frac{d\vec{A}}{dt} \times \vec{B} + \vec{A} \times \frac{d\vec{B}}{dt}$$

Torque can be expressed as a vector cross product

$$\vec{\tau} = \vec{r} \times \vec{F}$$

where $\vec{\tau}$ is the torque about a point O, $\vec{F}$ is the force acting on the object, and $\vec{r}$ is the displacement vector of the point of application of the force $\vec{F}$ from point O. Since the torque depends on the displacement vector from O, the torque will depend on the choice of coordinate system.

Example 11-2-A: Calculating the cross product. Determine the cross product of vector $\vec{A} = 3\hat{i} + 2\hat{j}$ and vector $\vec{B} = -2\hat{i} + \hat{j}$.

Approach: Since the Cartesian components of the two vectors are provided, the easiest approach is to apply the definition of the cross product in terms of the Cartesian coordinates.

Solution:

$$\vec{A} \times \vec{B} = \left(A_y B_z - A_z B_y\right)\hat{i} + \left(A_z B_x - A_x B_z\right)\hat{j} + \left(A_x B_y - A_y B_x\right)\hat{k} =$$

$$= \left\{(2)(0) - (0)(1)\right\}\hat{i} + \left\{(0)(-2) - (3)(0)\right\}\hat{j} + \left\{(3)(1) - (2)(-2)\right\}\hat{k} = 7\hat{k}$$

The direction of the cross product should not be a surprise. The two vectors $\vec{A}$ and $\vec{B}$ are located in the xy plane. Since the result of the cross product is perpendicular to the plane defined by these two vectors, the vector product must be perpendicular to the xy plane and thus parallel or anti-parallel to the z axis.

Example 11-2-B: The dependence of the torque on the choice of coordinate system. A force $\vec{F} = 3\hat{i} + \hat{j} - 2\hat{k}$ acts at a position $\vec{r} = -\hat{i} + 2\hat{j} + 2\hat{k}$ relative to point O. What is the torque of this force about point O? What is the torque with respect to a point $O' = (-1, 2, 0)$.

Approach: The problem provides the force and the position vectors in terms of their Cartesian coordinates, and the easiest approach is to apply the definition of the cross product in terms of the Cartesian coordinates. Since the position vector connecting our reference point to the point of application of the force depends on our choice of reference point, the torque will also depend on our choice of reference point.

Solution:

$$\vec{\tau} = \vec{r} \times \vec{F} = \left(r_y F_z - r_z F_y\right)\hat{i} + \left(r_z F_x - r_x F_z\right)\hat{j} + \left(r_x F_y - r_y F_x\right)\hat{k} =$$

$$= \left\{(2)(-2) - (2)(1)\right\}\hat{i} + \left\{(2)(3) - (-1)(-2)\right\}\hat{j} + \left\{(-1)(1) - (2)(3)\right\}\hat{k} = -6\hat{i} + 4\hat{j} - 7\hat{k}$$

When we use a different reference point, we need to recalculate the position vector from the new reference point to the point at which the force is applied before evaluating the cross product. The position vector from the new reference point to the point on which the force is acting is $\vec{r} = \left(-\hat{i} + 2\hat{j} + 2\hat{k}\right) - \left(-\hat{i} + 2\hat{j}\right) = 2\hat{k}$. The torque with respect to O' is thus equal to

$$\vec{\tau} = \vec{r} \times \vec{F} = \left(r_y F_z - r_z F_y\right)\hat{i} + \left(r_z F_x - r_x F_z\right)\hat{j} + \left(r_x F_y - r_y F_x\right)\hat{k} =$$

$$= \left\{(0)(-2) - (2)(1)\right\}\hat{i} + \left\{(2)(3) - (0)(-2)\right\}\hat{j} + \left\{(0)(1) - (0)(3)\right\}\hat{k} = -2\hat{i} + 6\hat{j}$$

Section 11-3. Angular Momentum of a Particle

The angular momentum of a particle about a point O is defined as

$$\vec{l} = \vec{r} \times \vec{p}$$

where $\vec{l}$ is the angular momentum of the particle about point O, $\vec{r}$ is the displacement of the particle from point O, and $\vec{p}$ is the linear momentum of the particle. Newton's second law can be written in terms of the torque on a particle about point O and the angular momentum of the particle about the same point:

$$\sum \vec{\tau} = \frac{d\vec{l}}{dt}$$

Example 11-3-A: Angular momentum associated with linear motion. A particle of mass $m = 2$ kg travels with constant velocity $\vec{v} = -3\hat{i}$ along a path that passes a distance of 2 m from point O. What is the angular momentum of the particle about point O as a function of time?

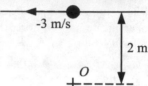

Approach: In order to determine the angular momentum of the particle as function of time we need to determine its linear momentum and its position as function of time. We will assume that time $t = 0$ is defined as the time at which the particle is closest to O (the time shown in the Figure.) The coordinate system of this problem has its x axis along the direction of motion of the particle and the y axis perpendicular to it.

Solution: The position of the particle as function of time is equal to

$$\vec{r}(t) = -3t\,\hat{i} + 2\,\hat{j}$$

At time $t = 0$, the particle is located at $(0, 2)$ which is the position of closest approach. The linear momentum of the particle is equal to

$$\vec{p}(t) = m\vec{v}(t) = (2)(-3\hat{i}) = -6\,\hat{i}$$

Applying the definition of the angular momentum of a particle we obtain

$$\vec{l} = \vec{r} \times \vec{p} = \begin{vmatrix} \hat{i} & \hat{j} & \hat{k} \\ r_x & r_y & r_z \\ p_x & p_y & p_z \end{vmatrix} = \begin{vmatrix} \hat{i} & \hat{j} & \hat{k} \\ -3t & 2 & 0 \\ -6 & 0 & 0 \end{vmatrix} = 12\,\hat{k}$$

Even though $\vec{r}$ is time dependent, the angular momentum is constant. This is a surprise since the force on the particle, and therefore the torque on the particle, is zero and angular momentum is thus conserved.

Section 11-4. Angular Momentum and Torque for a System of Particles; General Motion

For measurements made in an inertial reference frame, the angular momenta of a system of particles about a given point can be added together. The net torque about the same point acting on this system is the vector sum of the torques associated with the individual particles. In this sum, the torques resulting from the forces between the particles of the system add to zero as a consequence of Newton's third law. The result of the summation thus only depends on the torques associated with external forces, the so-called external torques:

$$\sum \vec{\tau} = \sum \vec{\tau}_{ext} = \frac{d}{dt} \sum \vec{l} = \frac{d\vec{L}}{dt}$$

where $\vec{\tau}_{ext}$ is a torque associated with external forces and $\vec{L}$ is the total angular momentum of the system. If the point about which the torques and angular momenta are calculated is the center of the mass of the system, then

$$\sum \vec{\tau}_{CM,ext} = \frac{d}{dt} \sum \vec{l}_{CM} = \frac{d\vec{L}_{CM}}{dt}$$

This relation applies even to noninertial or accelerating reference frames. The *CM* subscript means the quantities are calculated about the center of mass of the system.

Section 11-5. Angular Momentum and Torque for a Rigid Object

If the relationship between torque and angular momentum of a system of particles is applied to a rigid extended body, we can write

$$\vec{\mathbf{L}} = I\,\vec{\omega}$$

if the rotation axis of the body is also a symmetry axis of the body. This implies that the angular momentum vector $\vec{\mathbf{L}}$ and the angular velocity vector $\vec{\omega}$ of the body are pointing in the same direction. If the rotation axis is not a symmetry axis of the body, than the angular momentum vector of the body is not necessarily pointing in the same direction as the rotation axis of the body and the relation between the angular momentum vector and the angular velocity vector is more complicated. In this case,

$$\sum \tau_{axis} = \frac{dL_\omega}{dt} = \frac{d}{dt}\left(I\omega\right) = I\frac{d\omega}{dt} = I\alpha$$

where τ_{axis} is the component of torque along the rotation axis, I is the moment of inertia of the body about the rotation axis, L_ω is the component of the angular momentum along the rotation axis, and ω is the angular velocity about the rotation axis. The components of the torque in directions perpendicular to the rotation axis are in general non-zero.

When the rotation axis of an object is not aligned with a symmetry axis of the object, torque needs to be applied to the object to keep it rotating at constant angular velocity. This is a result of the fact that the angular momentum vector will rotate around the axis of rotation as the body changes orientation. Thus, $d\vec{\mathbf{L}}/dt$ is not equal to zero; the magnitude of the angular momentum vector may be constant, but its orientation in space is not. Since $d\vec{\mathbf{L}}/dt$ is related to the external torque acting on the object, we conclude that the external torque must be non-zero.

Example 11-5-A. Rotational motion of a cylinder. A uniform cylindrical spool of mass $M = 0.187$ kg and radius $r = 2.56$ cm has thread wrapped around it. The spool is free to rotate about its cylindrical axis. Attached to the end of the thread is a mass $m = 0.0542$ kg. What is the angular acceleration of the spool?

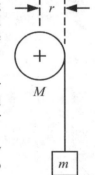

Approach: Since the cylinder will rotate about its symmetry axis, we will calculate its angular momentum and the external torques with respect to this axis. When the cylinder carries out rotational motion, mass m will carry out linear motion. If we assume that the thread does not slip, then the linear velocity of mass m must be the same as the linear velocity of a point on the rim of the cylinder.

To solve this problem we can take two different approaches. In solution 1 we consider the cylinder by itself and determine how it responds to external forces/torques. In solution 2 we consider our system to consist of the cylinder, the thread, and mass m and examine how it responds to external forces/torques.

Solution 1: The only non-zero external torque acting on the cylinder is the torque associated with the tension T in the thread:

$$\vec{\tau}_{ext} = -rT\,\hat{\mathbf{k}}$$

where the unit vector $\hat{\mathbf{k}}$ is directed out of the plane of the page. The direction of the external torque can be determined using the right-hand rule. As a result of the external torque, the cylinder will start to rotate, and the magnitude of its angular acceleration is equal to

$$|\vec{\alpha}| = \frac{|\vec{\tau}_{ext}|}{I} = \frac{rT}{\frac{1}{2}Mr^2} = 2\frac{T}{Mr}$$

Assuming that the thread does not slip, the linear acceleration of mass m must be equal to the linear acceleration of a point on the rim of the cylinder. Thus:

$$|\vec{\mathbf{a}}_m| = |\vec{\alpha}|r = 2\frac{T}{M}$$

The net force on mass m is the difference between the gravitational force on mass m and the tension in the thread. This net force is related to the acceleration of mass m:

$$\left|\vec{\mathbf{a}}_m\right| = \frac{mg - T}{m} = g - \frac{T}{m}$$

Using these last two equations we can determine the tension T:

$$2\frac{T}{M} = g - \frac{T}{m} \quad \rightarrow \quad T = \frac{g}{\dfrac{2}{M} + \dfrac{1}{m}} = \frac{gmM}{2m + M}$$

This result can now be used to determine the angular acceleration of the cylinder:

$$\left|\vec{\alpha}\right| = 2\frac{T}{Mr} = 2\frac{\left(\dfrac{gmM}{2m + M}\right)}{Mr} = \frac{gm}{\left(m + \dfrac{1}{2}M\right)r} = \frac{(9.80)(0.0542)}{\left[(0.0542) + \dfrac{1}{2}(0.187)\right](0.0256)} = 140 \text{ s}^{-2}$$

Solution 2: Now consider a system that consists of the spool, the thread, and mass m. With this choice of system, the tension T is an internal force, and it thus does not change the angular momentum of the system.

In this solution, we will calculate the torque and angular momentum about the fixed axis of the spool. The external forces acting on the system are the gravitational force on the spool, the force exerted by the axis on the spool, and the gravitational force on mass m. The torque associated with the first two forces is zero, since the arms of these forces with respect to the rotation axis are zero. The torque associated with the third force is

$$\left|\vec{\tau}_{ext}\right| = \left|r \times F_m\right| = r_{\perp}F = rmg$$

The total angular momentum of the system is the angular momentum of the spool plus the angular momentum of mass m. The angular momentum of the spool and the angular momentum of mass m are directed in the same direction, and the magnitude of the angular momentum of the system is the linear sum of the angular momenta of the spool and of mass m:

$$L = L_M + L_m = \tfrac{1}{2}Mr^2\omega + mvr$$

Because the mass is tied to the thread, the linear speed of the mass is related to the angular speed of the spool by

$$v = \omega r$$

Applying Newton's second law for rotational motion we obtain

$$\left|\vec{\tau}_{ext}\right| = \frac{dL}{dt} = \frac{d}{dt}\left(\frac{1}{2}Mr^2\omega + mvr\right) = \frac{d}{dt}\left[\left(\frac{1}{2}M + m\right)r^2\omega\right] = \left(\frac{1}{2}M + m\right)r^2\alpha$$

Using our expression for the external torque on the system we can now determine the angular acceleration of the spool:

$$\alpha = \frac{\left|\vec{\tau}_{ext}\right|}{\left(\dfrac{1}{2}M + m\right)r^2} = \frac{rmg}{\left(\dfrac{1}{2}M + m\right)r^2} = \frac{mg}{\left(\dfrac{1}{2}M + m\right)r} = 140 \text{ s}^{-2}$$

which is of course the same as the result obtained previously.

Section 11-6. Conservation of Angular Momentum

For a system of objects with no net external torque acting on the system, the total angular momentum is conserved. This is a statement of the **law of conservation of angular momentum.**

Example 11-6-A. Conservation of angular momentum on a merry-go-round. A child of mass $M = 32.8$ kg stands on a playground merry-go-round that is at rest, holding an object of mass $m = 1.45$ kg. The child is standing a distance $r = 1.34$ m from the axis of rotation of the merry-go-round. The moment of inertia of the merry-go-round about its axis is $I = 86.4$ kg·m^2. The child throws the object to a friend with a speed of $v = 4.65$ m/s in a direction that is $\theta = 66.4°$ from

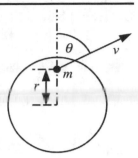

the direction directly away from the axis of rotation of the merry-go-round as shown in the Figure. What will be the angular velocity of the merry-go-round after the child releases the object?

Approach: The best approach to solve this problem is to consider the merry-go-round, the child, and the object as our system. With this choice of system, the force between the child and object is an internal force and will not change the angular momentum of the system. We will only consider the component of the angular momentum parallel to the rotation axis (the vertical direction) and this component is not affected by the gravitational forces since the torques associated with these forces are directed in the horizontal direction.

Solution: Since the system is initially at rest, its angular momentum is equal to 0. After the ball is released, both the object and the child/merry-go-round will have non-zero angular momenta. The final angular momentum of the ball is equal to

$$\vec{l}_{object} = \vec{r} \times \vec{p} = \left(r\,\hat{j}\right) \times mv\left(\cos\theta\,\hat{i} + \sin\theta\,\hat{j}\right) = -mrv\cos\theta\,\hat{k}$$

The merry-go-round and child rotate together after the object is released. Their angular momentum, $\vec{L}$, will be the total of their moments of inertia about the merry-go-round axis, multiplied by their common angular velocity:

$$\vec{L} = \left(I_{merry-go-round} + I_{child}\right)\vec{\omega} = \left(I_{merry-go-round} + Mr^2\right)\omega\,\hat{k}$$

Conservation of angular momentum requires that

$$0 = \vec{l}_{object} + \vec{L} = -mrv\cos\theta\,\hat{k} + \left(I_{merry-go-round} + Mr^2\right)\omega\,\hat{k} = \left[-mrv\cos\theta + \left(I_{merry-go-round} + Mr^2\right)\omega\right]\hat{k}$$

This relation can be used to determine the angular velocity of the merry-go-round:

$$\omega = \frac{mrv\cos\theta}{\left(I_{merry-go-round} + Mr^2\right)} = -\frac{(1.45)(1.34)(4.65)\cos 66.4°}{86.4 + (32.8)(1.34)^2} = 0.0249 \text{ rad/s}$$

Section 11-7. The Spinning Top

A slightly inclined toy top is an example of a system for which the torque on the system is perpendicular to the angular momentum vector of the system at any given instant. This produces a motion of the angular momentum vector called **precession**. The magnitude of the angular momentum vector does not change, but its direction precesses around the vertical direction with a constant angular speed. For a toy top, the angular speed of the precession, Ω, is given by

$$\Omega = \frac{Mgr}{L}$$

where M is the mass of the top, g is the gravitational acceleration, r is the distance between the center of mass of the top and its contact point with the surface, and L is the magnitude of the angular momentum of the top.

Section 11-8. Rotating Frames of Reference; Inertial Forces

A frame of reference in which Newton's laws of motion do not hold is called a **noninertial reference frame**. Typical noninertial reference frames are those that are accelerating relative to inertial reference frames and rotating reference frames. In order to describe motion in noninertial reference frames **fictitious forces** or **pseudoforces** are sometimes introduced. We need to recognize that these forces are not real forces, but are only introduced in an attempt to describe motion in a noninertial reference frame.

A good example of a pseudo force is the apparent force acting on our body while we are going around a corner in a car at high speed. We "feel a force" that pushes our body away from the inside of the curve but there is no real force acting in this direction. Our bodies are trying to follow a straight line, but our accelerated reference frame (the car is accelerating because it is changing direction) causes us to invent/experience a pseudoforce in order to explain why our bodies move relative to the reference frame toward the outside of the curve.

Section 11-9. The Coriolis Effect

A good example of a pseudoforce that occurs for observers in rotating reference frames is the **Coriolis force**. The acceleration of an object associated with this pseudoforce is called the **Coriolis acceleration**. The magnitude of the Coriolis acceleration, $a_{Coriolis}$, is given by

$$a_{Coriolis} = 2\omega v_{\perp}$$

where ω is the angular velocity of the noninertial reference frame relative to an inertial reference frame and $v_{\perp}$ is the component of the object's velocity perpendicular to the direction of the angular velocity vector of the noninertial reference frame.

Example 11-9-A. Coriolis acceleration of a ball on a merry-go-round. A child rides on a playground merry-go-round that is making one rotation every $T = 3.24$ s. The child throws a ball with a speed of $v = 8.22$ m/s, radially outward from the axis of rotation of the merry-go-round. The ball is released a distance $r = 1.16$ m from the axis of rotation. What is the magnitude of the initial acceleration of the released ball, observed by the child on the merry-go-round? Ignore the relatively small contribution from the Earth being a rotating reference frame.

Approach: The observations made by the child on the merry-go-round are made in a noninertial reference frame. Once the child releases the ball, it will carry out linear motion in the direction in which it was released, but to the child it appears as if the ball has an acceleration equal to the Coriolis acceleration. In addition to this pseudo acceleration, the ball will accelerate due to the real gravitational force.

Solution: The child will observe that the motion of the ball is due to a combination of the Coriolis acceleration and the gravitational acceleration. The Coriolis acceleration vector is directed parallel to the horizontal plane in a direction tangential to the circular path of points on the merry-go-round, but opposite to the direction of rotation. The magnitude of the Coriolis acceleration is

$$a_{Coriolis} = 2\omega v_{\perp} = 2\frac{2\pi}{T}v_{\perp} = \frac{4\pi}{(3.24)}(8.22) = 31.9\,\text{m/s}^2$$

The gravitational acceleration is directed vertically downward, perpendicular to $a_{Coriolis}$. The magnitude of the observed acceleration will be

$$a = \sqrt{a_{Coriolis}^2 + g^2} = \sqrt{(31.9)^2 + (9.80)^2} = 33.4\,\text{m/s}^2$$

Practice Quiz

1. Two vectors have non-zero magnitudes. Under which of the following conditions do you know that the vectors are perpendicular to each other?
   a) The vector sum of the two vectors is zero.
   b) The vector difference of the two vectors is zero.
   c) The cross product of the two vectors is zero.
   d) The scalar product of the two vectors is zero.

2. Which direction does the angular momentum vector of the Earth point?
   a) Directly upward at your current location.
   b) Directly upward at the North Pole.
   c) Directly upward at the South Pole.
   d) Toward the east.

3. When you turn a screw with right-hand threads (this is the standard type that is tightened by turning clockwise), what is the direction of the torque vector that is required to tighten the screw?
   a) In the direction the screw will advance.
   b) Opposite the direction the screw will advance.
   c) To the right as you look at the head of the screw.
   d) To the left as you look at the head of the screw.

4. Under what conditions does the equation $\sum \vec{\tau} = d\vec{L} / dt$ apply for a system of particles?
   a) Under all conditions.
   b) Only if the system consists of a single particle.
   c) Only if the center of mass of the system is at rest.
   d) If the quantities are calculated in an inertial reference frame or about the center of mass.

5. In the Southern Hemisphere, which direction is the horizontal component of the Coriolis acceleration for an object moving straight north?
   a) West.
   b) East.
   c) North.
   d) South.

6. What is the magnitude of the vector cross product of two vectors of magnitude 1 that add to zero?
   a) 1
   b) 2
   c) Square root of 2
   d) 0

7. A particle of mass m moves along a straight line with an acceleration that is non-zero. Where can an axis be located such that the angular momentum of the particle is not constant?
   a) At any point on the path of the particle.
   b) At a point that is instantaneously at the location of the particle.
   c) At any point not on the path of the particle.
   d) There are no such points.

8. An object is spinning freely about a fixed axis when a small piece of its surface is ejected. The remainder will continue to spin with its initial angular velocity if the ejected piece leaves the surface with a velocity
   a) that is directly away from the axis of rotation.
   b) that is zero.
   c) that is equal to the velocity of the center of mass of the initial object.
   d) that is equal to its velocity on the surface of the object at the moment it is ejected.

9. What happens to the precession rate of a spinning top as the angular velocity of the top about its axis decreases?
   a) The precession rate remains constant.
   b) The precession rate decreases.
   c) The precession rate increases.
   d) It depends on the angle of tilt of the top.

10. Near the surface of the Earth, where will the Coriolis acceleration be greatest for an object that is in free fall?
    a) Near either of the poles.
    b) Near the equator.
    c) At a latitude of 45°.
    d) Same at all latitudes.

11. Calculate the cross product of a vector $\vec{A} = 3\hat{i} + \hat{j} - 4\hat{k}$ and a vector $\vec{B} = -\hat{i} + \hat{j} + \hat{k}$.

12. Determine the torque about point O for a force $\vec{F} = 2\hat{i} + 3\hat{j}$ that acts at a point displaced from O by the vector $\vec{r} = \hat{i} - 2\hat{j}$.

13. An asteroid of mass 14×10^6 kg strikes the Earth traveling tangential to the Earth at a latitude of $40°$ traveling south at a speed of 12 km/s. What will the tilt of the Earth's rotation axis be from its original direction? Treat the Earth as a uniform-density sphere.

14. The turbine of a jet engine rotates at 12,000 rev/min. The two bearings that it is mounted on can withstand a 100 N force perpendicular to the axis of rotation. The bearings have a moment arm of 13.8 cm for torque calculated about the center of mass of the turbine. What is the maximum component of angular momentum perpendicular to the axis of rotation that the bearings can support?

15. Determine the precession rate of a top that is spinning at a rate of 34 rev/s. The top has a 1.6 cm radius of gyration. The center of mass of the top is 3.2 cm above the surface it is spinning on.

Responses to Select End-of-Chapter Questions

1. (*a*) With more people at the equator, more mass would be farther from the axis of rotation, and the moment of inertia of the Earth would increase. Due to conservation of angular momentum, the Earth's angular velocity would decrease. The length of the day would increase.

7. The cross product remains the same. $\vec{V}_1 \times \vec{V}_2 = \left(-\vec{V}_1\right) \times \left(-\vec{V}_2\right)$

13. A force directed to the left will produce a torque that will cause the axis of the rotating wheel to move directly upward.

19. The Sun will pull on the bulge closer to it more than it pulls on the opposite bulge, due to the inverse-square law of gravity. These forces, and those from the Moon, create a torque which causes the precession of the axis of rotation of the Earth. The precession is about an axis perpendicular to the plane of the orbit. During the equinox, no torque exists, since the forces on the bulges lie along a line.

Solutions to Select End-of-Chapter Problems

1. The angular momentum is given by Eq. 11-1.

$$L = I\omega = MR^2\omega = (0.210\,\text{kg})(1.35\,\text{m})^2(10.4\,\text{rad/s}) = \boxed{3.98\,\text{kg}\cdot\text{m}^2/\text{s}}$$

7. (*a*) For the daily rotation about its axis, treat the Earth as a uniform sphere, with an angular frequency of one revolution per day.

$$L_{\text{daily}} = I\omega_{\text{daily}} = \left(\tfrac{2}{5} MR_{\text{Earth}}^2\right)\omega_{\text{daily}}$$

$$= \tfrac{2}{5}\left(6.0\times10^{24}\,\text{kg}\right)\left(6.4\times10^6\,\text{m}\right)^2\left[\left(\frac{2\pi\ \text{rad}}{1\ \text{day}}\right)\left(\frac{1\ \text{day}}{86,400\ \text{s}}\right)\right] = \boxed{7.1\times10^{33}\ \text{kg}\cdot\text{m}^2/\text{s}}$$

(*b*) For the yearly revolution about the Sun, treat the Earth as a particle, with an angular frequency of one revolution per year.

$$L_{\text{daily}} = I\omega_{\text{daily}} = \left(MR_{\substack{\text{Sun-}\\\text{Earth}}}^2\right)\omega_{\text{daily}}$$

$$= \left(6.0\times10^{24}\,\text{kg}\right)\left(1.5\times10^{11}\,\text{m}\right)^2\left[\left(\frac{2\pi\ \text{rad}}{365\ \text{day}}\right)\left(\frac{1\ \text{day}}{86,400\ \text{s}}\right)\right] = \boxed{2.7\times10^{40}\ \text{kg}\cdot\text{m}^2/\text{s}}$$

13. The angular momentum of the merry-go-round and people combination will be conserved because there are no external torques on the combination. This situation is a totally inelastic collision, in which the final angular velocity is the same for both the merry-go-round and the people. Subscript 1 represents before the collision, and subscript 2 represents after the collision. The people have no initial angular momentum.

$$I_i = I_f \quad \rightarrow \quad I_1 \omega_1 = I_2 \omega_2 \quad \rightarrow$$

$$\omega_2 = \omega_1 \frac{I_1}{I_2} = \omega_1 \frac{I_{\text{m-g-r}}}{I_{\text{m-g-r}} + I_{\text{people}}} = \omega_1 \left[\frac{I_{\text{m-g-r}}}{I_{\text{m-g-r}} + 4 M_{\text{person}} R^2} \right]$$

$$= \left(0.80 \, \text{rad/s} \right) \left[\frac{1760 \, \text{kg} \cdot \text{m}^2}{1760 \, \text{kg} \cdot \text{m}^2 + 4 \left(65 \, \text{kg} \right) \left(2.1 \, \text{m} \right)^2} \right] = \boxed{0.48 \, \text{rad/s}}$$

If the people jump off the merry-go-round radially, then they exert no torque on the merry-go-round, and thus cannot change the angular momentum of the merry-go-round. The merry-go-round would continue to rotate at $\boxed{0.80 \, \text{rad/s}}$.

19. The angular momentum of the person–turntable system will be conserved. Call the direction of the person's motion the positive rotation direction. Relative to the ground, the person's speed will be $v + v_T$ where v is the person's speed relative to the turntable, and v_T is the speed of the rim of the turntable with respect to the ground. The turntable's angular speed is $\omega_T = v_T/R$, and the person's angular speed relative to the ground is $\omega_P = (v + v_T)/R = v/R + \omega_T$. The person is treated as a point particle for calculation of the moment of inertia.

$$L_i = L_f \quad \rightarrow \quad 0 = I_T \omega_T + I_P \omega_P = I_T \omega_T + mR^2 \left(\omega_T + \frac{v}{R} \right) \quad \rightarrow$$

$$\omega_T = -\frac{mRv}{I_T + mR^2} = -\frac{\left(65 \, \text{kg} \right) \left(3.25 \, \text{m} \right) \left(3.8 \, \text{m/s} \right)}{1850 \, \text{kg} \cdot \text{m}^2 + \left(65 \, \text{kg} \right) \left(3.25 \, \text{m} \right)^2} = \boxed{-0.32 \, \text{rad/s}}$$

25. We choose coordinates so that the plane in which the particle rotates is the x-y plane, and so the angular velocity is in the z direction. The object is rotating in a circle of radius $r \sin \theta$, where θ is the angle between the position vector and the axis of rotation. Since the object is rigid and rotates about a fixed axis, the linear and angular velocities of the particle are related by $v = \omega r \sin \theta$. The magnitude of the tangential acceleration is $a_{\text{tan}} = \alpha r \sin \theta$. The radial acceleration is given by

$$a_R = \frac{v^2}{r \sin \theta} = v \frac{v}{r \sin \theta} = v \omega.$$

We assume the object is gaining speed. See the diagram showing the various vectors involved.

The velocity and tangential acceleration are parallel to each other, and the angular velocity and angular acceleration are parallel to each other. The radial acceleration is perpendicular to the velocity, and the velocity is perpendicular to the angular velocity.

We see from the diagram that, using the right-hand rule, the direction of $\vec{a}_R$ is in the direction of $\vec{u} \times \vec{v}$. Also, since $\vec{u}$ and $\vec{v}$ are perpendicular, we have $\left| \vec{u} \times \vec{v} \right| = \omega v$, which from above is $v \omega = a_R$. Since both the magnitude and direction check out, we have $\boxed{\vec{a}_R = \vec{u} \times \vec{v}}$.

We also see from the diagram that, using the right-hand rule, the direction of $\vec{a}_{\text{tan}}$ is in the direction of $\vec{a} \times \vec{r}$. The magnitude of $\vec{a} \times \vec{r}$ is $\left| \vec{a} \times \vec{r} \right| = \alpha r \sin \theta$, which from above is $\alpha r \sin \theta = a_{\text{tan}}$. Since both the magnitude and direction check out, we have $\boxed{\vec{a}_{\text{tan}} = \vec{a} \times \vec{r}}$.

31. Calculate the three "triple products" as requested.

$$\vec{A} \times \vec{B} = \begin{vmatrix} \hat{i} & \hat{j} & \hat{k} \\ A_x & A_y & A_z \\ B_x & B_y & B_z \end{vmatrix} = \boxed{\hat{i}\left(A_y B_z - A_z B_y\right) + \hat{j}\left(A_z B_x - A_x B_z\right) + \hat{k}\left(A_x B_y - A_y B_x\right)}$$

$$\vec{B} \times \vec{C} = \begin{vmatrix} \hat{i} & \hat{j} & \hat{k} \\ B_x & B_y & B_z \\ C_x & C_y & C_z \end{vmatrix} = \boxed{\hat{i}\left(B_y C_z - B_z C_y\right) + \hat{j}\left(B_z C_x - B_x C_z\right) + \hat{k}\left(B_x C_y - B_y C_x\right)}$$

$$\vec{C} \times \vec{A} = \begin{vmatrix} \hat{i} & \hat{j} & \hat{k} \\ C_x & C_y & C_z \\ A_x & A_y & A_z \end{vmatrix} = \boxed{\hat{i}\left(C_y A_z - C_z A_y\right) + \hat{j}\left(C_z A_x - C_x A_z\right) + \hat{k}\left(C_x A_y - C_y A_x\right)}$$

$$\vec{A} \cdot \left(\vec{B} \times \vec{C}\right) = \left(A_x\hat{i} + A_y\hat{j} + A_z\hat{k}\right) \cdot \left[\hat{i}\left(B_y C_z - B_z C_y\right) + \hat{j}\left(B_z C_x - B_x C_z\right) + \hat{k}\left(B_x C_y - B_y C_x\right)\right]$$
$$= A_x\left(B_y C_z - B_z C_y\right) + A_y\left(B_z C_x - B_x C_z\right) + A_z\left(B_x C_y - B_y C_x\right)$$
$$= A_x B_y C_z - A_x B_z C_y + A_y B_z C_x - A_y B_x C_z + A_z B_x C_y - A_z B_y C_x$$

$$\vec{B} \cdot \left(\vec{C} \times \vec{A}\right) = \left(B_x\hat{i} + B_y\hat{j} + B_z\hat{k}\right) \cdot \left[\hat{i}\left(C_y A_z - C_z A_y\right) + \hat{j}\left(C_z A_x - C_x A_z\right) + \hat{k}\left(C_x A_y - C_y A_x\right)\right]$$
$$= B_x\left(C_y A_z - C_z A_y\right) + B_y\left(C_z A_x - C_x A_z\right) + B_z\left(C_x A_y - C_y A_x\right)$$
$$= B_x C_y A_z - B_x C_z A_y + B_y C_z A_x - B_y C_x A_z + B_z C_x A_y - B_z C_y A_x$$

$$\vec{C} \cdot \left(\vec{A} \times \vec{B}\right) = \left(C_x\hat{i} + C_y\hat{j} + C_z\hat{k}\right) \cdot \left[\hat{i}\left(A_y B_z - A_z B_y\right) + \hat{j}\left(A_z B_x - A_x B_z\right) + \hat{k}\left(A_x B_y - A_y B_x\right)\right]$$
$$= C_x\left(A_y B_z - A_z B_y\right) + C_y\left(A_z B_x - A_x B_z\right) + C_z\left(A_x B_y - A_y B_x\right)$$
$$= C_x A_y B_z - C_x A_z B_y + C_y A_z B_x - C_y A_x B_z + C_z A_x B_y - C_z A_y B_x$$

A comparison of three results shows that they are all the same.

37. Use Eq. 11-6 to calculate the angular momentum.

$$\vec{L} = \vec{r} \times \vec{p} = m\left(\vec{r} \times \vec{v}\right) = \left(3.8\,\text{kg}\right) \begin{vmatrix} \hat{i} & \hat{j} & \hat{k} \\ 1.0 & 2.0 & 3.0 \\ -5.0 & 2.8 & -3.1 \end{vmatrix} \text{m}^2/\text{s}$$

$$= \left(3.8\right)\left(-14.6\hat{i} - 11.9\hat{j} + 12.8\hat{k}\right)\text{kg}\cdot\text{m}^2/\text{s} = \boxed{\left(-55\hat{i} - 45\hat{j} + 49\hat{k}\right)\text{kg}\cdot\text{m}^2/\text{s}}$$

43. We follow the notation and derivation of Eq. 11-9b. Start with the general definition of angular momentum:

$$\vec{L} = \sum_i \vec{r}_i \times \vec{p}_i$$

Then express position and velocity with respect to the center of mass.

$\vec{r}_i = \vec{r}_{CM} + \vec{r}_i^{\,*}$, where $\vec{r}_i^{\,*}$ is the position of the ith particle with respect to the center of mass

$\vec{v}_i = \vec{v}_{CM} + \vec{v}_i^{\,*}$, which comes from differentiating the above relationship for position

$$\vec{L} = \sum_i \vec{r}_i \times \vec{p}_i = \sum_i \vec{r}_i \times m_i \vec{v}_i = \sum_i \left(\vec{r}_{CM} + \vec{r}_i^* \right) \times m_i \left(\vec{v}_{CM} + \vec{v}_i^* \right)$$

$$= \sum_i m_i \vec{r}_{CM} \times \vec{v}_{CM} + \sum_i m_i \vec{r}_{CM} \times \vec{v}_i^* + \sum_i m_i \vec{r}_i^* \times \vec{v}_{CM} + \sum_i m_i \vec{r}_i^* \times \vec{v}_i^*$$

Note that the center of mass quantities are not dependent on the summation subscript, and so they may be taken outside the summation process.

$$\vec{L} = \left(\vec{r}_{CM} \times \vec{v}_{CM} \right) \sum_i m_i + \vec{r}_{CM} \times \sum_i m_i \vec{v}_i^* + \left(\sum_i m_i \vec{r}_i^* \right) \times \vec{v}_{CM} + \sum_i m_i \vec{r}_i^* \times \vec{v}_i^*$$

In the first term, $\sum_i m_i = M$. In the second term, we have the following.

$$\sum_i m_i \vec{v}_i^* = \sum_i m_i \left(\vec{v}_i - \vec{v}_{CM} \right) = \sum_i m_i \vec{v}_i - \sum_i m_i \vec{v}_{CM} = \sum_i m_i \vec{v}_i - M \vec{v}_{CM} = 0$$

This is true from the definition of center of mass velocity:

$$\vec{v}_{CM} = \frac{1}{M} \sum_i m_i \vec{v}_i$$

Likewise, in the third term, we have the following.

$$\sum_i m_i \vec{r}_i^* = \sum_i m_i \left(\vec{r}_i - \vec{r}_{CM} \right) = \sum_i m_i \vec{r}_i - \sum_i m_i \vec{r}_{CM} = \sum_i m_i \vec{r}_i - M \vec{r}_{CM} = 0$$

This is true from the definition of center of mass:

$$\vec{r}_{CM} = \frac{1}{M} \sum_i m_i \vec{r}_i$$

Thus $\vec{L} = M \left(\vec{r}_{CM} \times \vec{v}_{CM} \right) + \sum_i m_i \vec{r}_i^* \times \vec{v}_i^* = \boxed{\vec{L}^* + \left(\vec{r}_{CM} \times M\vec{v}_{CM} \right)}$ as desired.

49. The angular momentum of the Earth–meteorite system is conserved in the collision. The Earth is spinning counterclockwise as viewed in the diagram. We take that direction as the positive direction for rotation about the Earth's axis, and so the initial angular momentum of the meteorite is negative.

$$L_{\text{initial}} = L_{\text{final}} \quad \rightarrow \quad I_{\text{Earth}} \omega_0 - m R_E v \sin 45° = \left(I_{\text{Earth}} + I_{\text{meteorite}} \right) \omega \quad \rightarrow$$

$$\omega = \frac{I_{\text{Earth}} \omega_0 - m R_E v \sin 45°}{\left(I_{\text{Earth}} + I_{\text{meteorite}} \right)} = \frac{\frac{2}{5} M_E R_E^2 \omega_0 - m R_E v \sin 45°}{\left(\frac{2}{5} M_E R_E^2 + m R_E^2 \right)}$$

$$\frac{\omega}{\omega_0} = \frac{\frac{2}{5} M_E R_E^2 - m R_E \dfrac{v}{\omega_0} \dfrac{1}{\sqrt{2}}}{R_E^2 \left(\frac{2}{5} M_E + m \right)} = \frac{R_E^2 \left(\frac{2}{5} M_E - \dfrac{mv}{\sqrt{2}\omega_0 R_E} \right)}{R_E^2 \left(m + \frac{2}{5} M_E \right)} = \frac{\left(\frac{2}{5} M_E - \dfrac{mv}{\sqrt{2}\omega_0 R_E} \right)}{\left(m + \frac{2}{5} M_E \right)}$$

$$\frac{\Delta\omega}{\omega_0} = \frac{\omega - \omega_0}{\omega_0} = \frac{\omega}{\omega_0} - 1 = \frac{\left(\frac{2}{5} M_E - \dfrac{mv}{\sqrt{2}\omega_0 R_E} \right)}{\left(m + \frac{2}{5} M_E \right)} - 1 = \frac{-\left(\dfrac{v}{\sqrt{2}\omega_0 R_E} + 1 \right)}{\left(1 + \frac{2}{5} \dfrac{M_E}{m} \right)}$$

$$= -\frac{\left(\dfrac{2.2 \times 10^4 \text{ m/s}}{\sqrt{2}\left(\dfrac{2\pi}{86,400} \text{ rad/s} \right)\left(6.38 \times 10^6 \text{ m} \right)} + 1 \right)}{\left(1 + \frac{2}{5} \dfrac{5.97 \times 10^{24} \text{ kg}}{5.8 \times 10^{10} \text{ kg}} \right)} = -8.387 \times 10^{-13} \approx \boxed{-8.4 \times 10^{-13}}$$

55. Use Eq. 11-13c for the precessional angular velocity.

$$\Omega = \frac{Mgr}{I\omega} = \frac{Mg\left(\frac{1}{2}l_{axle}\right)}{\frac{1}{2}Mr^2_{wheel}\omega} = \frac{gl_{axle}}{r^2_{wheel}\omega} = \frac{\left(9.80\,\text{m/s}^2\right)\left(0.25\,\text{m}\right)}{\left(0.060\,\text{m}\right)^2\left(85\,\text{rad/s}\right)} = \boxed{8.0\,\text{rad/s}} \quad (1.3\,\text{rev/s})$$

61. The footnote on page 302 gives the Coriolis acceleration as $\vec{\mathbf{a}}_{Cor} = 2\vec{\mathbf{u}} \times \vec{\mathbf{v}}$. The angular velocity vector is parallel to the axis of rotation of the Earth. For the Coriolis acceleration to be 0, then, the velocity must be parallel to the axis of rotation of the Earth. At the equator this means moving either due north or due south.

67. (*a*) See the free-body diagram for the vehicle, tilted up on 2 wheels, on the verge of rolling over. The center of the curve is to the left in the diagram, and so the center of mass is accelerating to the left. The force of gravity acts through the center of mass, and so causes no torque about the center of mass, but the normal force and friction cause opposing torques about the center of mass. The amount of tilt is exaggerated. Write Newton's second laws for the horizontal and vertical directions and for torques, taking left, up, and counterclockwise as positive.

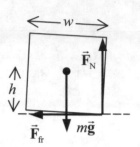

$$\sum F_{vertical} = F_N - Mg = 0 \quad \rightarrow \quad F_N = Mg$$

$$\sum F_{horizontal} = F_{fr} = M\frac{v_C^2}{R}$$

$$\sum \tau = F_N\left(\tfrac{1}{2}w\right) - F_{fr}h = 0 \quad \rightarrow \quad F_N\left(\tfrac{1}{2}w\right) = F_{fr}h$$

$$Mg\left(\tfrac{1}{2}w\right) = M\frac{v_C^2}{R}h \quad \rightarrow \quad \boxed{v_C = \sqrt{Rg\left(\frac{w}{2h}\right)}}$$

(*b*) From the above result, we see that $R = \dfrac{v_C^2}{g}\dfrac{2h}{w} = \dfrac{v_C^2}{g\left(SSF\right)}$.

$$\frac{R_{car}}{R_{SUV}} = \frac{\dfrac{v_C^2}{g\left(SSF\right)_{car}}}{\dfrac{v_C^2}{g\left(SSF\right)_{SUV}}} = \frac{\left(SSF\right)_{SUV}}{\left(SSF\right)_{car}} = \frac{1.05}{1.40} = \boxed{0.750}$$

73. (*a*) The angular momentum delivered to the waterwheel is that lost by the water.

$$\Delta L_{wheel} = -\Delta L_{water} = L_{initial\atop water} - L_{final\atop water} = mv_1R - mv_2R \quad \rightarrow$$

$$\frac{\Delta L_{wheel}}{\Delta t} = \frac{mv_1R - mv_2R}{\Delta t} = \frac{mR}{\Delta t}\left(v_1 - v_2\right) = \left(85\,\text{kg/s}\right)\left(3.0\,\text{m}\right)\left(3.2\,\text{m/s}\right) = 816\,\text{kg}\cdot\text{m}^2/\text{s}^2$$

$$\approx \boxed{820\,\text{kg}\cdot\text{m}^2/\text{s}^2}$$

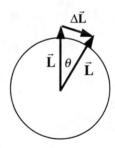

(*b*) The torque is the rate of change of angular momentum, from Eq. 11-9.

$$\tau_{on\atop wheel} = \frac{\Delta L_{wheel}}{\Delta t} = 816\,\text{kg}\cdot\text{m}^2/\text{s}^2 = 816\,\text{m}\cdot\text{N} \approx \boxed{820\,\text{m}\cdot\text{N}}$$

(*c*) Power is given by Eq. 10-21, $P = \tau\omega$.

$$P = \tau\omega = \left(816\,\text{m}\cdot\text{N}\right)\left(\frac{2\pi\,\text{rev}}{5.5\,\text{s}}\right) = \boxed{930\,\text{W}}$$

79. (a) During the jump (while airborne), the only force on the skater is gravity, which acts through the skater's center of mass. Accordingly, there is no torque about the center of mass, and so angular momentum is conserved during the jump.

(b) For a single axel, the skater must have 1.5 total revolutions. The number of revolutions during each phase of the motion is the rotational frequency times the elapsed time. Note that the rate of rotation is the same for both occurrences of the "open" position.

$$(1.2\,\text{rev/s})(0.10\,\text{s}) + f_{\text{single}}(0.50\,\text{s}) + (1.2\,\text{rev/s})(0.10\,\text{s}) = 1.5\,\text{rev} \;\rightarrow$$

$$f_{\text{single}} = \frac{1.5\,\text{rev} - 2(1.2\,\text{rev/s})(0.10\,\text{s})}{(0.50\,\text{s})} = 2.52\,\text{rev/s} \approx \boxed{2.5\,\text{rev/s}}$$

The calculation is similar for the triple axel.

$$(1.2\,\text{rev/s})(0.10\,\text{s}) + f_{\text{triple}}(0.50\,\text{s}) + (1.2\,\text{rev/s})(0.10\,\text{s}) = 3.5\,\text{rev} \;\rightarrow$$

$$f_{\text{triple}} = \frac{3.5\,\text{rev} - 2(1.2\,\text{rev/s})(0.10\,\text{s})}{(0.50\,\text{s})} = 6.52\,\text{rev/s} \approx \boxed{6.5\,\text{rev/s}}$$

(c) Apply angular momentum conservation to relate the moments of inertia.

$$L_{\substack{\text{single}\\\text{open}}} = L_{\substack{\text{single}\\\text{closed}}} \;\rightarrow\; I_{\substack{\text{single}\\\text{open}}}\omega_{\substack{\text{single}\\\text{open}}} = I_{\substack{\text{single}\\\text{closed}}}\omega_{\substack{\text{single}\\\text{closed}}} \;\rightarrow$$

$$\frac{I_{\substack{\text{single}\\\text{closed}}}}{I_{\substack{\text{single}\\\text{open}}}} = \frac{\omega_{\substack{\text{single}\\\text{open}}}}{\omega_{\substack{\text{single}\\\text{closed}}}} = \frac{f_{\substack{\text{single}\\\text{open}}}}{f_{\substack{\text{single}\\\text{closed}}}} = \frac{1.2\,\text{rev/s}}{2.52\,\text{rev/s}} = 0.476 \approx \boxed{\tfrac{1}{2}}$$

Thus the single axel moment of inertia must be reduced by a factor of about 2.

For the triple axel, the calculation is similar.

$$\frac{I_{\substack{\text{triple}\\\text{closed}}}}{I_{\substack{\text{triple}\\\text{open}}}} = \frac{f_{\substack{\text{single}\\\text{open}}}}{f_{\substack{\text{single}\\\text{closed}}}} = \frac{1.2\,\text{rev/s}}{6.52\,\text{rev/s}} = 0.184 \approx \boxed{\tfrac{1}{5}}$$

Thus the triple axel moment of inertia must be reduced by a factor of about 5.

Chapter 12: Static Equilibrium; Elasticity and Fracture

Chapter Overview and Objectives

In this chapter, the requirements for equilibrium and stability are defined. The properties of solid materials exposed to external forces are described and the concepts of stress and strain are introduced. Hooke's law is applied to different types of stress. Several different types of structures are discussed that take advantage of the greater strength of materials under compressive stress.

After completing this chapter you should:
- Know the conditions for static equilibrium.
- Know how to solve statics problems.
- Know the different classifications of stability and what leads to those conditions.
- Know how to relate strain and stress in the Hooke's law limit.
- Know the names of the different types of stresses that can be applied to solids.

Summary of Equations

Conditions for equilibrium:	$\sum \vec{F} = 0$	(Section 12-1)
	$\sum \vec{\tau} = 0$	(Section 12-1)
General definition of stress:	$\text{stress} = \dfrac{\text{force}}{\text{area}} = \dfrac{F}{A}$	(Section 12-4)
General definition of strain:	$\text{strain} = \dfrac{\text{distortion}}{\text{length}} = \dfrac{\Delta L}{L_0}$	(Section 12-4)
Hooke's law for compressive or tensile stress:	$\dfrac{F}{A} = E \dfrac{\Delta L}{L_0}$	(Section 12-4)
Hooke's law for shear stress:	$\dfrac{F}{A} = G \dfrac{\Delta L}{L_0}$	(Section 12-4)
Hooke's law for isotropic pressure:	$\Delta P = -B \dfrac{\Delta V}{V_0}$	(Section 12-4)

Chapter Summary

Section 12-1. The Conditions for Equilibrium

A body with no linear and rotational acceleration must have no net force and no net torque acting on it. A body in this condition is said to be in **equilibrium.**
The absence of linear motion implies that the sum of the forces acting on it must be zero:

$$\sum \vec{F} = 0$$

This condition is called the **first condition for equilibrium**. It can be rewritten in terms of the Cartesian components of the forces:

$$\sum F_x = 0 \qquad \sum F_y = 0 \qquad \sum F_z = 0$$

The absence of rotational motion implies that the net torque acting on the object must be zero:

$$\sum \vec{\tau} = 0$$

This condition is called the **second condition for equilibrium**. It can be rewritten in terms of the Cartesian components of the torques:

$$\sum \tau_x = 0 \qquad \sum \tau_y = 0 \qquad \sum \tau_z = 0$$

Most of the problems we encounter in this chapter are restricted to cases where all the forces act in a single plane. Assuming this plane is the x-y plane, the six conditions for equilibrium reduce to the following three conditions:

$$\sum F_x = 0 \qquad \sum F_y = 0 \qquad \sum \tau_z = 0$$

Example 12-1-A: The physics of a seesaw. Two children sit on a seesaw. One child has a mass of $m_1 = 36$ kg and sits a distance of $d_1 = 1.3$ m from the fulcrum. The second child has a mass of $m_2 = 41$ kg. How far from the fulcrum must the second child sit to balance the seesaw? What is the magnitude of the force that the fulcrum exerts on the seesaw?

Approach: Since the seesaw is in equilibrium, the conditions for equilibrium must be satisfied. A schematic of the problem is shown in the Figure on the right. In this solution we have assumed that the mass of the seesaw is 0 kg. We also have assumed that the seesaw is oriented horizontally. There are two unknown parameters in this problem: the force exerted by the fulcrum on the seesaw, $F_{fulcrum}$, and the distance of the second child from the fulcrum, d_2. Using the two conditions for equilibrium we can determine these parameters. The coordinate system we will be using in this problem has its x axis directed horizontally (positive x corresponds to displacement to the right) and its y axis vertically (positive y corresponds to upwards displacement).

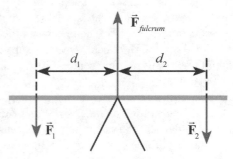

Solution: Each child exerts a force on the seesaw equal to his/her weight. These forces are directed in the vertical direction. The only other force acting on the seesaw is the force exerted by the fulcrum. This force must also be directed in the vertical direction, but opposite to the direction of the weight of the children.
Since all forces all lie in the same plane, the following three conditions for equilibrium must be satisfied:

$$\sum F_x = 0 \qquad \sum F_y = 0 \qquad \sum \tau_z = 0$$

The condition on the x components of the forces is satisfied since all forces are directed along the y axis. The condition on the y components of the forces is:

$$-F_1 + F_{fulcrum} - F_2 = 0 \quad \Rightarrow \quad F_{fulcrum} = F_1 + F_2 = \left(m_1 + m_2\right)g$$

The condition on the sum of the z components of the torques can be written in terms of the forces once we have selected a rotation axis. We will use an axis through the point of contact of the fulcrum with the seesaw, directed perpendicular to the page. With this choice of axis, we must require that

$$F_1 d_1 + F_{fulcrum} 0 - F_2 d_2 = 0 \quad \Rightarrow \quad d_2 = \frac{F_1 d_1}{F_2} = \frac{m_1 d_1}{m_2} = 1.1 \text{ m}$$

The axis we use to evaluate the torque can be chosen at will. In this solution, we have used an axis through the fulcrum. This choice removes the dependence on the force exerted by the fulcrum (since the arm of this force with respect to the axis is 0 m). However, any other axis would have produced the same result for d_2. Consider for example an axis going through child 1.

The condition on the torque is now

$$F_1 0 + F_{fulcrum} d_1 - F_2 (d_1 + d_2) = 0 \quad \Rightarrow \quad d_2 = \frac{F_{fulcrum} d_1 - F_2 d_1}{F_2} = \frac{(m_1 + m_2) g d_1 - m_2 g d_1}{m_2 g} = \frac{m_1 d_1}{m_2} = 1.1 \text{ m}$$

As expected, the value obtained for d_2 using this axis is the same as the value of d_2 obtained using the previous axis.

Section 12-2. Solving Statics Problems

The following procedure should be used to solve statics problems:
1. Consider one body at a time. Draw a free-body diagram for the body under consideration, showing each force that acts on that body. Draw the forces so that they act at the correct point(s) of application.
2. Choose a Cartesian coordinate system that is convenient. Often it is helpful to choose a coordinate system that is aligned with the direction of as many of the forces as possible. If all forces act in a plane, choose the x and y axes to lie in this plane. Resolve each of the forces into its Cartesian components.
3. Write down the first condition for equilibrium, and use symbols to represent the unknown parameters.
4. Write down the second condition for equilibrium. If all forces act in a plane, choose an axis perpendicular to that plane to calculate torque. The equations to solve are usually simplified by picking the origin at the point of application of one of the unknown forces. Write down the torques, being careful to use the proper signs.
5. Solve the system of equations for the unknown parameters.

Example 12-2-A: Irregular-shaped objects in equilibrium. An irregularly-shaped sign is supported from the ceiling by two cables as shown in the Figure on the right. The angles between the cables and the ceiling are $\theta_1 = 60°$ and $\theta_2 = 45°$. The tension in the left cable is $T_1 = 40$ N. What is the tension T_2 in the right cable? What is the mass of the sign? What can you say about the location of the center of mass of the sign?

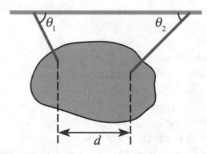

Approach: We will solve this problem by applying the five steps outlined above. The known quantities in this problem are the direction of the tension in the cables and the magnitude of the tension in one of the cables. The unknown are the tension in the second cable, the gravitational force of the sign, and the location of the center-of-mass of the sign.

Solution: Consider each of the steps discussed in this section.
1. *Consider one body at a time. Draw a free-body diagram for the body under consideration, showing each force that acts on the body. Draw the forces so that they act at the correct point(s) of application.*
In this problem we only have one body, and the forces acting on this body are shown in the free-body diagram on the right.
2. *Choose a Cartesian coordinate system that is convenient. Often it is helpful to choose a coordinate system that is aligned with the direction of as many of the forces as possible. If all forces act in a plane, choose the x and y axes to lie in this plane. Resolve each of the forces into its Cartesian components.*
In the solution of this problem we will use an xy coordinate system where the x axis is parallel to the horizontal direction and the y axis is parallel to the vertical

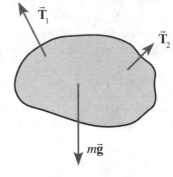

direction. Using the angles provided we can determine the x and y components of the 3 forces that act on the sign:

$$\vec{T}_1 = -T_1 \cos\theta_1 \hat{i} + T_1 \sin\theta_1 \hat{j}$$

$$\vec{T}_2 = T_2 \cos\theta_2 \hat{i} + T_2 \sin\theta_2 \hat{j}$$

$$\vec{W} = -mg\hat{j}$$

3. *Write down the first condition for equilibrium, and use symbols to represent the unknown parameters.*
Since all forces are lying in the xy plane, the conditions for equilibrium are

$$\sum F_x = 0 \quad \text{and} \quad \sum F_y = 0$$

Using the components of the forces obtained in step 2, we obtain the following two conditions for equilibrium:

$$\sum F_x = -T_1 \cos\theta_1 + T_2 \cos\theta_2 = 0 \quad \text{and} \quad \sum F_y = T_1 \sin\theta_1 + T_2 \sin\theta_2 - mg = 0$$

The unknown parameters are the mass m and the tension T_2. Since there are two equations with two unknown, we can determine these unknown parameters.
4. *Write down the second condition for equilibrium. If all forces act in a plane, choose an axis perpendicular to that plane to calculate torque. The equations to solve are usually simplified by picking the origin at the point of application of one of the unknown forces. Write down the torques, being careful to use the proper signs.*

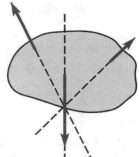

In step 3 we have shown that we can determine the unknown parameters by applying the force conditions for equilibrium. The condition that the net torque around an axis parallel to the z axis does not provide any additional constraints on the mass m and the tension T_2. However, this condition puts a constraint on the location of the center of mass of the object. Consider the diagram on the right. The dashed lines drawn through the forces exerted by the strings on the sign intersect. Now consider the torque with respect to an axis through this intersection point. The torques due to the tensions are equal to 0 Nm, since the arm of each tension force with respect to this rotation axis is 0 m. Since the net torque with respect to this axis must be zero, the torque associated with the gravitational force with respect to this axis must thus also be equal to 0 Nm. Since the gravitational force is non-zero, the arm of this force must be equal to 0 m. The center of mass of the sign must thus be located on the vertical dashed line, going through the rotation axis.
5. *Solve the system of equations for the unknowns.*
The two equations obtained in step 3 can be solved easily:

$$T_2 = \frac{T_1 \cos\theta_1}{\cos\theta_2} = 28.3\,\text{N} \quad \text{and} \quad m = \frac{T_1 \sin\theta_1 + T_2 \sin\theta_2}{g} = 5.6\,\text{kg}$$

Example 12-2-B: Equilibrium of a table. Consider the table with three legs, shown in the diagram. The table top has a uniform density and a mass of 60 kg. What is the force exerted on each leg of the table?

Approach: We will solve this problem by applying the five steps outlined above. The unknown parameters are the forces exerted by the three legs on the table top.

Solution: Consider each of the steps discussed in this section.
1. *Consider one body at a time. Draw a free-body diagram for the body under consideration, showing each force that acts on the body. Draw the forces so that they act at the correct point(s) of application.*
There are four forces acting on the table: three forces are exerted by the legs on the table top and one force is the gravitational force, acting on the center of mass of the table top. These four forces do not act in a single plane, and we thus have to consider the full set of equilibrium requirements. The four forces are shown in the free-body diagram on the right. Note that we have determined the location of the center of mass of the table top by locating the intersection point of the dashed lines shown on the surface of the table top.
2. *Choose a Cartesian coordinate system that is convenient. Often it is helpful to choose a coordinate system that is aligned with the direction of as many of the forces as possible. If all forces act in a plane, choose the x and y axes to lie in this plane. Resolve each of the forces into its Cartesian components.*

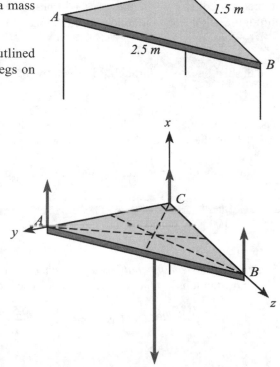

We have chosen our coordinate system such that the x axis points in the vertical direction and the y and z axes are directed along CA and CB, respectively. The four forces, resolved into their Cartesian components, can be written as

$$\vec{F}_A = F_A \hat{i}$$

$$\vec{F}_B = F_B \hat{i}$$

$$\vec{F}_C = F_C \hat{i}$$

$$\vec{W} = -mg\hat{i}$$

3. *Write down the first condition for equilibrium, and use symbols to represent the unknown parameters.*
Since all forces are parallel to the x axis, the equilibrium conditions for the y and z components of the forces are automatically satisfied. The equilibrium condition for the x components of the forces is

$$\sum F_x = F_A + F_B + F_C - mg = 0$$

There are three unknown parameters in this equation, and we can thus not solve it. Two additional equations are required to determine the unknown forces. These two additional equations can be obtained by using the torque requirements for equilibrium (step 4).

4. *Write down the second condition for equilibrium. If all forces act in a plane, choose an axis perpendicular to that plane to calculate torque. The equations to solve are usually simplified by picking the origin at the point of application of one of the unknown forces. Write down the torques, being careful to use the proper signs.*
The second condition for equilibrium requires that

$$\sum \tau_x = \sum \tau_y = \sum \tau_z = 0$$

By choosing the rotation axis to coincide with either the y or the z axis, we eliminate contributions due to two of the three unknown forces. In order to evaluate the torque due to weight of the table top, we need to determine the position of the center of mass of the table top. The Figure on the right shows a top view of the table top. The position of the center of mass can be found by calculating the intersection of two of the three dashed lines. The dashed line going through C is given by the following equation:

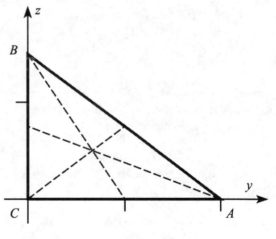

$$y = 0.75x$$

The dashed line going through B is given by the following equation:

$$y = -1.5x + 1.5$$

The intersection point of these two lines is located at (0.67, 0.5).
The arm of the weight of the table top with respect to the y axis is equal to 0.5 m; the arm with respect to the z axis is equal to 0.67 m.
The torques with respect to the y and z axes are equal to:

$$\sum \tau_y = 1.5 F_B - 0.5 mg = 0$$

$$\sum \tau_z = -2.0 F_A + 0.67 mg = 0$$

5. *Solve the system of equations for the unknowns.*
The two equations obtained in step 4 can be used to determine the forces due to legs A and B:

$$F_A = \frac{0.67 mg}{2.0} = 197 \text{ N} \quad \text{and} \quad F_B = \frac{0.5 mg}{1.5} = 196 \text{ N}$$

Using the equation obtained in step 3 we can calculate the force due to leg C:

$$F_C = mg - F_A + F_B = 195 \text{ N}$$

Section 12-3. Stability and Balance

Objects can be in three different conditions of **stability** when in equilibrium. The type of stability is determined by the nature of the forces and torques that act on the body when it is displaced slightly from its equilibrium position and/or orientation. If the forces and torques act to return the object to its equilibrium position and orientation, the object is in **stable equilibrium**. If the forces and torques act to push the object further from its equilibrium position or orientation, the object is in **unstable equilibrium**. If there are no forces and torques acting on the object when it is displaced from its equilibrium position and orientation, the object is in **neutral equilibrium**.

In many cases, we are interested in ensuring that objects are in stable equilibrium.

Section 12-4. Elasticity; Stress and Strain

Up to this point, bodies have been treated as being rigid. However, all materials change their size and/or shape when forces are applied to them. In this section, we look at some simple models of the change in size and shape of materials under applied forces.

In order to separate the properties of a type of material from the size and shape of an object, we introduce the concepts of stress and strain. **Stress** is defined as the force per unit area:

$$\text{stress} = \frac{\text{force}}{\text{area}} = \frac{F}{A}$$

where F is the magnitude of the force applied to the surface of area A. In response to a given stress, a piece of material will change its size by an amount characterized by its **strain**. Strain is defined as the distance of distortion, ΔL, divided by the length of the object, perpendicular to the surface on which the force is applied, L_0:

$$\text{strain} = \frac{\text{distortion}}{\text{length}} = \frac{\Delta L}{L_0}$$

In the limit of small distortions, the stress is proportional to the strain:

$$\text{stress} \propto \text{strain}$$

The constant of proportionality is called an **elastic modulus**. The elastic modulus is a property of the material. It does not depend on the geometry of the object. Hooke's law for springs is an instance of this general rule. In all cases, once the strain becomes large enough, the relationship between stress and strain changes. If the strain is below the **elastic limit**, the object will return to its original undistorted size and shape once the stress is removed. If the material is strained beyond the elastic limit, the material undergoes **plastic deformation**, in which it undergoes internal rearrangement and does not return to its original size and/or shape when the stress is removed. For even greater stress, the material will reach its **ultimate strength** and break apart.

Stress can be characterized by the direction of the forces applied to a given surface. The following three types of stress are considered for rigid objects:

1. **Tension stress:** The force is directed outward, normal to the surface under stress. The strain is an elongation of the volume element. We write

$$\frac{F}{A} = E \frac{\Delta L}{L_0}$$

where E is called the elastic or **Young's modulus**.

2. **Compression stress:** The force is directed inward, normal to the surface under stress. The strain is a decrease in length of the volume element. We write

$$\frac{F}{A} = E \frac{\Delta L}{L_0}$$

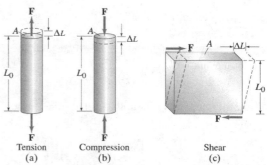
Tension (a)   Compression (b)   Shear (c)

where again E is the elastic modulus or Young's modulus.

3. **Shear stress:** The force is directed parallel to the surface under stress. The strain is the relative displacement of the top surface from the bottom surface, parallel to the plane of the surfaces. We write

$$\frac{F}{A} = G\frac{\Delta L}{L_0}$$

where G is called the **shear modulus**.

If a uniform force per unit area acts perpendicular on all surfaces, we say that the object is under **isotropic pressure**. The strain, in this case, is the fractional volume change of the object, $\Delta V/V_0$. The uniform force per unit area is called the pressure, P. Hooke's law in this case is written in terms of the change in pressure, ΔP, on the object. This measures the added stress to the object. The relationship between the stress and the strain is

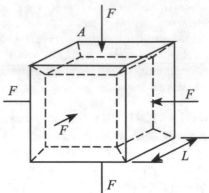

$$\Delta P = -B\frac{\Delta V}{V_0}$$

where B is called the **bulk modulus**. The minus sign indicates that the volume decreases when the pressure increases.

Example 12-4-A: Tension in a Cable. A steel cable of length $L = 500$ m (length without stress) is used to suspend a mass $m = 200$ kg from a cliff. The diameter of the cable is $d = 0.5$". How much is the cable stretched?

Approach: In order to solve this problem we must recognize that the force within the cable is non-uniform, assuming we do not ignore the mass of the cable. In order to find the final length of the cable, we determine the position dependence of the force inside the cable, and use it to calculate the position-dependent change in its length.

Solution: The coordinate system we will use to solve this problem has its x axis parallel to the cable (see Figure on the right). The origin of the coordinate system is located at the position of the mass m.

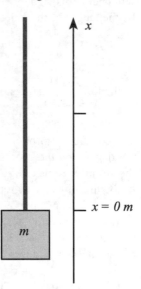

Consider a slice of the cable of thickness dx, located at position x. The weight of the cable below this point is equal to

$$W(x) = \rho V g = \rho A x g$$

where ρ is the density of the steel cable and A is its cross-sectional area. The total downward force on this section of the cable is the weight of the cable and the weight of mass m:

$$F(x) = W(x) + mg = (\rho A x + m)g$$

Hooke's law can now be used to determine the change in the length of this slice of the cable:

$$dL = \frac{1}{E}\frac{F(x)}{A}L = \frac{1}{E}\left(\rho x + \frac{m}{A}\right)gL$$

Since the length of the slice of the cable we are currently considering is dx, we can rewrite the change in the length as

$$dL = \frac{1}{E}\left(\rho x + \frac{m}{A}\right)g dx$$

To find the change in length of the entire cable, we integrate over the original length of the cable:

$$\Delta L = \int dL = \int_0^L \frac{1}{E}\left(\rho x + \frac{m}{A}\right)g dx = \left[\frac{1}{2}\frac{\rho g x^2}{E} + \frac{mg}{EA}x\right]_0^L = \frac{g}{E}\left(\frac{1}{2}\rho L^2 + \frac{m}{A}L\right) = 0.048\,\text{m}$$

In this calculation we have used the Young's modulus from Table 12-1 and assumed that the density of steel is 7,860 kg/m^3.

Section 12-5. Fracture

If the stress exceeds a limit called the **ultimate strength**, the material breaks apart. The magnitude of the ultimate strength is dependent on which type of stress is applied to the material.

Example 12-5-A: Maximum cable length. What is the longest steel cable, hanging vertically, that will support its own weight?

Approach: A material will break when the applied tension exceeds the tensile strength. When a wire is hanging vertically, it will experience a tension due to its own weight. The longer the wire, the larger its weight, and the larger its tension. By making the wire long enough, we will always be able to ensure that the tension exceeds the tensile strength.

Solution: As we discussed in Example 12-4-A, the tension in the cable will be position dependent. The tension will be the largest at the top of the cable. Assuming that the cable has a cross-sectional area A and length L we find that the weight W of the cable is

$$W = mg = \rho Vg = \rho ALg$$

The tensile stress at the top of the cable will be

$$\frac{F}{A} = \frac{\rho ALg}{A} = \rho Lg$$

The cable will break at the top when the tensile stress exceeds the tensile strength. The length required to make this happen can be found by solving the previous equation for L:

$$L = \frac{1}{\rho g}\left(\frac{F}{A}\right)_{max}$$

The tensile strength of the steel cable can be found in Table 12-2 and is 500×10^6 N/m^2. Using this tensile strength and the density used in Example 12-5-A we find the maximum length to be $L = 6.5 \times 10^3$ m.

Section 12-6. Trusses and Bridges

A **truss** is a structure that is composed of members that are always connected into triangles. Multiple triangles are used to form arbitrary shape trusses. The points at which beams are joined by pins are called **joints**. The model of a pin that is usually applied to problems is that the forces on a member at a pin cannot provide any net torque to the member about an origin centered on the pin.

Section 12-7. Arches and Domes

Arches and **domes** are used in construction to take advantage of the greater strength of materials to withstand compressive strain than tensile or shear strain.

Practice Quiz

1. Which of the following objects is in neutral equilibrium?
    a)   A chair resting on the floor
    b)   A pencil standing on its point
    c)   A sphere resting atop another sphere
    d)   A sphere resting on a flat horizontal surface

2. What is the ordering, from smallest to largest, of the proportional limit, the ultimate strength, and the elastic limit of a given material?
    a) Proportional limit, ultimate strength, elastic limit
    b) Ultimate strength, proportional limit, elastic limit
    c) Ultimate strength, elastic limit, proportional limit
    d) Proportional limit, elastic limit, ultimate strength

3. The typical behavior of a solid that has reached its proportional limit is that the strain for a given increase in stress is
    a) Greater than before the proportional limit is reached
    b) Less than before the proportional limit is reached
    c) The same as before the proportional limit is reached
    d) Unlimited, as this the point at which the material fractures

4. A block with no applied stress is shown on the left in the Figure. The same block with stress applied is shown on the right in the Figure. What type of stress has been applied to the block?
    a) Shear and compression
    b) Pressure and tension
    c) Pressure and shear
    d) Shear and tension

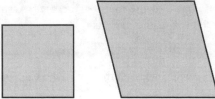

5. What is the purpose of prestressing concrete?
    a) To test to see if the concrete is strong enough for the intended load
    b) To compress the concrete so that it fits in the space provided
    c) To avoid tensile stress on the concrete
    d) To stretch the concrete out, so it is more flexible

6. What structural purpose does an architectural arch serve?
    a) None; it only has aesthetic value.
    b) To force the loads in the material to be mostly compressive
    c) To provide a well-controlled fracture point upon failure
    d) To place the structure into a condition of neutral equilibrium

7. An object is stretched and returns to its original shape when released. The maximum stress that could have been applied for this to be true is
    a) The proportional limit
    b) The elastic limit
    c) The ultimate strength
    d) Zero

8. A pencil standing vertically on its tip is an example of
    a) Non-equilibrium
    b) Stable equilibrium
    c) Unstable equilibrium
    d) Neutral equilibrium

9. What are the predominant types of strain that occur in a bent beam?
    a) Shear and pressure
    b) Tension and shear
    c) Tension and compression
    d) Compression and shear

10. What statement is necessarily true about a system that has no net torque and no net force acting on it?
    a) The system is in static equilibrium.
    b) The system is in stable static equilibrium.
    c) The system is not in equilibrium.
    d) The system is not undergoing translational or angular acceleration.

11. In the diagram, a shelf supported by a wall and a wire is shown. The shelf is 40 cm wide and has a mass of 2.4 kg. The block sitting on the shelf is 30 cm from the wall and has a mass of 1.8 kg. The wire is attached at the outer end of the shelf and to the wall a distance 40 cm above the shelf. Determine the tension in the wire. The mass of the wire is negligible.

12. Two children sit on the left side of a seesaw. One has a mass of 24 kg and sits 1.3 m from the fulcrum. The second has a mass of 31 kg and sits 1.6 m from the fulcrum. What is the mass of a person that must sit on the right side of the seesaw a distance 1.8 m from the fulcrum in order to balance the seesaw?

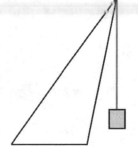

13. The triangular structure shown on the right has a uniform density and a mass of 32 kg. Determine the maximum mass that can be supported from the upper vertex of the triangle, as shown, before the entire structure is unstable. The base of the triangle is 1.0 m in length, the upper side of the triangle is 2.4 m in length, and the remaining side is 2.0 m in length.

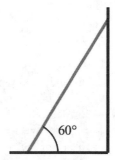

14. A concrete column has a uniform cross-section that is 20 in by 32 in and is 10 ft tall. What is the compression of the column when it carries a load of 300 tons?

15. A ladder leans against the wall at a 60° angle as shown. The ladder is 8.0 ft long and has a mass of 12.5 kg. The ladder starts to slip when a 88-kg person is three-fourths of the way up the ladder. What is the coefficient of static friction between the ladder and floor? Ignore friction with the wall.

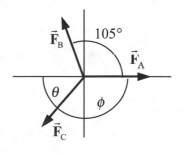

60°

Responses to Select End-of-Chapter Questions

1. Equilibrium requires both the net force and net torque on an object to be zero. One example is a meter stick with equal and opposite forces acting at opposite ends. The net force is zero but the net torque is not zero, because the forces are not co-linear. The meter stick will rotate about its center.

7. The ladder is more likely to slip when a person stands near the top of the ladder. The torque produced by the weight of the person about the bottom of the ladder increases as the person climbs the ladder, because the lever arm increases.

13. When you do a sit-up, you generate a torque with your abdominal muscles to rotate the upper part of your body off the floor while keeping the lower part of your body on the floor. The weight of your legs helps produce the torque about your hips. When your legs are stretched out, they have a longer lever arm, and so produce a larger torque, than when they are bent at the knee. When your knees are bent, your abdominal muscles must work harder to do the sit-up.

Solutions to Select End-of-Chapter Problems

1. If the tree is not accelerating, then the net force in all directions is 0.

$$\sum F_x = F_A + F_B \cos 105° + F_{Cx} = 0 \rightarrow F_{Cx} = -F_A - F_B \cos 105° = -262.1 \, \text{N}$$

$$\sum F_y = F_B \sin 105° + F_{Cy} = 0 \quad \rightarrow \quad F_{Cy} = -F_B \sin 105° = -458.8 \, \text{N}$$

$$F_C = \sqrt{F_{Cx}^2 + F_{Cy}^2} = \sqrt{(-262.1 \, \text{N})^2 + (-458.8 \, \text{N})^2} = 528.4 \, \text{N} \approx \boxed{528 \, \text{N}}$$

$$\theta = \tan^{-1} \frac{F_{Cy}}{F_{Cx}} = \tan^{-1} \frac{-458.8 \, \text{N}}{-262.1 \, \text{N}} = 60.3° \, , \, \phi = 180° - 60.3° = \boxed{120°}$$

And so $\vec{F}_C$ is 528 N, at an angle of 120° clockwise from $\vec{F}_A$. The angle has 3 sig. fig.

7. Since the backpack is midway between the two trees, the angles in the diagram are equal. Write Newton's second law for the vertical direction for the point at which the backpack is attached to the cord, with the weight of the backpack being the downward vertical force. The angle is determined by the distance between the trees and the amount of sag at the midpoint, as illustrated in the second diagram.

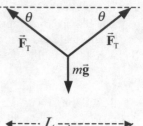

(a) $\theta = \tan^{-1}\dfrac{y}{L/2} = \tan^{-1}\dfrac{1.5\text{ m}}{3.3\text{ m}} = 24.4°$

$$\sum F_y = 2F_T\sin\theta_1 - mg = 0 \rightarrow$$

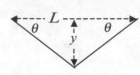

$$F_T = \dfrac{mg}{2\sin\theta_1} = \dfrac{(19\text{ kg})(9.80\text{ m/s}^2)}{2\sin 24.4°} = 225.4\text{ N} \approx \boxed{230\text{ N}}$$

(b) $\theta = \tan^{-1}\dfrac{y}{L/2} = \tan^{-1}\dfrac{0.15\text{ m}}{3.3\text{ m}} = 2.60°$

$$F_T = \dfrac{mg}{2\sin\theta_1} = \dfrac{(19\text{ kg})(9.80\text{ m/s}^2)}{2\sin 2.60°} = 2052\text{ N} \approx \boxed{2100\text{ N}}$$

13. The table is symmetric, so the person can sit near either edge and the same distance will result. We assume that the person (mass M) is on the right side of the table, and that the table (mass m) is on the verge of tipping, so that the left leg is on the verge of lifting off the floor. There will then be no normal force between the left leg of the table and the floor. Calculate torques about the right leg of the table, so that the normal force between the table and the floor causes no torque. Counterclockwise torques are taken to be positive. The conditions of equilibrium for the table are used to find the person's location.

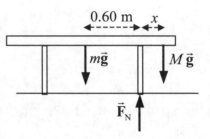

$$\sum\tau = mg(0.60\text{ m}) - Mgx = 0 \rightarrow x = (0.60\text{ m})\dfrac{m}{M} = (0.60\text{ m})\dfrac{24.0\text{ kg}}{66.0\text{ kg}} = 0.218\text{ m}$$

Thus the distance from the edge of the table is 0.50 m - 0.218 m = $\boxed{0.28\text{ m}}$.

19. There will be a normal force upwards at the ball of the foot, equal to the person's weight ($F_N = mg$). Calculate torques about a point on the floor directly below the leg bone (and so in line with the leg bone force, $\vec{F}_B$). Since the foot is in equilibrium, the sum of the torques will be zero. Take counterclockwise torques as positive.

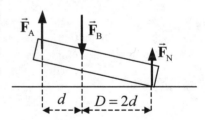

$$\sum\tau = F_N(2d) - F_A d = 0 \rightarrow$$
$$F_A = 2F_N = 2mg = 2(72\text{ kg})(9.80\text{ m/s}^2) = \boxed{1400\text{ N}}$$

The net force in the y direction must be zero. Use that to find F_B.

$$\sum F_y = F_N + F_A - F_B = 0 \rightarrow F_B = F_N + F_A = 2mg + mg = 3mg = \boxed{2100\text{ N}}$$

25. Because the board is firmly set against the ground, the top of the board would move upwards as the door opened. Thus the frictional force on the board at the door must be down. We also assume that the static frictional force is a maximum, and so is given by $F_{fr} = \mu F_N = \mu F_{push}$. Take torques about the point A in the free-body diagram, where the board rests on the ground. The board is of length l.

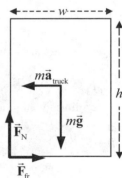

$$\sum \tau = F_{push} l \sin\theta - mg\left(\tfrac{1}{2}l\right)\cos\theta - F_{fr} l \cos\theta = 0 \quad \rightarrow$$

$$F_{push} l \sin\theta - mg\left(\tfrac{1}{2}l\right)\cos\theta - \mu F_{push} l \cos\theta = 0 \quad \rightarrow$$

$$F_{push} = \frac{mg}{2\left(\tan\theta - \mu\right)} = \frac{mg}{2\left(\tan\theta - \mu\right)} = \frac{(62.0\,\text{kg})(9.80\,\text{m/s}^2)}{2(\tan 45° - 0.45)} = 552.4\,\text{N} \approx \boxed{550\,\text{N}}$$

31. We assume the truck is accelerating to the right. We want the refrigerator to not tip in the non-inertial reference frame of the truck. Accordingly, to analyze the refrigerator in the non-inertial reference frame, we must add a pseudoforce in the opposite direction of the actual acceleration. The free-body diagram is for a side view of the refrigerator, just ready to tip so that the normal force and frictional force are at the lower back corner of the refrigerator. The center of mass is in the geometric center of the refrigerator. Write the conditions for equilibrium, taking torques about an axis through the center of mass, perpendicular to the plane of the paper. The normal force and frictional force cause no torque about that axis.

$$\sum F_{horiz} = F_{fr} - ma_{truck} = 0 \quad \rightarrow \quad F_{fr} = ma_{truck}$$

$$\sum F_{vert} = F_N - mg = 0 \quad \rightarrow \quad F_N = mg$$

$$\sum \tau = F_N\left(\tfrac{1}{2}w\right) - F_{fr}\left(\tfrac{1}{2}h\right) = 0 \quad \rightarrow \quad \frac{F_N}{F_{fr}} = \frac{h}{w}$$

$$\frac{F_N}{F_{fr}} = \frac{h}{w} = \frac{mg}{ma_{truck}} \quad \rightarrow \quad a_{truck} = g\frac{w}{h} = (9.80\,\text{m/s}^2)\frac{1.0\,\text{m}}{1.9\,\text{m}} = \boxed{5.2\,\text{m/s}^2}$$

37. (a) $\text{Stress} = \dfrac{F}{A} = \dfrac{mg}{A} = \dfrac{(1700\,\text{kg})(9.80\,\text{m/s}^2)}{0.012\,\text{m}^2} = 1.388 \times 10^6\,\text{N/m}^2 \approx \boxed{1.4 \times 10^6\,\text{N/m}^2}$

(b) $\text{Strain} = \dfrac{\text{Stress}}{\text{Young's Modulus}} = \dfrac{1.388 \times 10^6\,\text{N/m}^2}{200 \times 10^9\,\text{N/m}^2} = 6.94 \times 10^{-6} \approx \boxed{6.9 \times 10^{-6}}$

(c) $\Delta l = (\text{Strain})(l_0) = (6.94 \times 10^{-6})(9.50\,\text{m}) = 6.593 \times 10^{-5}\,\text{m} \approx \boxed{6.6 \times 10^{-5}\,\text{m}}$

43. (a) The maximum tension can be found from the ultimate tensile strength of the material.

$$\text{Tensile Strength} = \frac{F_{max}}{A} \quad \rightarrow$$

$$F_{max} = (\text{Tensile Strength})A = (500 \times 10^6\,\text{N/m}^2)\pi(5.00 \times 10^{-4}\,\text{m})^2 = \boxed{393\,\text{N}}$$

(b) To prevent breakage, thicker strings should be used, which will increase the cross-sectional area of the strings, and thus increase the maximum force. Breakage occurs because when the strings are hit by the ball, they stretch, increasing the tension. The strings are reasonably tight in the normal racket configuration, so when the tension is increased by a particularly hard hit, the tension may exceed the maximum force.

49. (*a*) The three forces on the truss as a whole are the tension force at point B, the load at point E, and the force at point A. Since the truss is in equilibrium, these three forces must add to be 0 and must cause no net torque. Take torques about point A, calling clockwise torques positive. Each member is 3.0 m in length.

$$\sum \tau = F_T (3.0\,\text{m}) \sin 60^\circ - Mg(6.0\,\text{m}) = 0 \rightarrow$$

$$F_T = \frac{Mg(6.0\,\text{m})}{(3.0\,\text{m}) \sin 60^\circ} = \frac{(66.0\,\text{kN})(6.0\,\text{m})}{(3.0\,\text{m}) \sin 60^\circ} = 152\,\text{kN} \approx \boxed{150\,\text{kN}}$$

The components of $\vec{F}_A$ are found from the force equilibrium equations, and then the magnitude and direction can be found.

$$\sum F_{\text{horiz}} = F_T - F_{A\,\text{horiz}} = 0 \rightarrow F_{A\,\text{horiz}} = F_T = 152\,\text{kN}$$

$$\sum F_{\text{vert}} = F_{A\,\text{vert}} - Mg = 0 \rightarrow F_{A\,\text{vert}} = Mg = 66.0\,\text{kN}$$

$$F_A = \sqrt{F_{A\,\text{horiz}}^2 + F_{A\,\text{vert}}^2} = \sqrt{(152\,\text{kN})^2 + (66.0\,\text{kN})^2} = 166\,\text{kN} \approx \boxed{170\,\text{kN}}$$

$$\theta_A = \tan^{-1} \frac{F_{A\,\text{vert}}}{F_{A\,\text{horiz}}} = \tan^{-1} \frac{66.0\,\text{kN}}{152\,\text{kN}} = 23.47^\circ \approx \boxed{23^\circ \text{ above AC}}$$

(*b*) Analyze the forces on the pin at point E. See the second free-body diagram. Write equilibrium equations for the horizontal and vertical directions.

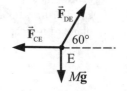

$$\sum F_{\text{vert}} = F_{DE} \sin 60^\circ - Mg = 0 \rightarrow$$

$$F_{DE} = \frac{Mg}{\sin 60^\circ} = \frac{66.0\,\text{kN}}{\sin 60^\circ} = 76.2\,\text{kN} \approx \boxed{76\,\text{kN, in tension}}$$

$$\sum F_{\text{horiz}} = F_{DE} \cos 60^\circ - F_{CE} = 0 \rightarrow$$

$$F_{CE} = F_{DE} \cos 60^\circ = (76.2\,\text{kN}) \cos 60^\circ = 38.1\,\text{kN} \approx \boxed{38\,\text{kN, in compression}}$$

Analyze the forces on the pin at point D. See the third free-body diagram. Write equilibrium equations for the horizontal and vertical directions.

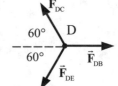

$$\sum F_{\text{vert}} = F_{DC} \sin 60^\circ - F_{DE} \sin 60^\circ = 0 \rightarrow$$

$$F_{DC} = F_{DE} = 76.2\,\text{kN} \approx \boxed{76\,\text{kN, in compression}}$$

$$\sum F_{\text{horiz}} = F_{DB} - F_{DE} \cos 60^\circ - F_{DC} \cos 60^\circ = 0 \rightarrow$$

$$F_{DB} = (F_{DE} + F_{DC}) \cos 60^\circ = 2(76.2\,\text{kN}) \cos 60^\circ = 76.2\,\text{kN}$$

$$\approx \boxed{76\,\text{kN, in tension}}$$

Analyze the forces on the pin at point C. See the fourth free-body diagram. Write equilibrium equations for the horizontal and vertical directions.

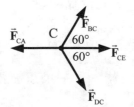

$$\sum F_{\text{vert}} = F_{BC} \sin 60^\circ - F_{DC} \sin 60^\circ = 0 \rightarrow$$

$$F_{BC} = F_{DC} = 76.2\,\text{kN} \approx \boxed{76\,\text{kN, in tension}}$$

$$\sum F_{\text{horiz}} = F_{CE} + F_{BC} \cos 60^\circ + F_{DC} \cos 60^\circ - F_{CA} = 0 \rightarrow$$

$$F_{CA} = F_{CE} + (F_{BC} + F_{DC}) \cos 60^\circ = 38.1\,\text{kN} + 2(76.2\,\text{kN}) \cos 60^\circ$$

$$= 114.3\,\text{kN} \approx \boxed{114\,\text{kN, in compression}}$$

Analyze the forces on the pin at point B. See the fifth free-body diagram. Write equilibrium equations for the horizontal and vertical directions.

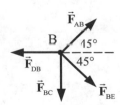

$$\sum F_{vert} = F_{AB} \sin 60° - F_{BC} \sin 60° = 0 \;\;\rightarrow$$

$$F_{AB} = F_{BC} = 76.2 \, \text{kN} \approx \boxed{76 \, \text{kN, in compression}}$$

$$\sum F_{horiz} = F_T - F_{BC} \cos 60° - F_{AB} \cos 60° - F_{DB} = 0 \;\;\rightarrow$$

$$F_T = \left(F_{BC} + F_{AB} \right) \cos 60° + F_{DB} = 2 \left(76.2 \, \text{kN} \right) \cos 60° + 76.2 \, \text{kN} = 152 \, \text{kN}$$

This final result confirms the earlier calculation, so the results are consistent. We could also analyze point A to check for consistency.

55. We first show a free-body diagram for the entire structure. All acute angles in the structure are 45°. Write the conditions for equilibrium for the entire truss by considering vertical forces and the torques about point A. Let clockwise torques be positive.

$$\sum F_{vert} = F_1 + F_2 - 5F = 0$$

$$\sum \tau = Fa + F\left(2a\right) + F\left(3a\right) + F\left(4a\right) - F_2\left(4a\right) = 0$$

$$F_2 = F \frac{10a}{4a} = 2.5F \;\; ; \;\; F_1 = 5F - F_2 = 2.5F$$

Note that the forces at the ends each support half of the load. Analyze the forces on the pin at point A. See the second free-body diagram. Write equilibrium equations for the horizontal and vertical directions.

$$\sum F_{vert} = F_1 - F - F_{AB} \sin 45° = 0 \;\;\rightarrow$$

$$F_{AB} = \frac{F_1 - F}{\sin 45°} = \frac{\frac{3}{2}F}{\frac{1}{2}\sqrt{2}} = \boxed{\frac{3F}{\sqrt{2}}}, \text{ in compression}$$

$$\sum F_{horiz} = F_{AC} - F_{AB} \cos 45° = 0 \;\;\rightarrow\;\; F_{AC} = F_{AB} \cos 45° = \frac{3F}{\sqrt{2}} \frac{\sqrt{2}}{2} = \boxed{\frac{3}{2}F}, \text{ in tension}$$

Analyze the forces on the pin at point C. See the third free-body diagram. Write equilibrium equations for the horizontal and vertical directions.

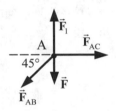

$$\sum F_{vert} = F_{BC} - F = 0 \;\;\rightarrow\;\; F_{BC} = \boxed{F, \text{ tension}}$$

$$\sum F_{horiz} = F_{CE} - F_{AC} = 0 \;\;\rightarrow\;\; F_{CE} = F_{AC} = \boxed{\frac{3}{2}F, \text{ in tension}}$$

Analyze the forces on the pin at point B. See the fourth free-body diagram. Write equilibrium equations for the horizontal and vertical directions.

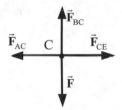

$$\sum F_{vert} = F_{AB} \sin 45° - F_{BE} \sin 45° - F_{BC} = 0 \;\;\rightarrow$$

$$F_{BE} = F_{AB} - \frac{F_{BC}}{\sin 45°} = \frac{3F}{\sqrt{2}} - \frac{F}{\frac{1}{2}\sqrt{2}} = \boxed{\frac{F}{\sqrt{2}}}, \text{ tension}$$

$$\sum F_{horiz} = F_{AB} \cos 45° + F_{BE} \cos 45° - F_{DB} = 0 \;\;\rightarrow$$

$$F_{DB} = \left(F_{AB} + F_{BE} \right) \cos 45° = \left(\frac{3F}{\sqrt{2}} + \frac{F}{\sqrt{2}} \right) \frac{\sqrt{2}}{2} = \boxed{2F, \text{ in compression}}$$

Analyze the forces on the pin at point D. See the fifth free-body diagram. Write equilibrium equations for the vertical direction.

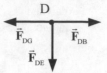

$$\sum F_{vert} = -F_{DE} \quad \rightarrow \quad F_{DE} = \boxed{0}$$

All of the other forces can be found from the equilibrium of the structure.

$$F_{DG} = F_{DB} = \boxed{2F, \text{ in compression}} \,, \quad F_{GE} = F_{BE} = \boxed{\frac{F}{\sqrt{2}}, \text{ tension}} \,,$$

$$F_{EH} = F_{CE} = \boxed{\tfrac{3}{2}F, \text{ in tension}} \,, \quad F_{GH} = F_{BC} = \boxed{F, \text{ tension}} \,, \quad F_{HJ} = F_{AC} = \boxed{\tfrac{3}{2}F, \text{ in tension}} \,,$$

$$F_{GJ} = F_{AB} = \boxed{\frac{3F}{\sqrt{2}}, \text{ in compression}}$$

61. (a) The weight of the shelf exerts a downward force and a clockwise torque about the point where the shelf touches the wall. Thus there must be an upward force and a counterclockwise torque exerted by the slot for the shelf to be in equilibrium. Since any force exerted by the slot will have a short lever arm relative to the point where the shelf touches the wall, the upward force must be larger than the gravity force. Accordingly, there then must be a downward force exerted by the slot at its left edge, exerting no torque, but balancing the vertical forces.

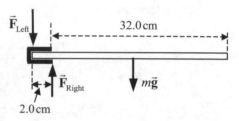

(b) Calculate the values of the three forces by first taking torques about the left end of the shelf, with the net torque being zero, and then sum the vertical forces, with the sum being zero.

$$\sum \tau = F_{Right}\left(2.0 \times 10^{-2}\,\text{m}\right) - mg\left(17.0 \times 10^{-2}\,\text{m}\right) = 0 \quad \rightarrow$$

$$F_{Right} = \left(6.6\,\text{kg}\right)\left(9.80\,\text{m/s}^2\right)\left(\frac{17.0 \times 10^{-2}\,\text{m}}{2.0 \times 10^{-2}\,\text{m}}\right) = 549.8\,\text{N} \approx \boxed{550\,\text{N}}$$

$$\sum F_y = F_{Right} - F_{Left} - mg \quad \rightarrow$$

$$F_{Left} = F_{Right} - mg = 549.8\,\text{N} - \left(6.6\,\text{kg}\right)\left(9.80\,\text{m/s}^2\right) = \boxed{490\,\text{N}}$$

$$mg = \left(6.6\,\text{kg}\right)\left(9.80\,\text{m/s}^2\right) = \boxed{65\,\text{N}}$$

(c) The torque exerted by the support about the left end of the rod is

$$\tau = F_{Right}\left(2.0 \times 10^{-2}\,\text{m}\right) = \left(549.8\,\text{N}\right)\left(2.0 \times 10^{-2}\,\text{m}\right) = \boxed{11\,\text{m·N}}$$

67. The airplane is in equilibrium, and so the net force in each direction and the net torque are all equal to zero. First write Newton's second law for both the horizontal and vertical directions, to find the values of the forces.

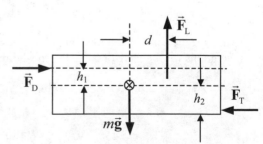

$$\sum F_x = F_D - F_T = 0 \quad \rightarrow \quad F_D = F_T = \boxed{5.0 \times 10^5\,\text{N}}$$

$$\sum F_y = F_L - mg = 0$$

$$F_L = mg = \left(7.7 \times 10^4\,\text{kg}\right)\left(9.80\,\text{m/s}^2\right) = 7.546 \times 10^5\,\text{N}$$

Calculate the torques about the CM, calling counterclockwise torques positive.

$$\sum \tau = F_L d - F_D h_1 - F_T h_2 = 0 \quad \rightarrow \quad h_1 = \frac{F_L d - F_T h_2}{F_D} = \frac{\left(7.546 \times 10^5\,\text{N}\right)\left(3.2\,\text{m}\right) - \left(5.0 \times 10^5\,\text{N}\right)\left(1.6\,\text{m}\right)}{\left(5.0 \times 10^5\,\text{N}\right)} = \boxed{3.2\,\text{m}}$$

73. The force on the sphere from each plane is a normal force, and so is perpendicular to the plane at the point of contact. Use Newton's second law in both the horizontal and vertical directions to determine the magnitudes of the forces.

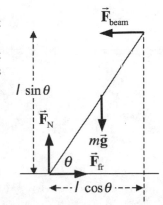

$$\sum F_x = F_L \sin\theta_L - F_R \sin\theta_R = 0 \quad \rightarrow \quad F_R = F_L \frac{\sin\theta_L}{\sin\theta_R} = F_L \frac{\sin 67°}{\sin 32°}$$

$$\sum F_y = F_L \cos\theta_L + F_R \cos\theta_R - mg = 0 \quad \rightarrow \quad F_L\left(\cos 67° + \frac{\sin 67°}{\sin 32°}\cos 32°\right) = mg$$

$$F_L = \frac{mg}{\left(\cos 67° + \dfrac{\sin 67°}{\sin 32°}\cos 32°\right)} = \frac{(23\,\text{kg})(9.80\,\text{m/s}^2)}{\left(\cos 67° + \dfrac{\sin 67°}{\sin 32°}\cos 32°\right)} = 120.9\,\text{N} \approx \boxed{120\,\text{N}}$$

$$F_R = F_L \frac{\sin 67°}{\sin 32°} = (120.9\,\text{N})\frac{\sin 67°}{\sin 32°} = 210.0\,\text{N} \approx \boxed{210\,\text{N}}$$

79. The tension in the string when it breaks is found from the ultimate strength of nylon under tension, from Table 12-2.

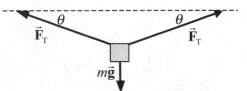

$$\frac{F_T}{A} = \text{Tensile Strength} \quad \rightarrow$$

$$F_T = A\left(\text{Tensile Strength}\right)$$

$$= \pi\left[\tfrac{1}{2}(1.15\times 10^{-3}\,\text{m})\right]^2 (500\times 10^6\,\text{N/m}^2) = 519.3\,\text{N}$$

From the force diagram for the box, we calculate the angle of the rope relative to the horizontal from Newton's second law in the vertical direction. Note that since the tension is the same throughout the string, the angles must be the same so that the object does not accelerate horizontally.

$$\sum F_y = 2F_T \sin\theta - mg = 0 \quad \rightarrow$$

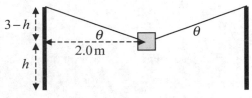

$$\theta = \sin^{-1}\frac{mg}{2F_T} = \sin^{-1}\frac{(25\,\text{kg})(9.80\,\text{m/s}^2)}{2(519.3\,\text{N})} = 13.64°$$

To find the height above the ground, consider the second diagram.

$$\tan\theta = \frac{3.00\,\text{m} - h}{2.00\,\text{m}} \quad \rightarrow \quad h = 3.00\,\text{m} - 2.00\,\text{m}\left(\tan\theta\right) = 3.00\,\text{m} - 2.00\,\text{m}\left(\tan 13.64°\right) = \boxed{2.5\,\text{m}}$$

85. Draw a free-body diagram for one of the beams. By Newton's third law, if the right beam pushes down on the left beam, then the left beam pushes up on the right beam. But the geometry is symmetric for the two beams, and so the beam contact force must be horizontal. For the beam to be in equilibrium, $F_N = mg$ and so $F_{fr} = \mu_s F_N = \mu mg$ is the maximum friction force. Take torques about the top of the beam, so that $\vec{F}_{beam}$ exerts no torque. Let clockwise torques be positive.

$$\sum \tau = F_N l \cos\theta - mg\left(\tfrac{1}{2}l\right)\cos\theta - F_{fr} l \sin\theta = 0 \quad \rightarrow$$

$$\theta = \tan^{-1}\frac{1}{2\mu_s} = \tan^{-1}\frac{1}{2(0.5)} = \boxed{45°}$$

91. Consider the free-body diagram for the box. The box is assumed to be in equilibrium, but just on the verge of both sliding and tipping. Since it is on the verge of sliding, the static frictional force is at its maximum value. Use the equations of equilibrium. Take torques about the lower right corner where the box touches the floor, and take clockwise torques as positive. We also assume that the box is just barely tipped up on its corner, so that the forces are still parallel and perpendicular to the edges of the box.

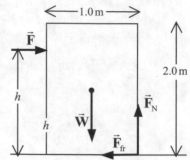

$$\sum F_y = F_N - W = 0 \quad \rightarrow \quad F_N = W$$

$$\sum F_x = F - F_{fr} = 0 \quad \rightarrow \quad F = F_{fr} = \mu W = (0.60)(250\,\text{N}) = \boxed{150\,\text{N}}$$

$$\sum \tau = Fh - W(0.5\,\text{m}) = 0 \quad \rightarrow \quad h = (0.5\,\text{m})\frac{W}{F} = (0.5\,\text{m})\frac{250\,\text{N}}{150\,\text{N}} = \boxed{0.83\,\text{m}}$$

97. (a) The stress is given by F/A, the applied force divided by the cross-sectional area, and the strain is given by $\Delta l/l_0$, the elongation over the original length.

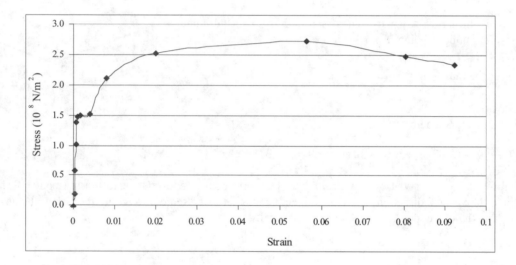

(b) The elastic region is shown in the graph.

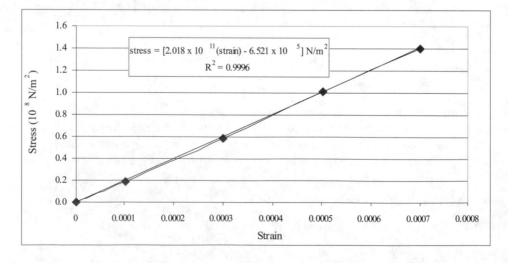

The slope of the stress vs. strain graph is the elastic modulus, and is $\boxed{2.02 \times 10^{11}\,\text{N/m}^2}$.

Chapter 13: Fluids

Chapter Overview and Objectives

In this chapter, the physics of fluids is introduced. The concepts of pressure, density, buoyancy, flow rate, viscosity, surface tension, and capillarity are presented, and the relationships between depth and pressure of a fluid at rest are discussed. The types of instruments used to measure pressure are described and Archimedes' principle and Bernoulli's equation are introduced.

After completing this chapter you should:
- Know the definition of density.
- Know the definition of pressure.
- Know the relationship between pressure and depth within a fluid at rest.
- Know the difference between absolute pressure and gauge pressure.
- Know Pascal's principle and be able to apply it to problems.
- Know how to determine pressure using the height of a column of fluid supported by a pressure difference.
- Know what buoyancy is and how to use Archimedes' principle to calculate the buoyant force.
- Know what the equation of continuity is.
- Know Bernoulli's equation and how to apply it to fluids in motion.
- Know what viscosity is.
- Know Poiseuille's equation and how to apply it to problems.
- Know what surface tension is.
- Know what a pump is.

Summary of Equations

Definition of density:	$\rho = \dfrac{m}{V}$	(Section 13-2)
Definition of pressure:	$P = \dfrac{F}{A}$	(Section 13-3)
Dependence of pressure on depth within fluid:	$P(h) = P(0) + \rho g h$	(Section 13-3)
	$P(h) = P(0) + \int_0^h \rho(y) g \, dy$	(Section 13-3)
Definition of gauge pressure:	$P = P_0 + P_G$	(Section 13-4)
Buoyant force on object displacing fluid:	$F_B = \rho g V_{fluid\,displaced}$	(Section 13-4)
Equation of continuity:	$\rho_1 A_1 v_1 = \rho_2 A_2 v_2$	(Section 13-8)
Equation of continuity for a constant density fluid:	$A_1 v_1 = A_2 v_2$	(Section 13-8)
Bernoulli's equation:	$P_1 + \frac{1}{2}\rho v_1^2 + \rho g y_1 = P_2 + \frac{1}{2}\rho v_2^2 + \rho g y_2$	(Section 13-9)
Torricelli's theorem:	$v_1 = \sqrt{2g(y_2 - y_1)}$	(Section 13-10)

Viscous force:
$$\frac{F}{A} = \eta \frac{dv}{dx}$$
(Section 13-11)

Poiseuille's equation:
$$Q = \frac{\pi R^4 \left(P_1 - P_2\right)}{8\eta L}$$
(Section 13-12)

Definition of surface tension:
$$\gamma = \frac{F}{L}$$
(Section 13-13)

Chapter Summary

Section 13-1. Phases of Matter

Ordinary matter has three common phases:
- **Solid phase:** maintains fixed shape and fixed size.
- **Liquid phase:** maintains fixed size but its shape adjusts itself to the shape of the container.
- **Gas phase:** has neither a fixed size nor a fixed shape and adjusts its shape and size to the shape and size of the container.

Section 13-2. Density and Specific Gravity

The **density** ρ of a material is defined as the ratio of its mass m and its volume V:

$$\rho = \frac{m}{V}$$

The **specific gravity** of a substance is the ratio of the density of the material to the density of water at 4.00 °C. The density of water at 4.00 °C is 1000 kg/m^3 or 1.000 g/cm^3.

Example 13-2-A. Mass of ice. Ice cubes are sold in ten-pound bags. You need 1000 ice cubes for your party. Each ice cube is a rectangular prism of dimension 2 cm × 2 cm × 3 cm. How many bags of ice do you need?

Approach: In this problem you know the volume of ice you need. Using the known density of ice, you can determine the total mass of the ice you need. Since the ice cubes are sold in ten-pound bags, you now know how many bags to get.

Solution: The volume of each ice cube is its length L times its width W times its height H:

$$V_{ice\,cube} = LWH = \left(0.02\right)\left(0.02\right)\left(0.03\right) = 1.2 \times 10^{-5}\,\text{m}^3$$

Since you need 1000 ice cubes for your party, the total volume of ice you need is $V_{ice} = 1000 V_{ice\,cube} = 1.2 \times 10^{-2}$ m^3. The density of ice is 917 kg/m^3 and the total mass of ice required for the party is

$$M_{ice} = \rho_{ice}\,V_{ice} = \left(917\,\text{kg}/\text{m}^3\right)\left(1.2 \times 10^{-2}\,\text{m}^3\right) = 11\,\text{kg}$$

The weight of one bag of ice is 10 lbs. Its mass (10/2.2) kg = 4.5 kg. Since you need 11 kg of ice, you will need to purchase three bags of ice (assuming you cannot buy a fraction of a bag).

Section 13-3. Pressure in Fluids

The pressure P exerted by a fluid on a surface area A is defined as the ratio of the normal force F and the area A:

$$P = \frac{F}{A}$$

The SI unit for pressure is N/m^2. A commonly used unit for pressure is the **Pascal** (Pa); 1 Pascal = 1 N/m^2. Another commonly used unit is lbs/in^2.

The pressure in a fluid is isotropic; it is the same in all directions. In a fluid at rest, the force on a given surface is always perpendicular to the surface. The pressure is the same at all positions at the same depth within the fluid. The pressure in a fluid increases with depth, because the pressure at a given depth must support the weight of the fluid above it. The pressure at a depth h below the surface is the pressure at the top surface, P_0, plus the weight per unit area of the fluid above the depth h:

$$P(h) = P(0) + \rho g h$$

where ρ is the density of the fluid and g is the gravitational acceleration. To derive this equation, we have assumed that the density of the liquid does not depend on pressure; this is a reasonable assumption for most liquids but not for gases. For systems with a density that depends strongly on pressure, the following relationship between depth and pressure must be used:

$$P(h) = P(0) + \int_0^h \rho(y) g \, dy$$

where $\rho(y)$ is the density as a function of depth y within the fluid or gas.

Example 13-3-A. Systems with pressure-dependent densities. A gas has a temperature gradient such that pressure and density have the following relationship:

$$\rho = cP^2$$

The pressure at the top of a column of gas of height h is P_0 and the density at this location is ρ_0. What are the pressure and the density at the bottom of this column?

Approach: Since the density has a strong pressure dependence, we cannot use the approximation that the density is constant to solve this problem. In addition, since the problem does not provide us with an expression for the density as function of depth, we cannot use the expression for $P(h)$ in terms of $\rho(y)$ as discussed above.

Solution: The constant c in the expression of the density is determined by the conditions at the top of the gas column:

$$\rho_0 = cP_0^2 \quad \Rightarrow \quad c = \frac{\rho_0}{P_0^2} \quad \Rightarrow \quad \rho(P) = \frac{\rho_0}{P_0^2} P^2$$

Now consider a small slice of the gas column, at depth y and with thickness dy. The difference in pressure dP between the top and the bottom of this slice is equal to

$$dP = P_{top} - P_{bottom} = -\rho g dy = -\frac{\rho_0}{P_0^2} P^2 g dy$$

This differential equation can be rewritten as

$$\frac{1}{P^2} dP = -\frac{\rho_0}{P_0^2} g dy$$

After integrating both sides we obtain

$$\int_{P_0}^{P(h)} \frac{1}{P^2} dP = \left(\frac{1}{P_0} - \frac{1}{P(h)} \right) = -\int_0^h \frac{\rho_0}{P_0^2} g \, dy = -\frac{\rho_0}{P_0^2} gh$$

The pressure at the bottom of the column can be determined by solving this equation for $P(h)$:

$$P(h) = \frac{1}{\frac{\rho_0}{P_0^2} gh + \frac{1}{P_0}} = \frac{P_0^2}{\rho_0 gh + P_0}$$

Section 13-4. Atmospheric Pressure and Gauge Pressure

The pressure of the air at sea level has a mean value of 1.013×10^5 N/m². This value is used to define 1 **atmosphere** of pressure, a commonly used unit of pressure. A **bar** is another unit of pressure convenient for measuring atmospheric pressure. One bar is equal to 1.00×10^5 N/m² and one bar is thus slightly less than one atmosphere.

The pressure we have defined previously is sometimes referred to as **absolute pressure**. This is in contrast to gauge pressure. **Gauge pressure** is the difference between absolute pressure and atmospheric pressure:

$$P_{gauge} = P - P_{atmosphere}$$

Most mechanical pressure gauges, such as compressed gas gauges, measure gauge pressure rather than absolute pressure. When using these gauges we have to take into consideration that the atmospheric pressure varies and corrections for these variations must be made when this type of gauge is being used to measure absolute pressures.

Example 13-4-A. Using air pressure to provide support. A child sits on an inner tube of a tire in such a way that the inner tube keeps the child off the floor. The area of contact of the inner tube with the floor is 6.8 square inches. The gauge pressure in the inner tube is 7.2 lbs/in². What is the mass of the child? Assume the weight of the inner tube is negligible.

Approach: Since the inner tube is made of flexible rubber, we can assume that no support is provided by the inner tube itself. The force required to support the child must thus be associated with the pressure difference between the air inside the inner tube and the air outside the inner tube. This pressure difference is the gauge pressure.

If we define our system to consist of the inner tube and the child, we conclude that the net external force acting on it must be zero. The external force has two components that must be equal in magnitude but directed in opposite directions: the gravitational force acting on the child and the normal force exerted by the floor on the inner tube. The normal force can be determined from the gauge pressure and the area of contact between the tire and the floor.

Solution: The force exerted by the ground on the inner tube must be equal in magnitude to the force exerted by the inner tube on the ground. This latter force is equal to the product of the gauge pressure of the air inside the inner tube and the contact area between the tube and the ground. The weight of the child must thus be equal to

$$W_{child} = P_{gauge} A = (7.2)(6.8) = 49 \text{ lbs}$$

Since a one kilogram mass has a weight of 2.2 pounds we conclude that the mass of the child is equal to 49/2.2 = 22 kg.

Section 13-5. Pascal's Principle

Pascal's principle states that *if an external pressure is applied to a confined fluid, the pressure at every point within the fluid increases by that amount.* Pascal's principle has many important applications such as hydraulic lifts and hydraulic brakes.

Section 13-6. Measurement of Pressure; Gauges and the Barometer

There are several different instruments that are used to measure fluid pressure. One common device is a **manometer**. A manometer uses the height of a column of fluid, supported by the pressure difference at two surfaces of the liquid, to determine this pressure difference. If the pressure at one surface is known (e.g., volume above the surface is evacuated in which case the pressure is 0, or the volume is open to the atmosphere in which case the pressure would be about one atm.) the height of the column can be used to determine the pressure at the other surface. A typical manometer is shown in the diagram on the right. The pressure at the lower surface of the liquid, P_2, is related to P_1 by

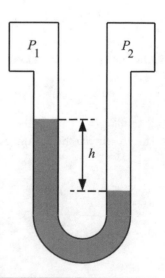

$$P_2 = P_1 + \rho g h$$

where ρ is the density of the liquid in the manometer and h is the difference in the vertical position of the two surfaces. Pressures are often expressed in terms of the height of the column of liquid supported (e.g., mm of mercury).

A **barometer** is a manometer with one end closed with negligible pressure ($P_1 = 0$ atm) and the other end open to atmospheric pressure. The barometer is used to measure atmospheric pressure. If the height of the liquid column is h, then the pressure $P_2 = \rho g h$.

Example-13-6-A. The mercury barometer. The atmospheric pressure on Earth is 760 mm of mercury (the height of the column of mercury in a barometer is 760 mm). If a mercury barometer were brought to the moon to measure the air pressure inside a spacecraft, what would the height of the mercury column be if the air pressure inside the spacecraft is the same as Earth's atmospheric pressure?

Approach: The operation of a barometer depends on the density of the liquid and the gravitational acceleration. Since the gravitational acceleration on the surface of the moon differs from the gravitational acceleration on the surface of the Earth, the height of the column of mercury will differ at the two locations, even though the measured pressure is the same.

Solution: The measured pressure in both locations is the same. This requires that

$$\rho g_{moon} h_{moon} = \rho g_{earth} h_{earth}$$

The height of the column on the moon is thus equal to

$$h_{moon} = \frac{g_{earth} h_{earth}}{g_{moon}} = \frac{(9.80)(0.760)}{(1.62)} = 4.60 \, \text{m}$$

The gravitational acceleration on the surface of the moon was obtained from the mass of the moon and its radius:

$$g_{moon} = G \frac{M_{moon}}{R_{moon}^{\,2}} = (6.67 \times 10^{-11}) \frac{(7.35 \times 10^{22})}{(1.74 \times 10^6)} = 1.62 \, \text{m/s}^2$$

A mercury barometer would be somewhat impractical for measuring pressures as large as atmospheric pressure on the moon because of its required height.

Section 13-7. Buoyancy and Archimedes' Principle

An object either completely or partially immersed in a fluid is acted on by an upward force of a magnitude equal to the weight of the fluid displaced by the object.

$$F_B = \rho g V_{fluid\ displaced}$$

This is a statement of **Archimedes' principle**. The upward force on the object, F_B, is called the **buoyant force**. The weight of the displaced fluid is written in terms of the density of the fluid ρ, the acceleration of gravity g, and the volume of the fluid displaced by the object $V_{fluid\ displaced}$.

Example 13-7-A: Maximum load of a boat. A raft is made from wood and has an average density $\rho_{raft} = 378 \, \text{kg/m}^3$. The volume of the raft is $V_{raft} = 3.87 \, \text{m}^3$. What is the maximum load that can be placed on the raft before its top surface reaches the surface of the water?

Approach: When the boat carries its maximum load, the buoyant force exerted by the water on the boat will be equal to the gravitational force acting on the boat and its load. Since we know the volume of the boat, we can determine the maximum buoyant force and thus determine the total gravitational force due to the boat and its load.

Solution: The buoyant force acting on the boat, when it carries its maximum load, is determined by the volume of the water displaced. The problem provides us with the information required to calculate this force:

$$F_B = \rho_{water} V_{raft} g$$

The gravitational force acing on the boat and its load is equal to

$$F_{grav} = \left(m_{raft} + m_{load} \right) g = \left(\rho_{raft} V_{raft} + m_{load} \right) g$$

In this last expression we have assumed that the submerged volume is the entire volume of the raft.
If the system is in equilibrium, the magnitude of the buoyant force must be equal to the magnitude of the gravitational force. This requires that

$$\rho_{water} V_{raft} g = \left(\rho_{raft} V_{raft} + m_{load} \right) g$$

This equation contains only one unknown, m_{load}, which can now be determined:

$$m_{load} = \left(\rho_{water} - \rho_{raft} \right) V_{raft} = (1000 - 378)(3.87) = 2.41 \times 10^3 \text{ kg}$$

Section 13-8. Fluids in Motion; Flow Rate and the Equation of Continuity

Fluid dynamics is the study of moving fluids. The study of moving water is an important example of fluid dynamics and is frequently referred to as **hydrodynamics**.
There are two main types of fluid flow:
- **laminar** flow which occurs when the fluid flows in smooth layers.
- **turbulent** flow which occurs when there are eddies or whirlpools in the flow.
Consider the flow of a fluid through a pipe of varying diameter. When the velocity distribution of the fluid is independent of time, the flow is called **steady-state flow**; the velocity may still be position dependent, and the fluid may thus accelerate, but the velocity at a given position remains constant. In steady-state flow, the flow rates ($kg/m^2/s$) must be the same everywhere. This requires that

$$\rho_1 A_1 v_1 = \rho_2 A_2 v_2$$

where ρ is the density of the fluid, A is the cross-sectional area of the flow tube, and v is the speed of the velocity of the fluid. The subscripts refer to different locations along the flow tube. This expression is called the **equation of continuity**. If the fluid has a constant density, $\rho_1 = \rho_2$, then the equation of continuity can be rewritten as

$$A_1 v_1 = A_2 v_2$$

This approximation is a good approximation for most liquids and gases in those situations in which the speeds involved are small compared to the speed of sound in the gas.

Section 13-9. Bernoulli's Equation

In the particular case of an incompressible fluid undergoing steady-state laminar flow through a flow tube, there is a simple relationship, called **Bernoulli's equation,** that relates the speed of the fluid, the pressure in the fluid, and the height of the fluid within a given flow tube. Bernoulli's equation can be derived from the work-energy principle and is written as

$$P_1 + \tfrac{1}{2}\rho v_1^2 + \rho g y_1 = P_2 + \tfrac{1}{2}\rho v_2^2 + \rho g y_2$$

where P is the pressure, ρ is the density, v is the speed, g is the acceleration of gravity, and y is the vertical position of the fluid. The subscripts refer to different locations along the flow tube.

Example 13-9-A. Applying Bernoulli's equation to water flow in a fire hose. A fire hose sprays water at a rate $R = 82$ gallons per minute through a nozzle of diameter $d_{out} = 2.2$ cm. The hose has a diameter of $d_{hose} = 6.3$ cm. What is the pressure in the hose lying on the ground before the water reaches the nozzle held by a fireman at a height $h = 1.4$ m above the ground?

Approach: This problem is a good example of an application of Bernoulli's equation. However, in order to determine the pressure inside the hose we need to know the values of all other variables in this equation. Based on the information required to apply Bernoulli's equation to the current problem we need to determine the velocity of the water when it leaves the hose and its velocity in the section of the hose that lies on the ground.

Solution: Consider that the water leaves the hose with a velocity v. The volume that passes through the nozzle each second is $(1/4)\pi d_{out}^2 v$. The problem states that this rate is R.

Using this rate and the diameter of the nozzle, as specified in the problem, we can calculate the velocity of the water leaving the hose:

$$v = \frac{R}{\frac{1}{4}\pi d_{out}^2} = \frac{(82)(6.3 \times 10^{-5})}{\frac{1}{4}\pi(0.022)^2} = 13.6 \text{ m/s}$$

In this calculation we have converted the rate R from gal/min to m³/s: 1 gal/min = 3.78 x 10⁻³ m³/min = 6.3 x 10⁻⁵ m³/s. The speed of the fluid in the hose can be obtained using the equation of continuity:

$$A_{hose}v_{hose} = A_{noz}v_{noz} \quad \Rightarrow \quad v_{hose} = \frac{A_{noz}v_{noz}}{A_{hose}} = \frac{(0.022)^2(13.6)}{(0.063)^2} = 1.66 \text{ m/s}$$

We can now use Bernoulli's equation to relate the pressure in the hose to the known quantities:

$$P_{hose} + \tfrac{1}{2}\rho v_{hose}^2 + \rho g y_{hose} = P_{noz} + \tfrac{1}{2}\rho v_{noz}^2 + \rho g y_{noz} \quad \Rightarrow$$

$$P_{hose} = P_{noz} + \rho\left[\tfrac{1}{2}\left(v_{noz}^2 - v_{hose}^2\right) + g\left(y_{noz} - y_{hose}\right)\right] \quad \Rightarrow$$

$$P_{hose} = \left(1.01 \times 10^5\right) + \left(1000\right)\left\{\tfrac{1}{2}\left[(13.6)^2 - (1.66)^2\right] + (9.80)(1.4)\right\} = 2.1 \times 10^5 \text{ N/m}^2 = 2.1 \text{ atm}$$

Section 13-10. Application of Bernoulli's Principle: From Torricelli to Sailboats, Airfoils, and TIA

Consider the special case of a fluid in which the pressure is at positions 1 and 2. If the velocity of the fluid is zero at point 2, Bernoulli's equation can be simplified to

$$\tfrac{1}{2}\rho v_1^2 + \rho g y_1 = \rho g y_2$$

If this relationship is solved for the velocity of the fluid at point 1, we obtain

$$v_1 = \sqrt{2g(y_2 - y_1)}$$

This is known as **Torricelli's theorem**. We see that the speed of the fluid at position 1 is the same as the speed of an object undergoing free fall through a distance $y_2 - y_1$. Torricelli's equation is a good approximation to the situation where a fluid, contained in a large container, leaves the container through an opening whose diameter is small compared to the diameter of the container. In that case, the velocity inside the container, far from the small exit opening, is small compared to the velocity of the fluid moving through the opening.

Example 13-10-A. Measuring the flow rate of a liquid. A venturi tube is used to measure the flow rate of gasoline through a fuel line. The venturi tube contains a restriction, as shown in the Figure on the right, and the flow rate can be determined from the measured pressure at two locations along the fuel line. Consider a venturi tube whose large diameter is $d_1 = 1.22$ cm and small diameter is $d_2 = 0.26$ cm. The gauge pressure measured at position 1 is $P_1 = 2.45 \times 10^4$ N/m² and the pressure at position 2 is $P_2 = 1.34 \times 10^4$ N/m². What is the flow rate of gasoline through this tube? Note: the density of gasoline is 680 kg/m³.

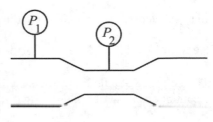

Approach: The flow rate through the tube is equal to Av where we can use either the area/speed in the section with the larger diameter or the area/speed in the section with the smaller diameter. This gives us a relation between the velocity at position 1 and the velocity at position 2. A second relation between these velocities can be obtained from Bernoulli's equation. With two equations and two unknown, we can determine both velocities and thus the flow rate.

Solution: Since the flow rate is the same in both sections of the venturi tube we conclude that

$$A_1 v_1 = A_2 v_2$$

The velocity of the fluid at location 2 can thus be specified in terms of the velocity at location 1:

$$v_2 = \frac{A_1}{A_2} v_1$$

Since the average height of the liquid in both positions is the same and it is reasonable to assume that the density of the liquid in both positions is the same, we can simplify Bernoulli's equation to

$$P_1 + \tfrac{1}{2}\rho v_1^2 = P_2 + \tfrac{1}{2}\rho v_2^2$$

Using our expression for the velocity of the liquid at position 2 in terms of the velocity of the liquid at position 1 we can rewrite this equation as

$$P_1 + \tfrac{1}{2}\rho v_1^2 = P_2 + \tfrac{1}{2}\rho \left(\frac{A_1}{A_2}\right)^2 v_1^2$$

The velocity v_1 is the only unknown in this equation and can now thus be determined:

$$v_1 = \sqrt{\frac{2(P_1 - P_2)}{\rho\left[\left(\frac{A_1}{A_2}\right)^2 - 1\right]}} = \sqrt{\frac{2\left(2.45 \times 10^4 - 1.34 \times 10^4\right)}{(680)\left[\left(\frac{(1.22)^2}{(0.26)^2}\right)^2 - 1\right]}} = 0.26\,\text{m/s}$$

Using this velocity we can now determine the flow rate Q:

$$Q = A_1 v_1 = \frac{1}{4}\pi d_1^2 v_1 = \frac{1}{4}\pi\left(1.22 \times 10^{-2}\right)^2 (0.26) = 3.0 \times 10^{-5}\,\text{m}^3/\text{s}$$

Section 13-11. Viscosity

Viscosity is the frictional force that exists between adjacent layers of a fluid or gas that move with different velocities. Viscosity is characterized in terms of the shear stress, F/A, that exists between layers of fluid or gas, and is found to be proportional to the spatial derivative (or gradient) of the velocity of the fluid in the direction perpendicular to the direction of flow:

$$\frac{F}{A} = \eta\,\frac{dv}{dx}$$

In this equation, η is the **coefficient of viscosity**.

Section 13-12. Flow in Tubes: Poiseuille's Equation, Blood Flow

The viscosity of a liquid restricts the flow rates of fluids or gases in tubes. A pressure differential is required to maintain a continuous flow of a liquid or gas through a pipe. The rate of flow Q is described by a relation known as **Poiseuille's equation**. Q is found to be proportional to the pressure difference between the ends of the pipe, proportional to the square of the cross-sectional area, and inversely proportional to the length of the pipe:

$$Q = \frac{\pi R^4 \left(P_1 - P_2\right)}{8\eta L}$$

where $P_1 - P_2$ is the pressure difference between the ends of the pipe, R is its radius, L is its length, and η is the viscosity of the fluid. Note that the flow rate is proportional to the fourth power of R; a small change in the radius of the pipe will have a big impact on the flow rate.

Example 13-12-A. Draining a tank. A tank of diameter D is filled with water to an initial depth H. A pipe with inner radius r much smaller than D and length l is used to drain the water horizontally out of the bottom of the tank. How long does it take to drain the water out of the tank? Assume the flow through the pipe is laminar.

Approach: Since the flow is laminar flow we can calculate the flow rate using Poiseuille's equation. Using this flow rate and the initial volume of water in the tank we can calculate the time required to drain it. Since the pressure at the input of the drain pipe will decrease when the tank drains, the flow rate will be time dependent.

Solution: The tank system is shown schematically in the Figure to the right. The rate of flow through the drain pipe depends on the pressure difference between the ends of the pipe. The pressure at the exit of the pipe, P_2, will be independent of time and equal to the atmospheric pressure. Because the radius of the pipe is small compared to the diameter of the tank, the velocity of the water in the tank is approximately zero and there is thus no viscous pressure drop in the tank. The pressure at the entrance to the pipe is equal to the sum of the atmospheric pressure and the pressure associated with the column of water. As a consequence, the pressure will decrease as the depth of the water in the tank decreases. For a water depth h the pressure is

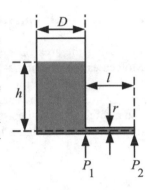

$$P_1(h) = P_A + \rho g h$$

The height of the water in the tank depends on the amount of water drained from the tank:

$$h(t) = \frac{V_{\text{in tank}}}{A_{\text{tank}}} = \frac{V_0 - V_{\text{drained}}}{A_{\text{tank}}} = \frac{HA_{\text{tank}} - \int_0^t Q\,dt'}{A_{\text{tank}}} = H - \frac{\int_0^t Q\,dt'}{A_{\text{tank}}}$$

In this equation, $Q(t)$ is the rate at which the water flows out of the tank and A_{tank} is its cross-sectional area. This rate can be obtained by using Poiseuille's equation:

$$Q(t) = \frac{\pi r^4 (P_1 - P_2)}{8\eta l} = \frac{\pi r^4}{8\eta l}\left(P_A + \rho g h(t) - P_A\right) = \frac{\pi r^4 \rho g}{8\eta l} h(t)$$

If we put this expression for $Q(t)$ into our equation for $h(t)$ we obtain for $h(t)$:

$$h(t) = H - \frac{\pi r^4 \rho g}{8\eta l A_{\text{tank}}} \int_0^t h(t')\,dt'$$

By taking the derivative of both sides of this equation with respect to t we get

$$\frac{dh(t)}{dt} = -\frac{\pi r^4 \rho g}{8\eta l A_{\text{tank}}} h(t) \quad \text{or} \quad \frac{dh(t)}{h(t)} = -\frac{\pi r^4 \rho g}{8\eta l A_{\text{tank}}} dt$$

Integrating each side of this expression we obtain

$$\ln h(t) = c - \frac{\pi r^4 \rho g}{8\eta l A_{\text{tank}}} t$$

where c is an integration constant. This equation can be rewritten as

$$h(t) = e^{c - \frac{\pi r^4 \rho g}{8\eta l A_{\text{tank}}} t} = e^c e^{-\frac{\pi r^4 \rho g}{8\eta l A_{\text{tank}}} t}$$

The constant c must be adjusted to match the initial conditions. If time $t = 0$ is defined as the time at which we start to drain the tank, then $h(0) = H$ and $e^c = H$. The time dependence of the height of the liquid in the tank can thus be written as

$$h(t) = H e^{-\alpha t}$$

where

$$\alpha = \frac{\pi r^4 \rho g}{8 \eta l A_{\text{tank}}}$$

Because the height of the water in the tank decreases exponentially, there never is a time at which the tank is completely empty. The height of the tank decreases by a factor $1/e$ during a time interval $1/\alpha$.

Section 13-13. Surface Tension and Capillarity

Increasing the area of the surface of a liquid requires moving molecules of the liquid from its interior to its surface. Molecules of the liquid at the surface do not have as many molecules surrounding them as do molecules in the interior of the liquid. Because most molecules of the liquid have a strong attraction for the other molecules in the liquid, it takes energy to increase the surface area. This means that work must be done on the liquid as the size of its surface increases and a force can thus be associated with an increase of the surface area. The **surface tension** γ is defined as the force per unit surface length that is directed perpendicular to any "imaginary" line on the surface:

$$\gamma = \frac{F}{L}$$

When you look up the value of the surface tension of a liquid, it will usually be the surface tension associated with an air-liquid interface, unless otherwise specified. If the air is replaced with a different gas, the surface tension may change. The surface tension associated with the boundary between water and glass causes **capillary action**. Capillary action causes water to rise within a small-diameter glass tube.

Section 13-14. Pumps, and the Heart

A pump is a device that is used to maintain a pressure difference in a fluid. The pressure difference is often used to create a flow of the fluid. The human heart is an example of a pump. It creates a pressure difference in the blood to cause the blood to move through the blood vessels. A pump is characterized by the flow rate through the pump as a function of the pressure difference maintained by the pump.

Practice Quiz

1. An object floats on water so that of one-half of its volume is above the surface of the water. Next, the object is placed in a container that contains half water and half oil (the oil floats on the water). The bottom of the object floats at the boundary between the water and the oil. What statement can be made about the fraction of the volume of the object that is now above the surface of the oil?
    a) The fraction of the volume above the surface of the oil is still 50%.
    b) The fraction of the volume above the surface of the oil is greater than 50%.
    c) The fraction of the volume above the surface of the oil is less than 50%.
    d) More information about the oil is needed to answer the question.

2. A pressure gauge on a compressed air tank reads -6.0 lbs/in^2. How can the pressure be negative?
    a) The gauge must be broken; negative pressure is impossible.
    b) Someone put the gauge on the tank backwards.
    c) Gauge pressure can be negative, if the absolute pressure is less than atmospheric pressure.
    d) The air in the tank is burning.

3. Which has the greatest mass: air, water, or iron?
    a) Air.
    b) Water.
    c) Iron.
    d) None; mass is not a property of a type of material.

4. A plunger is often used to clean out blocked drainpipes in homes. The blockage is usually some distance away from where the plunger applies force to the water that is blocked from flowing down the pipe. Which principle explains why the force applied on the water at one position helps to increase the pressure at the location of the blockage to clear the pipe?
   a) Bernoulli's principle.
   b) Pascal's principle.
   c) Archimedes' principle.
   d) Poiseuille's principle.

5. A pump provides a pressure difference to pump water through some pipes of fixed length. Which arrangement of pipes provides the greatest flow rate of water?
   a) One pipe of diameter D.
   b) Four pipes of diameter D.
   c) Ten pipes of diameter D.
   d) One pipe of diameter $2D$.

6. Water is at rest inside a U-shaped tube. The water in the two legs of the U is at identical height. Oil with a density less than the density of water is poured into one leg of the U. What happens to the water level in the other leg?
   a) The water level drops.
   b) The water level stays the same.
   c) The water level rises to be at the same level as the top of the oil.
   d) The water level rises, but will stay below the top of the oil.

7. A compressed air tank has an internal gauge pressure P_0 when the atmospheric pressure is P_A. The atmospheric pressure then drops to a pressure that is 99% of the original atmospheric pressure. What happens to the gauge pressure on the compressed air tank?
   a) The gauge pressure drops to $0.99\,P_0$.
   b) The gauge pressure drops to $P_0 - 0.01P_A$.
   c) The gauge pressure rises to $1.01P_0$.
   d) The gauge pressure rises to $P_0 + 0.01P_A$.

8. On a freshly cleaned and waxed car, water beads up. On a dirty car the water spreads out and wets the surface. Which statement is consistent with this observation?
   a) The surface tension between the water and the air is smaller than the cohesive forces between the water and the wax.
   b) The cohesive force of the wax for the dirt is larger than the cohesive force for the water.
   c) The cohesive force of the water for the water is stronger than the cohesive force of the water for the wax.
   d) The surface tension of the wax is larger than the surface tension of the dirt.

9. A small raft of density 500 kg/m^3 floats by itself on water. A person of density 980 kg/m^3 floats on water. However, if the person stands on the raft, the raft sinks below the surface of the water. Why?
   a) If you add the density of the person to the density of the raft, the result is greater than the density of water.
   b) If you multiply the density of the person by the density of the raft, the result is greater than the density of water.
   c) The volume of the person is smaller than the volume of the raft.
   d) The volume of the person does not displace any water unless the raft sinks.

10. A pipe carries water a certain distance up a hill. A faucet on the pipe at the top of the hill can control the flow rate or stop the flow all together. The pipe has been under-designed and could burst under certain conditions. Under what conditions and where is the pipe most likely to burst?
   a) The pipe is most likely to burst at the top when the water is flowing fastest.
   b) The pipe is most likely to burst at the top when the water is stopped.
   c) The pipe is most likely to burst at the bottom when the water is flowing fastest.
   d) The pipe is most likely to burst at the bottom when the water is stopped.

11. A wooden block floats in water with 32% of its volume above the surface. What is the density of the wood?

12. A pump transfers water from one container to another through a hose that is 38.4 m long. The pump provides a pressure difference of 5.00×10^4 N/m^2 from its intake to its output side. If the hose has a diameter 0.500 in, how long will it take to pump 14,000 gallons of water? Assume that the water is at 20° C and that the flow is laminar.

13. The water main to a home is 1.44 m below ground level and has an internal diameter of 1.25 in. Water leaves a tap in the home through a ½-in. internal diameter drain at a height of 1.08 m above ground level; the flow rate at this point is 1.54 gallons per minute. What is the absolute pressure in the water main? Ignore viscosity.

14. A small hydraulic jack is used to jack up a car to change its tire. The piston that pumps fluid into the chamber supporting the car has a diameter of 0.43 cm. The piston that supports the car has a diameter of 5.25 cm. What force is needed to push on the pumping piston to support the weight of a 724-kg car?

15. Ethyl alcohol is used to make a barometer. What height of column of ethyl alcohol is supported by atmospheric pressure?

Responses to Select End-of-Chapter Questions

1. No. If one material has a higher density than another, then the molecules of the first could be heavier than those of the second, or the molecules of the first could be more closely packed together than the molecules of the second.

7. Ice floats in water, so ice is less dense than water. When ice floats, it displaces a volume of water that is equal to the weight of the ice. Since ice is less dense than water, the volume of water displaced is smaller than the volume of the ice, and some of the ice extends above the top of the water. When the ice melts and turns back into water, it will fill a volume exactly equal to the original volume of water displaced. The water will not overflow the glass as the ice melts.

13. As the weather balloon rises into the upper atmosphere, atmospheric pressure on it decreases, allowing the balloon to expand as the gas inside it expands. If the balloon were filled to maximum capacity on the ground, then the balloon fabric would burst shortly after take-off, as the balloon fabric would be unable to expand any additional amount. Filling the balloon to a minimum value on take-off allows plenty of room for expansion as the balloon rises.

19. As a high-speed train travels, it pulls some of the surrounding air with it, due to the viscosity of the air. The moving air reduces the air pressure around the train (Bernoulli's principle), which in turn creates a force toward the train from the surrounding higher air pressure. This force is large enough that it could push a light-weight child toward the train.

Solutions to Select End-of-Chapter Problems

1. The mass is found from the density of granite (found in Table 13-1) and the volume of granite.

$$m = \rho V = \left(2.7 \times 10^3 \text{ kg/m}^3\right)\left(10^8 \text{ m}^3\right) = 2.7 \times 10^{11} \text{ kg} \approx \boxed{3 \times 10^{11} \text{ kg}}$$

7. (a) The density from the three-part model is found from the total mass divided by the total volume. Let subscript 1 represent the inner core, subscript 2 represent the outer core, and subscript 3 represent the mantle. The radii are then the outer boundaries of the labeled region.

$$\rho_{\substack{\text{three} \\ \text{layers}}} = \frac{m_1 + m_2 + m_3}{V_1 + V_2 + V_3} = \frac{\rho_1 m_1 + \rho_2 m_2 + \rho_3 m_3}{V_1 + V_2 + V_3} = \frac{\rho_1 \frac{4}{3}\pi r_1^3 + \rho_2 \frac{4}{3}\pi \left(r_2^3 - r_1^3\right) + \rho_3 \frac{4}{3}\pi \left(r_3^3 - r_2^3\right)}{\frac{4}{3}\pi r_1^3 + \frac{4}{3}\pi \left(r_2^3 - r_1^3\right) + \frac{4}{3}\pi \left(r_3^3 - r_2^3\right)} =$$

$$= \frac{\rho_1 r_1^3 + \rho_2 \left(r_2^3 - r_1^3\right) + \rho_3 \left(r_3^3 - r_2^3\right)}{r_3^3} = \frac{r_1^3 \left(\rho_1 - \rho_2\right) + r_2^3 \left(\rho_2 - \rho_3\right) + r_3^3 \rho_3}{r_3^3} =$$

$$= 5505 \text{ kg/m}^3 \approx \boxed{5500 \text{ kg/m}^3}$$

(b) $\rho_{\substack{\text{one} \\ \text{density}}} = \dfrac{M}{V} = \dfrac{M}{\frac{4}{3}\pi R^3} = \dfrac{5.98\times10^{24}\,\text{kg}}{\frac{4}{3}\pi\left(6371\times10^3\,\text{m}\right)^3} = 5521\,\text{kg/m}^3 \approx \boxed{5520\,\text{kg/m}^3}$

$\%\,\text{diff} = 100\left(\dfrac{\rho_{\substack{\text{one} \\ \text{density}}} - \rho_{\substack{\text{three} \\ \text{layers}}}}{\rho_{\substack{\text{three} \\ \text{layers}}}}\right) = 100\left(\dfrac{5521\,\text{kg/m}^3 - 5505\,\text{kg/m}^3}{5505\,\text{kg/m}^3}\right) = 0.2906 = \boxed{0.3\%}$

13. The force exerted by the gauge pressure will be equal to the weight of the vehicle.

$mg = PA = P\left(\pi r^2\right) \quad\rightarrow$

$m = \dfrac{P\pi r^2}{g} = \dfrac{\left(17.0\,\text{atm}\right)\left(\dfrac{1.013\times10^5\,\text{N/m}^2}{1\,\text{atm}}\right)\pi\left[\frac{1}{2}\left(0.225\,\text{m}\right)\right]^2}{\left(9.80\,\text{m/s}^2\right)} = \boxed{6990\,\text{kg}}$

19. We use the relationship developed in Example 13-5.

$P = P_0 e^{-(\rho_0 g/P_0)y} = \left(1.013\times10^5\,\text{N/m}^2\right)e^{-\left(1.25\times10^{-4}\,\text{m}^{-1}\right)\left(8850\,\text{m}\right)} = \boxed{3.35\times10^4\,\text{N/m}^2} \approx 0.331\,\text{atm}$

Note that if we used the constant density approximation, $P = P_0 + \rho gh$, a negative pressure would result.

25. Consider a layer of liquid of (small) height Δh, and ignore the pressure variation due to height in that layer. Take a cylindrical ring of water of height Δh, radius r, and thickness dr. See the diagram (the height is not shown). The volume of the ring of liquid is $(2\pi r\Delta h)\,dr$, and so has a mass of $dm = (2\pi\rho\, r\Delta h)\,dr$. That mass of water has a net centripetal force on it of magnitude

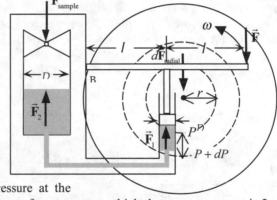

$dF_{\text{radial}} = \omega^2 r\left(dm\right) = \omega^2 r\rho\left(2\pi r\Delta h\right)dr$

That force comes from a pressure difference across the surface area of the liquid. Let the pressure at the inside surface be P, which causes an outward force, and the pressure at the outside surface be $P + dP$, which causes an inward force. The surface area over which these pressures act is $2\pi r\Delta h$, the "walls" of the cylindrical ring. Use Newton's second law.

$dF_{\substack{\text{radial}}} = dF_{\substack{\text{outer} \\ \text{wall}}} - dF_{\substack{\text{inner} \\ \text{wall}}} \quad\rightarrow\quad \omega^2 r\rho\left(2\pi r\Delta h\right)dr = \left(P + dP\right)2\pi r\Delta h - \left(P\right)2\pi r\Delta h \quad\rightarrow$

$dP = \omega^2 r\rho\, dr \quad\rightarrow\quad \displaystyle\int_{P_0}^{P} dP = \int_{0}^{r}\omega^2 r\rho\, dr \quad\rightarrow\quad P - P_0 = \tfrac{1}{2}\rho\omega^2 r^2 \quad\rightarrow\quad \boxed{P = P_0 + \tfrac{1}{2}\rho\omega^2 r^2}$

31. The apparent weight is the actual weight minus the buoyant force. The buoyant force is weight of a mass of water occupying the volume of the metal sample.

$m_{\text{apparent}}g = m_{\text{metal}}g - F_{\text{B}} = m_{\text{metal}}g - V_{\text{metal}}\rho_{\text{H}_2\text{O}}g = m_{\text{metal}}g - \dfrac{m_{\text{metal}}}{\rho_{\text{metal}}}\rho_{\text{H}_2\text{O}}g \quad\rightarrow\quad m_{\text{apparent}} = m_{\text{metal}} - \dfrac{m_{\text{metal}}}{\rho_{\text{metal}}}\rho_{\text{H}_2\text{O}} \quad\rightarrow$

$\rho_{\text{metal}} = \dfrac{m_{\text{metal}}}{\left(m_{\text{metal}} - m_{\text{apparent}}\right)}\rho_{\text{H}_2\text{O}} = \dfrac{63.5\,\text{g}}{\left(63.5\,\text{g} - 55.4\,\text{g}\right)}\left(1000\,\text{kg/m}^3\right) = 7840\,\text{kg/m}^3$

Based on the density value, the metal is probably $\boxed{\text{iron or steel}}$.

37. (*a*) The buoyant force on the object is equal to the weight of the fluid displaced. The force of gravity of the fluid can be considered to act at the center of gravity of the fluid (see Section 9-8). If the object were removed from the fluid and that space refilled with an equal volume of fluid, that fluid would be in equilibrium. Since there are only two forces on that volume of fluid, gravity and the buoyant force, they must be equal in magnitude <u>and</u> act at the same point. Otherwise they would be a couple (see Figure 12-4), exert a non-zero torque, and cause rotation of the fluid. Since the fluid does not rotate, we may conclude that the buoyant force acts at the center of gravity.

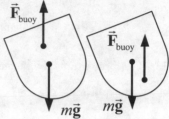

(*b*) From the diagram, if the center of buoyancy (the point where the buoyancy force acts) is above the center of gravity (the point where gravity acts) of the entire ship, when the ship tilts, the net torque about the center of mass will tend to reduce the tilt. If the center of buoyancy is below the center of gravity of the entire ship, when the ship tilts, the net torque about the center of mass will tend to increase the tilt. Stability is achieved when the center of buoyancy is above the center of gravity .

43. We apply the equation of continuity at constant density, Eq. 13-7b.

Flow rate out of duct = Flow rate into room

$$A_{duct}v_{duct} = \pi r^2 v_{duct} = \frac{V_{room}}{t_{\substack{to\ fill \\ room}}} \rightarrow v_{duct} = \frac{V_{room}}{\pi r^2 t_{\substack{to\ fill \\ room}}} = \frac{(8.2\,m)(5.0\,m)(3.5\,m)}{\pi(0.15\,m)^2 (12\,min)\left(\frac{60\,s}{1\,min}\right)} = \boxed{2.8\,m/s}$$

49. We assume that there is no appreciable height difference between the two sides of the roof. Then the net force on the roof due to the air is the difference in pressure on the two sides of the roof, times the area of the roof. The difference in pressure can be found from Bernoulli's equation.

$$P_{inside} + \tfrac{1}{2}\rho v_{inside}^2 + \rho g y_{inside} = P_{outside} + \tfrac{1}{2}\rho v_{outside}^2 + \rho g y_{outside} \rightarrow$$

$$P_{inside} - P_{outside} = \tfrac{1}{2}\rho_{air}v_{outside}^2 = \frac{F_{air}}{A_{roof}} \rightarrow$$

$$F_{air} = \tfrac{1}{2}\rho_{air}v_{outside}^2 A_{roof} = \tfrac{1}{2}(1.29\,kg/m^3)\left[(180\,km/h)\left(\frac{1\,m/s}{3.6\,km/h}\right)\right]^2 (6.2\,m)(12.4\,m)$$

$$= \boxed{1.2 \times 10^5\,N}$$

55. Apply both Bernoulli's equation and the equation of continuity between the two openings of the tank. Note that the pressure at each opening will be atmospheric pressure.

$$A_2 v_2 = A_1 v_1 \rightarrow v_2 = v_1 \frac{A_1}{A_2}$$

$$P_1 + \tfrac{1}{2}\rho v_1^2 + \rho g y_1 = P_2 + \tfrac{1}{2}\rho v_2^2 + \rho g y_2 \rightarrow v_1^2 - v_2^2 = 2g(y_2 - y_1) = 2gh$$

$$v_1^2 - \left(v_1 \frac{A_1}{A_2}\right) = 2gh \rightarrow v_1^2\left(1 - \frac{A_1^2}{A_2^2}\right) = 2gh \rightarrow \boxed{v_1 = \sqrt{\frac{2gh}{\left(1 - A_1^2/A_2^2\right)}}}$$

61. (*a*) Relate the conditions inside the rocket and just outside the exit orifice by means of Bernoulli's equation and the equation of continuity. We ignore any height difference between the two locations.

$$P_{in} + \tfrac{1}{2}\rho v_{in}^2 + \rho g y_{in} = P_{out} + \tfrac{1}{2}\rho v_{out}^2 + \rho g y_{out} \quad \rightarrow \quad P + \tfrac{1}{2}\rho v_{in}^2 = P_0 + \tfrac{1}{2}\rho v_{out}^2 \quad \rightarrow$$

$$\frac{2(P - P_0)}{\rho} = v_{out}^2 - v_{in}^2 = v_{out}^2\left[1 - \left(\frac{v_{in}}{v_{out}}\right)^2\right]$$

$$A_{in}v_{in} = A_{out}v_{out} \quad \rightarrow \quad Av_{in} = A_0 v_{out} \quad \rightarrow \quad \frac{v_{in}}{v_{out}} = \frac{A_0}{A} \ll 1 \quad \rightarrow$$

$$\frac{2(P - P_0)}{\rho} = v_{out}^2\left[1 - \left(\frac{v_{in}}{v_{out}}\right)^2\right] \approx v_{out}^2 \quad \rightarrow \quad \boxed{v_{out} = v = \sqrt{\frac{2(P - P_0)}{\rho}}}$$

(*b*) Thrust is defined in Section 9-10, by $F_{thrust} = v_{rel}\dfrac{dm}{dt}$, and is interpreted as the force on the rocket due to the ejection of mass.

$$F_{thrust} = v_{rel}\frac{dm}{dt} = v_{out}\frac{d(\rho V)}{dt} = v_{out}\rho\frac{dV}{dt} = v_{out}\rho(v_{out}A_{out}) = \rho v^2 A_0 = \rho\frac{2(P - P_0)}{\rho}A_0$$

$$= \boxed{2(P - P_0)A_0}$$

67. Use Poiseuille's equation to find the radius, and then double the radius to the diameter.

$$Q = \frac{\pi R^4(P_2 - P_1)}{8\eta l} \quad \rightarrow$$

$$d = 2R = 2\left[\frac{8\eta lQ}{\pi(P_2 - P_1)}\right]^{1/4} = 2\left[\frac{8(1.8\times10^{-5}\,\text{Pa}\bullet\text{s})(15.5\,\text{m})\left(\dfrac{8.0\times14.0\times4.0\,\text{m}^3}{720\,\text{s}}\right)}{\pi(0.71\times10^{-3}\,\text{atm})(1.013\times10^5\,\text{Pa/atm})}\right]^{1/4} = \boxed{0.10\,\text{m}}$$

73. In Figure 13-35, we have $\gamma = F/2l$. Use this relationship to calculate the force.

$$\gamma = F/2l \quad \rightarrow \quad F = 2\gamma l = 2(0.025\,\text{N/m})(0.245\,\text{m}) = \boxed{1.2\times10^{-2}\,\text{N}}$$

79. The pressures for parts (*a*) and (*b*) stated in this problem are gauge pressures, relative to atmospheric pressure. The pressure change due to depth in a fluid is given by $\Delta P = \rho g\,\Delta h$.

(*a*) $$\Delta h = \frac{\Delta P}{\rho g} = \frac{(55\,\text{mm-Hg})\left(\dfrac{133\,\text{N/m}^2}{1\,\text{mm-Hg}}\right)}{\left(1.00\dfrac{\text{g}}{\text{cm}^3} \times \dfrac{1\,\text{kg}}{1000\,\text{g}} \times \dfrac{10^6\,\text{cm}^3}{1\,\text{m}^3}\right)(9.80\,\text{m/s}^2)} = \boxed{0.75\,\text{m}}$$

(*b*) $$\Delta h = \frac{\Delta P}{\rho g} = \frac{(650\,\text{mm-H}_2\text{O})\left(\dfrac{9.81\,\text{N/m}^2}{1\,\text{mm-H}_2\text{O}}\right)}{\left(1.00\dfrac{\text{g}}{\text{cm}^3} \times \dfrac{1\,\text{kg}}{1000\,\text{g}} \times \dfrac{10^6\,\text{cm}^3}{1\,\text{m}^3}\right)(9.80\,\text{m/s}^2)} = \boxed{0.65\,\text{m}}$$

(*c*) For the fluid to just barely enter the vein, the fluid pressure must be the same as the blood pressure.

$$\Delta h = \frac{\Delta P}{\rho g} = \frac{(78\,\text{mm-Hg})\left(\dfrac{133\,\text{N}/\text{m}^2}{1\,\text{mm-Hg}}\right)}{\left(1.00\,\dfrac{\text{g}}{\text{cm}^3} \times \dfrac{1\,\text{kg}}{1000\,\text{g}} \times \dfrac{10^6\,\text{cm}^3}{1\,\text{m}^3}\right)(9.80\,\text{m}/\text{s}^2)} = 1.059\,\text{m} \approx \boxed{1.1\,\text{m}}$$

85. The pressure difference due to the lungs is the pressure change in the column of water.

$$\Delta P = \rho g \Delta h \quad \rightarrow \quad \Delta h = \frac{\Delta P}{\rho g} = \frac{(75\,\text{mm-Hg})\left(\dfrac{133\,\text{N}/\text{m}^2}{1\,\text{mm-Hg}}\right)}{(1.00 \times 10^3\,\text{kg}/\text{m}^3)(9.80\,\text{m}/\text{s}^2)} = 1.018\,\text{m} \approx \boxed{1.0\,\text{m}}$$

91. We assume that the air pressure is due to the weight of the atmosphere, with the area equal to the surface area of the Earth.

$$P = \frac{F}{A} \rightarrow F = PA = mg \rightarrow$$

$$m = \frac{PA}{g} = \frac{4\pi R_{\text{Earth}}^2 P}{g} = \frac{4\pi (6.38 \times 10^6\,\text{m})^2 (1.013 \times 10^5\,\text{N}/\text{m}^2)}{9.80\,\text{m}/\text{s}^2} = 5.29 \times 10^{18}\,\text{kg} \approx \boxed{5 \times 10^{18}\,\text{kg}}$$

97. The upward force due to air pressure on the bottom of the wing must be equal to the weight of the airplane plus the downward force due to air pressure on the top of the wing. Bernoulli's equation can be used to relate the forces due to air pressure. We assume that there is no appreciable height difference between the top and the bottom of the wing.

$$P_{\text{top}} A + mg = P_{\text{bottom}} A \quad \rightarrow \quad \left(P_{\text{bottom}} - P_{\text{top}}\right) = \frac{mg}{A}$$

$$P_0 + P_{\text{bottom}} + \tfrac{1}{2}\rho v_{\text{bottom}}^2 + \rho g y_{\text{bottom}} = P_0 + P_{\text{top}} + \tfrac{1}{2}\rho v_{\text{top}}^2 + \rho g y_{\text{top}}$$

$$v_{\text{top}}^2 = \frac{2\left(P_{\text{bottom}} - P_{\text{top}}\right)}{\rho} + v_{\text{bottom}}^2 \quad \rightarrow$$

$$v_{\text{top}} = \sqrt{\frac{2\left(P_{\text{bottom}} - P_{\text{top}}\right)}{\rho} + v_{\text{bottom}}^2} = \sqrt{\frac{2mg}{\rho A} + v_{\text{bottom}}^2} = \sqrt{\frac{2(1.7 \times 10^6\,\text{kg})(9.80\,\text{m}/\text{s}^2)}{(1.29\,\text{kg}/\text{m}^3)(1200\,\text{m}^2)} + (95\,\text{m}/\text{s})^2}$$

$$= 174.8\,\text{m}/\text{s} \approx \boxed{170\,\text{m}/\text{s}}$$

103. Use the definition of density and specific gravity, and then solve for the fat fraction, f.

$$m_{\text{fat}} = mf = V_{\text{fat}} \rho_{\text{fat}} \quad ; \quad m_{\substack{\text{fat} \\ \text{free}}} = m(1-f) = V_{\substack{\text{fat} \\ \text{free}}} \rho_{\substack{\text{fat} \\ \text{free}}}$$

$$\rho_{\text{body}} = X\rho_{\text{water}} = \frac{m_{\text{total}}}{V_{\text{total}}} = \frac{m_{\text{fat}} + m_{\substack{\text{fat} \\ \text{free}}}}{V_{\text{fat}} + V_{\substack{\text{fat} \\ \text{free}}}} = \frac{m}{\dfrac{mf}{\rho_{\text{fat}}} + \dfrac{m(1-f)}{\rho_{\substack{\text{fat} \\ \text{free}}}}} = \frac{1}{\dfrac{f}{\rho_{\text{fat}}} + \dfrac{(1-f)}{\rho_{\substack{\text{fat} \\ \text{free}}}}} \quad \rightarrow$$

$$f = \frac{\rho_{\text{fat}} \rho_{\substack{\text{fat} \\ \text{free}}}}{X\rho_{\text{water}}\left(\rho_{\substack{\text{fat} \\ \text{free}}} - \rho_{\text{fat}}\right)} - \frac{\rho_{\text{fat}}}{\left(\rho_{\substack{\text{fat} \\ \text{free}}} - \rho_{\text{fat}}\right)} = \frac{(0.90\,\text{g}/\text{cm}^3)(1.10\,\text{g}/\text{cm}^3)}{X(1.0\,\text{g}/\text{cm}^3)(0.20\,\text{g}/\text{cm}^3)} - \frac{(0.90\,\text{g}/\text{cm}^3)}{(0.20\,\text{g}/\text{cm}^3)}$$

$$= \frac{4.95}{X} - 4.5 \quad \rightarrow \quad \%\ \text{Body fat} = 100 f = 100\left(\frac{4.95}{X} - 4.5\right) = \boxed{\frac{495}{X} - 450}$$

Chapter 14: Oscillations

Chapter Overview and Objectives

In this chapter, vibrating or oscillatory motion is introduced. The details of a particular type of oscillatory motion, called simple harmonic motion, are described. Damped and driven harmonic oscillators are discussed.

After completing this chapter you should:
- Know what periodic motion is and the definitions of frequency, period, and angular frequency.
- Know what simple harmonic motion is and what forces lead to simple harmonic motion.
- Know the parameters that describe simple harmonic motion: amplitude, angular frequency, and phase angle.
- Know how to determine the period or frequency of a mass–spring system.
- Know how to determine the period or frequency of a simple pendulum, a physical pendulum, and a torsion pendulum.
- Know the solution of a damped harmonic oscillator and its characteristics.
- Know the steady-state solution of the driven harmonic oscillator and the characteristics of its solutions.

Summary of Equations

Relationships between period T and frequency f:
$$T = \frac{1}{f}$$
(Section 14-1)

Simple harmonic motion:
$$x(t) = A\cos\left(\sqrt{\frac{k}{m}}t + \phi\right) = A\cos(\omega t + \phi)$$
(Section 14-2)

Relationships between angular frequency ω, frequency f, and period T:
$$\omega = \frac{2\pi}{T} = 2\pi f$$
(Section 14-2)

Total energy of mass–spring system in simple harmonic motion:
$$E = \frac{1}{2}kA^2$$
(Section 14-3)

Period of motion of simple pendulum:
$$T = 2\pi\sqrt{\frac{L}{g}}$$
(Section 14-5)

Period of motion of physical pendulum:
$$T = 2\pi\sqrt{\frac{I}{mgh}}$$
(Section 14-6)

Angular frequency of torsion pendulum:
$$\omega = \sqrt{\frac{K}{I}}$$
(Section 14-6)

Motion of damped harmonic oscillator:
$$x(t) = Ae^{-\alpha t}\cos(\omega' t + \phi)$$
(Section 14-7)

Steady-state solution of driven harmonic oscillator: $\quad x(t) = \dfrac{F_0}{m\sqrt{\left(\omega^2 - \omega_0^2\right) + b^2\omega^2 / m^2}}\cos\left(\omega t + \phi_0\right)$

(Section 14-8)

Quality factor Q of a driven harmonic oscillator: $\qquad Q = \dfrac{m\omega_0}{b}$ $\qquad\qquad$ (Section 14-8)

Resonance width: $\qquad\qquad\qquad\qquad\qquad\qquad \Delta\omega \approx \dfrac{\omega_0}{Q}$ $\qquad\qquad$ (Section 14-8)

Chapter Summary

Section 14-1. Oscillations of a Spring

When motion repeats itself over and over, it is called **periodic motion**. The time for one repetition of the motion is called the **period** of the motion. The number of repetitions per unit time is called the **frequency** of the motion. The period T and frequency f are related by

$$T = \frac{1}{f}$$

An example of a system carrying out periodic motion is the mass-spring system; the mass oscillates back and forth about its **equilibrium position**. The equilibrium position of the mass is the position at which there is no net force acting on it (see Chapter 12). The difference in the position of the mass and its equilibrium position is called the **displacement** of the mass. The magnitude of the maximum displacement is called the **amplitude** of the periodic motion.

Section 14-2. Simple Harmonic Motion

A restoring force that has a magnitude proportional to the displacement of the object, such as the spring force $F = -kx$, results in a form of periodic motion that is called **simple harmonic motion**. Writing down Newton's second law for such a system we obtain:

$$F = ma \quad \Rightarrow \quad -kx = m\frac{d^2x}{dt^2} \quad \text{or} \quad \frac{d^2x}{dt^2} + \frac{k}{m}x = 0$$

The most general solution of this differential equation is

$$x(t) = A\cos\left(\sqrt{\frac{k}{m}}\,t + \phi\right) = A\cos\left(\omega t + \phi\right)$$

where ω is the **angular frequency**, A is the **amplitude**, and ϕ is the **phase angle**. The amplitude and the phase angle are determined by the **initial conditions** of the motion (e.g., the position and velocity at time $t = 0$). The relationships between the angular frequency ω, the frequency f, and the period T for simple harmonic motion are

$$\omega = \frac{2\pi}{T} = 2\pi f$$

Example 14-2-A: Motion of a mass on a spring. A mass $m = 2.44$ kg is attached to a spring with spring constant k. The motion of the mass is observed as function of time and found to be given by the following relation:

$$x(t) = A\cos\left(\omega t + \phi\right)$$

where $A = 3.32$ cm, $\omega = 2.29$ s^{-1}, and $\phi = 1.19$ rad. What is the spring constant of the spring? What is the position of the mass at time $t = 1.45$ s?

Approach: By comparing the observed motion of the mass with the predicted motion of a mass/spring system, we can determine the spring constant k.

Solution: A mass on a spring will carry out simple harmonic motion with an angular frequency $\omega = \sqrt{(k/m)}$. The observed angular frequency ω can thus be used to determine the unknown spring constant k:

$$k = m\omega^2 = (2.44)(2.29)^2 = 12.8 \, \text{N/m}$$

The position at $t = 1.45$ s can be found by evaluating the expression for $x(t = 1.45 \text{ s})$:

$$x(t = 1.45 \text{ s}) = (0.0332)\cos(2.29(1.45) + 1.19) = -0.000666 \, \text{m}$$

Note: make sure you realize that the argument of the cosine function is expressed in terms of radians, not in terms of degrees.

Example 14-2-B: Velocity and acceleration associated with harmonic motion. For the system in Example 14-2-A, determine the velocity and the acceleration of the mass as function of time. Also, determine the maximum speed and the maximum magnitude of the acceleration.

Approach: In Example 14-2-A, the observed time dependence of mass m is parameterized. Using the definition of the velocity and acceleration in terms of the position, we can now determine the maximum speed and acceleration.

Solution: The velocity of mass m can be obtained by differentiating the position with respect to time:

$$v(t) = \frac{dx(t)}{dt} = A\frac{d}{dt}\left[\cos(\omega t + \phi)\right] = -A\omega \sin(\omega t + \phi)$$

The acceleration of mass m can be obtained by differentiating the velocity with respect to time:

$$a(t) = \frac{dv(t)}{dt} = -A\omega \frac{d}{dt}\left[\sin(\omega t + \phi)\right] = -A\omega^2 \cos(\omega t + \phi) = -\omega^2 x(t)$$

Since the maximum and minimum values of the sin and cos functions is 1 and -1, the maximum speed and acceleration of mass m are

$$v_{max} = A\omega = (0.0332)(2.29) = 0.0760 \text{ m/s} \quad \text{and} \quad a_{max} = A\omega^2 = (0.0332)(2.29)^2 = 0.174 \text{ m/s}^2$$

Section 14-3. Energy in the Simple Harmonic Oscillator

The total energy of a mass and spring in simple harmonic motion is given by the sum of the kinetic and potential energies:

$$E = K + U = \frac{1}{2}mv^2 + \frac{1}{2}kx^2$$

When the object is at its maximum displacement from equilibrium, $x = \pm A$, its speed is zero. The total energy of the system is thus entirely in the form of potential energy of the spring and is equal to

$$E = \frac{1}{2}kA^2$$

We can come to the same conclusion when we use the expression for x and v, discussed in Examples 14-2-A and 14-2-B:

$$E = K + U = \frac{1}{2}mv^2 + \frac{1}{2}kx^2 = \frac{1}{2}mA^2\omega^2 \sin^2(\omega t + \phi) + \frac{1}{2}kA^2 \cos^2(\omega t + \phi) =$$

$$= \frac{1}{2}kA^2\left[\sin^2(\omega t + \phi) + \cos^2(\omega t + \phi)\right] = \frac{1}{2}kA^2$$

Example 14-3-A: Amplitude of a mass-spring system. Determine the amplitude A of a mass–spring system that has a speed of $v_1 = 2.34$ m/s when the displacement of the mass is $x_1 = 1.45$ cm. The spring constant is $k = 98.2$ N/m and the mass is $m = 287.2$ g.

Approach: This problem can be solved using the fact that the total energy of the system is constant. The information provided can be used to calculate the total energy when the displacement of mass m is $x_1 = 1.45$ cm. However, since this energy is constant, this energy must also be equal to the energy when mass m reaches its maximum amplitude where all the energy is in the form of potential energy of the spring.

Solution: The total energy of the system when mass m is located at $x_1 = 1.45$ cm is equal to

$$E = K + U = \frac{1}{2}mv_1^2 + \frac{1}{2}kx_1^2 = \frac{1}{2}(0.2872)(2.34)^2 + \frac{1}{2}(98.2)(0.0145)^2 = 0.797\,\text{J}$$

The total energy of the system is constant and is also equal to $\frac{1}{2}kA^2$. Using the total energy E we can thus determine the amplitude A:

$$A = \sqrt{\frac{2E}{k}} = \sqrt{\frac{2(0.797)}{98.2}} = 0.127\,\text{m}$$

Section 14-4. Simple Harmonic Motion Related to Uniform Circular Motion

The Cartesian components of a mass carrying out uniform circular motion in a plane defined by two Cartesian coordinate axes have a time dependence that is consistent with simple harmonic motion. If the center of the circular orbit coincides with the origin of the coordinate system, and if the angular position of the mass is given by $\theta = \omega t$, then the x and y coordinates of mass m are equal to

$$x(t) = r\cos\omega t$$

$$y(t) = r\sin\omega t$$

Here we have assumed that at time $t = 0$, the mass is located at $(r, 0)$. The x and y coordinates carry out a motion with identical amplitude and frequency; they only differ by a phase angle difference of $\pi/2$.

Section 14-5. The Simple Pendulum

A **simple pendulum** consists of a small mass hanging from the end of a string of length L. It is assumed that the mass is a point mass and that the string has negligible mass. The mass is free to swing back and forth with only the force of gravity and the string force acting on it; the path of the mass can be described as a circular segment. The motion of the mass can be described in terms of the angle θ of the string with the vertical direction; the equilibrium position is the position for which $\theta = 0$. The restoring force is the component of the gravitational force directed along this circular segment ($-mg\sin\theta$); the displacement along this segment is equal to $L\theta$. Newton's second law can be used to express the motion of the mass in terms of the angle θ:

$$F = ma \quad\Rightarrow\quad -mg\sin\theta = m\frac{d^2}{dt^2}(L\theta) = mL\frac{d^2\theta}{dt^2} \quad\text{or}\quad \frac{d^2\theta}{dt^2} + \frac{g}{L}\sin\theta = 0$$

This is a difficult equation to solve with the calculus you are expected to know at this point. However, when θ is small (less than 15°), and expressed in radians, we can use the following approximation:

$$\sin\theta \approx \theta$$

Using this approximation we can rewrite the equation of motion of the simple pendulum as:

$$\frac{d^2\theta}{dt^2} + \frac{g}{L}\theta = 0$$

which has the same form as the equation for simple-harmonic motion discussed in Section 14.2. The angular frequency, the frequency, and the period of the simple pendulum are

$$\omega = \sqrt{\frac{g}{L}} \qquad f = \frac{\omega}{2\pi} = \frac{1}{2\pi}\sqrt{\frac{g}{l}} \qquad T = \frac{1}{f} = 2\pi\sqrt{\frac{L}{g}}$$

Example 14-5-A. Using a simple pendulum to measure planetary masses. A planet has a radius $R = 4.75 \times 10^3$ km. A pendulum with a length $L = 1.00$ m has a period of $T = 2.1$ s on the planet's surface. What is the mass of the planet?

Approach: The period of the pendulum depends on the length of the pendulum and the gravitational acceleration at the location of the pendulum. The gravitational acceleration at this location depends on the mass of the planet and its radius. Based on the information provided, we can determine the gravitational acceleration at the location of the pendulum and use this information to calculate the mass of the planet.

Solution: The gravitational acceleration at the location of the pendulum is

$$g = \frac{4\pi^2 L}{T^2} = \frac{4\pi^2 (1.00)}{(2.1)^2} = 8.95 \, \text{m/s}^2$$

The gravitational acceleration at the surface of the planet is related to the mass and radius of the planet:

$$g = \frac{F}{m} = \frac{\left(\dfrac{GmM_{planet}}{R_{planet}^2}\right)}{m} = \frac{GM_{planet}}{R_{planet}^2} \Rightarrow M_{planet} = \frac{gR_{planet}^2}{G} = \frac{(8.95)\left(4.75\times10^6\right)^2}{\left(6.67\times10^{-11}\right)} = 3.02\times10^{24} \, \text{kg}$$

Section 14-6. The Physical Pendulum and the Torsion Pendulum

In our discussion of the simple pendulum in Section 14-5 we assumed that all its mass is located at a single point. The simple pendulum is a poor approximation for a realistic extended object that is suspended about a fixed axis that does not pass through its center of mass, and we thus need to develop a new approach to describe the oscillatory motion of this object. Such object is called a **physical pendulum**. The period T of a physical pendulum is given by

$$T = 2\pi\sqrt{\frac{I}{mgh}}$$

where I is the moment of inertia of the object about the axis of rotation, m is the mass of the object, and h is the distance of the axis of rotation from the center of mass of the object. We see that a measurement of the period of a physical pendulum can be used to determine its moment of inertia, which for many extended objects may be difficult to calculate directly.

The simple pendulum is a special case of the physical pendulum. For the simple pendulum, $I = mh^2$ and the corresponding period is thus equal to

$$T - 2\pi\sqrt{\frac{mh^2}{mgh}} = 2\pi\sqrt{\frac{h}{g}}$$

which is the period of a simple pendulum of length h.

An object suspended from a thin rod or wire will also oscillate back and forth in simple harmonic rotational motion if the rod or wire is twisted and released. This configuration is called a **torsion pendulum**. The twist angle θ of the rod or wire is proportional to the applied torque: $\tau = -K\theta$. The constant of proportionality K is called the **torsion constant**, and is an important property of the rod or wire. The angular frequency of the torsion pendulum is given by

$$\omega = \sqrt{\frac{K}{I}}$$

where I is the moment of inertia of the object about the twist axis of the system. In general, the moment of inertia is determined by a large circular mass that is connected to the end of the wire and is thus easy to determine. A measurement of the period of the pendulum can thus be used to determine the torsion constant of the rod or wire.

Example 14-6-A. The physical pendulum. A uniform rod of mass $m = 1.3$ kg and length $L = 59$ cm is suspended on a pivot at one end of the rod. At the other end of the rod, a small object with a mass of $M = 1.4$ kg is located. What is the period of the physical pendulum?

Approach: Since the mass of the rod is significant compared to the mass located at its end, we cannot apply the equations of motion of the simple pendulum to this system. Instead, we need to use the equations of motion of the physical pendulum.

Solution: The moment of inertia of the pendulum about the pivot point is equal to

$$I = I_{rod} + I_{mass} = \frac{1}{2}mL^2 + ML^2 = \frac{1}{2}\left(m + 2M\right)L^2$$

The distance h between the center of mass of the pendulum and the pivot point is equal to

$$h = \frac{\sum m_i y_i}{\sum m_i} = \frac{m\left(\frac{1}{2}L\right) + M\left(L\right)}{m + M} = \frac{\frac{1}{2}\left(m + 2M\right)L}{m + M}$$

The period of the physical pendulum is thus equal to

$$T = 2\pi\sqrt{\frac{I}{\left(m + M\right)gh}} = 2\pi\sqrt{\frac{\frac{1}{2}\left(m + 2M\right)L^2}{\left(m + M\right)g\left(\dfrac{\frac{1}{2}\left(m + 2M\right)L}{m + M}\right)}} = 2\pi\sqrt{\frac{L}{g}} = 1.5\,\text{s}$$

It is interesting to note that the period of the pendulum is the same as the period of a simple pendulum if all of its mass was located at the end of the rod. This could not have been anticipated.

Example 14-6-B. The torsion pendulum. A square plate that is $L = 12$ cm on a side has a mass $M = 1.05$ kg. It is suspended from a wire that requires a torque of $\tau_0 = 0.21$ Nm to twist it through an angle $\theta = 45°$. What is the period of this torsion pendulum?

Approach: To determine the period of the pendulum we need to calculate its moment of inertia I and the torsion constant K. We make the assumption that all the mass of the system is located in the square plate, and that the moment of inertia of the pendulum is thus equal to the moment of inertia of the square plate. The torsion constant K can be obtained from the torque required to twist the wire by an angle θ_0.

Solution: The moment of inertia of the pendulum is equal to the moment of inertia of the square plate with respect to its center. This moment of inertia can be obtained using the information provided in Figure 10-20 of the textbook.

$$I = \frac{1}{12}M\left(L^2 + L^2\right) = \frac{1}{6}ML^2$$

The torsion constant K of the wire is equal to

$$K = \frac{\tau_0}{\theta_0}$$

Note: we have to make sure that the angle is expressed in radians (not in degrees).

The period of the torsion pendulum can now be determined:

$$T = 2\pi\sqrt{\frac{I}{K}} = 2\pi\sqrt{\frac{\left(\frac{1}{6}ML^2\right)}{\left(\frac{\tau_0}{\theta_0}\right)}} = 2\pi\sqrt{\frac{1}{6}ML^2\left(\frac{\theta_0}{\tau_0}\right)} = 2\pi\sqrt{\frac{1}{6}(1.05)(0.12)^2\frac{\left(\frac{1}{4}\pi\right)}{(0.21)}} = 0.61\,\text{s}$$

Section 14-7. Damped Harmonic Motion

There is always some friction present in any macroscopic-sized system. Since the work associated with the frictional force is negative, the total mechanical energy of the system will decrease with time. For a system in simple harmonic motion, this means that the amplitude of the motion will decrease with time. The motion of such a system is called **damped harmonic motion**.

In general, the friction or damping force has a complicated mathematical dependence, which makes it difficult to solve Newton's second law and determine the motion of the system as function of time. One particular force for which an analytical solution can be obtained is a damping force that is proportional to the velocity of the object:

$$F_{damping} = -bv$$

With this additional force, Newton's second law can be rewritten as

$$F = -kx - bv = ma = m\frac{dx^2}{dt^2} \quad \text{or} \quad \frac{dx^2}{dt^2} + \frac{b}{m}\frac{dx}{dt} + \frac{k}{m}x = 0$$

The solution of this equation is

$$x(t) = Ae^{-\alpha t}\cos(\omega' t + \phi)$$

where

$$\alpha = \frac{b}{2m} \quad \text{and} \quad \omega' = \sqrt{\frac{k}{m} - \frac{b^2}{4m^2}}$$

A is the initial amplitude, ω' is the angular frequency, and ϕ is the phase angle. Their values are determined by the initial conditions (e.g., the position and velocity of the system at $t = 0$ s). The type of motion of the system depends on the angular frequency:

- If $b^2/4m^2 < k/m$ the angular frequency is a real number and the system will carry out damped oscillatory motion. This motion is called **underdamped** motion.
- If $b^2/4m^2 > k/m$ the angular frequency is not a real number and the system will return to its equilibrium position without oscillating. This motion is called **overdamped** motion.
- If $b^2/4m^2 = k/m$ the angular frequency is equal to 0 and the system returns to its equilibrium position in the shortest possible time. This motion is called **critically damped** motion.

Example 14-7-A: Amplitude of damped harmonic motion. How long does it take for the amplitude of a damped harmonic oscillator to reach one-half its original amplitude?

Approach: The amplitude of the damped harmonic motion is time dependent and equal to $A(t) = A_0e^{-\alpha t}$ where A_0 is the amplitude at time $t = 0$. The time at which the amplitude has reached half of its original value will only depend on the value of α.

Solution: The time t at which the amplitude reaches one-half of its original value is defined by requiring that

$$e^{-\alpha t} = \tfrac{1}{2} \quad \text{or} \quad t = -\frac{1}{\alpha}\ln\left(\frac{1}{2}\right) = \frac{1}{\alpha}\ln 2$$

Example 14-7-B: Kinetic energy of damped harmonic motion. How long does it take for the energy of a damped harmonic oscillator to reach one-half of its original value? Assume that the oscillator carries out underdamped motion.

Approach: If we do know the time dependence of the position of the oscillator, we can determine its velocity. At the times that the system crosses the equilibrium position, the potential energy of the system is 0 and all of the energy is in the form of kinetic energy. By examining the time dependence of the kinetic energy we can determine the time at which the energy of the oscillator reaches one-half of its original value.

Solution: The velocity of the oscillator can be determined by differentiating its position with respect to time:

$$v(t) = \frac{dx(t)}{dt} = -\alpha A e^{-\alpha t} \cos(\omega' t + \phi) - \omega' A e^{-\alpha t} \sin(\omega' t + \phi)$$

When the system crosses its equilibrium position ($x = 0$), the energy of the system will entirely be in the form of kinetic energy. The system is located at $x = 0$ when $\cos(\omega' t + \phi) = 0$. The velocity at these positions is equal to

$$v_{eq}(t) = \pm \omega' A e^{-\alpha t}$$

Here we have used the fact that when $\cos(\omega' t + \phi) = 0$, $\sin(\omega' t + \phi) = \pm 1$. The kinetic energy of the system when it crosses the equilibrium positions is thus equal to

$$K_{eq}(t) = \frac{1}{2} m v_{eq}^2(t) = \frac{1}{2} m \left(\omega' A e^{-\alpha t} \right)^2 = \frac{1}{2} m \left(\omega' A \right)^2 e^{-2\alpha t} = K_{eq}(t = 0) e^{-2\alpha t}$$

The energy of the system reaches half of its original value when the kinetic energy at its equilibrium position has been reduced by a factor of 2. This requires that

$$e^{-2\alpha t} = \frac{1}{2} \quad \text{or} \quad t = \frac{1}{2\alpha} \ln 2$$

Section 14-8. Forced Oscillations; Resonance

Sometimes an external force is applied to an oscillatory system. Consider a damped mass-spring system to which an external force is applied. If the time dependence of this external force has the following form

$$F_{external} = F_0 \cos \omega t$$

then Newton's second law becomes

$$F_{total} = F_{external} - kx - bv = ma = m\frac{d^2 x}{dt^2} \quad \Rightarrow \quad \frac{d^2 x}{dt^2} + \frac{b}{m}\frac{dx}{dt} + \frac{k}{m}x = \frac{F_0}{m} \cos \omega t$$

A **steady-state solution** of this equation is

$$x(t) = \frac{F_0}{m\sqrt{\left(\omega^2 - \omega_0^2\right)^2 + \frac{b^2 \omega^2}{m^2}}} \cos(\omega t + \phi_0)$$

where

$$\omega_0 = \sqrt{\frac{k}{m}} \quad \text{and} \quad \phi_0 = \tan^{-1}\left[\frac{\omega_0^2 - \omega^2}{\omega\left(\frac{b}{m}\right)}\right]$$

The frequency ω_0 is called the **natural frequency** of the system; it is the frequency that describes the motion of the system if the driving and damping forces are absent. We observe that when the driving force is present, the system will not oscillate with its natural frequency but with the frequency of the driving force.

The amplitude of the oscillations reaches a maximum value when the driving frequency ω becomes equal to the natural frequency ω_0. When the damping force is relatively small (small b), the amplitude of the steady-state solution becomes very large near resonance.

The **quality factor** Q of the system is a measurement of the ratio of the amplitude of oscillation of the system at resonance to the amplitude at low frequency:

$$Q = \frac{A(\omega_0)}{A(\omega \to 0)} = \frac{\dfrac{F_0}{m\sqrt{\left(\omega_0^2 - \omega_0^2\right)^2 + b^2 \omega_0^2 / m^2}}}{\dfrac{F_0}{m\sqrt{\left(0 - \omega_0^2\right)^2 + b^2 0^2 / m^2}}} = \frac{\dfrac{1}{b\omega_0 / m}}{\dfrac{1}{\omega_0^2}} = \frac{m\omega_0}{b}$$

The width of the resonance peak in the graph of amplitude versus frequency is usually measured by the full width at half maximum power. The power is proportional to the square of the amplitude of the oscillator. That implies the half maximum power is at amplitudes that are the maximum amplitude divided by the square root of 2. By this definition, the full width at half maximum power, $\Delta\omega$, is approximately

$$\Delta\omega \approx \frac{\omega_0}{Q}$$

when the damping is relatively small.

Practice Quiz

1. Which of the following changes in a mass–spring system does not increase the frequency of the system?
   a) An increase of the spring constant.
   b) A decrease of the mass.
   c) An increase of the stiffness of the spring.
   d) An increase of the amplitude of the motion.

2. When a simple harmonic oscillator reaches its maximum displacement, which of the following statements is true?
   a) The speed of the harmonic oscillator reaches its maximum speed.
   b) The acceleration of the harmonic oscillator becomes zero.
   c) The acceleration of the harmonic oscillator is pointing in the direction of the displacement.
   d) The acceleration of the harmonic oscillator is pointing in a direction opposite to the direction of the displacement.

3. If the amplitude of a simple harmonic oscillator doubles, what happens to the total energy of the system?
   a) The total energy doubles.
   b) The total energy triples.
   c) The total energy quadruples.
   d) The total energy becomes half as great.

4. Which set of quantities is sufficient for specifying the motion of a simple harmonic oscillator?
   a) Amplitude, period, and frequency
   b) Phase angle, period, and frequency
   c) Amplitude, phase angle, and period
   d) Spring constant, mass, and phase angle

5. A simple pendulum undergoes simple harmonic motion for small amplitudes. What happens to the period of the pendulum as the amplitude increases to large angles?
   a) Period remains the same.
   b) Period increases.
   c) Period decreases.
   d) It depends on the length of the pendulum.

6. A mass-spring system has a frequency f when the mass slides horizontally across a frictionless surface. If instead of sliding horizontally, the mass is suspended from the spring vertically, what will be the frequency of the mass-spring system?
    a) Depends on the gravitational acceleration
    b) Depends on how far the spring stretches to support the weight of the mass at equilibrium
    c) $2f$
    d) f

7. The amplitude of a damped harmonic oscillator decreases to half its initial amplitude in time t. How much longer does it take for the amplitude to reduce to one-fourth of its original amplitude?
    a) $t/2$
    b) $t/4$
    c) $t \ln 2$
    d) t

8. The acceleration of gravity at the surface of the moon is approximately one-sixth the acceleration of gravity at the surface of the earth. A pendulum at the surface of the earth has length L. On the surface of the moon, a pendulum with the same period will have a length
    a) $6L$
    b) $L/6$
    c) $36L$
    d) $L/\sqrt{6}$

9. Which of these systems would benefit most by having the system critically damped?
    a) A pendulum on a clock
    b) The suspension of an automobile
    c) A tuning fork
    d) A child on a swing

10. The steady-state amplitude of a driven harmonic oscillator is A. Which of the following options will definitely reduce the amplitude of the system?
    a) A reduction of the frequency of the driving force.
    b) An increase of the frequency of the driving force.
    c) An increase of the quality factor of the system.
    d) A decrease of the quality factor of the system.

11. A mass–spring system has a total energy of 18.4 J, oscillates with a period of 1.46 s, and has an amplitude of 4.6 cm. What is the spring constant and the mass of the system?

12. A torsion pendulum contains a wire with a torsion constant of 24.6 Nm/rad that supports a uniform rod at its center of mass. The rod has a length of 10.8 cm and a mass of 0.422 kg. What is the period of oscillation of the torsion pendulum?

13. The initial amplitude of a damped harmonic oscillator is 3.64 cm. After 24.5 s the amplitude is 2.99 cm. How long does it take before the amplitude becomes 1.00 cm?

14. Consider a driven harmonic oscillator. What is the limit of the phase angle when
    a) the driving frequency goes to zero?
    b) the driving frequency goes to infinity?
    c) the driving frequency goes to the natural frequency?

15. A driven harmonic oscillator consists of a spring with a spring constant $k = 145$ N/m attached to a mass $m = 0.388$ kg. The driving force has an amplitude $F_0 = 3.65$ N and a frequency $f_0 = 2.87$ Hz. The damping force is given by $F_{damping} = -bv$ where $b = 8.45$ Ns/m. What is the amplitude of the oscillation of the mass?

Responses to Select End-of-Chapter Questions

1. Examples are: a child's swing (SHM, for small oscillations), stereo speakers (complicated motion, the addition of many SHMs), the blade on a jigsaw (approximately SHM), the string on a guitar (complicated motion, the addition of many SHMs).

7. At high altitude, g is slightly smaller than it is at sea level. If g is smaller, then the period T of the pendulum clock will be longer, and the clock will run slow (or lose time).

13. When walking at a normal pace, about 1 s (timed). The faster you walk, the shorter the period. The shorter your legs, the shorter the period.

19. Yes. Rattles which occur only when driving at certain speeds are most likely resonance phenomena.

Solutions to Select End-of-Chapter Problems

1. The particle would travel four times the amplitude: from $x = A$ to $x = 0$ to $x = -A$ to $x = 0$ to $x = A$. So the total distance $= 4A = 4(0.18 \text{ m}) = \boxed{0.72 \text{ m}}$.

7. The maximum velocity is given by Eq. 14-9a.

$$v_{max} = \omega A = \frac{2\pi A}{T} = \frac{2\pi (0.15 \text{ m})}{7.0 \text{ s}} = \boxed{0.13 \text{ m/s}}$$

The maximum acceleration is given by Eq. 14-9b.

$$a_{max} = \omega^2 A = \frac{4\pi^2 A}{T^2} = \frac{4\pi^2 (0.15 \text{ m})}{(7.0 \text{ s})^2} = 0.1209 \text{ m/s}^2 \approx \boxed{0.12 \text{ m/s}^2}$$

$$\frac{a_{max}}{g} = \frac{0.1209 \text{ m/s}^2}{9.80 \text{ m/s}^2} = 1.2 \times 10^{-2} = \boxed{1.2\%}$$

13. (a) For A, the amplitude is $A_A = \boxed{2.5 \text{ m}}$. For B, the amplitude is $A_B = \boxed{3.5 \text{ m}}$.
   (b) For A, the frequency is 1 cycle every 4.0 seconds, so $f_A = \boxed{0.25 \text{ Hz}}$. For B, the frequency is 1 cycle every 2.0 seconds, so $f_B = \boxed{0.50 \text{ Hz}}$.
   (c) For C, the period is $T_A = \boxed{4.0 \text{ s}}$. For B, the period is $T_B = \boxed{2.0 \text{ s}}$.
   (d) Object A has a displacement of 0 when $t = 0$, so it is a sine function.

$$x_A = A_A \sin(2\pi f_A t) \quad \rightarrow \quad \boxed{x_A = (2.5 \text{ m}) \sin\left(\tfrac{1}{2}\pi t\right)}$$

Object B has a maximum displacement when $t = 0$, so it is a cosine function.

$$x_B = A_B \cos(2\pi f_B t) \quad \rightarrow \quad \boxed{x_B = (3.5 \text{ m}) \cos(\pi t)}$$

19. When the object is at rest, the magnitude of the spring force is equal to the force of gravity. This determines the spring constant. The period can then be found.

$$\sum F_{vertical} = kx_0 - mg \quad \rightarrow \quad k = \frac{mg}{x_0}$$

$$T = 2\pi \sqrt{\frac{m}{k}} = 2\pi \sqrt{\frac{m}{\dfrac{mg}{x_0}}} = 2\pi \sqrt{\frac{x_0}{g}} = 2\pi \sqrt{\frac{0.14 \text{ m}}{9.80 \text{ m/s}^2}} = \boxed{0.75 \text{ s}}$$

25. (*a*) If the block is displaced a distance x to the right in Figure 14-32a, then the length of spring # 1 will be increased by a distance x_1 and the length of spring # 2 will be increased by a distance x_2, where $x = x_1 + x_2$. The force on the block can be written $F = -k_{eff}\, x$. Because the springs are massless, they act similar to a rope under tension, and the same force F is exerted by each spring. Thus $F = -k_{eff}x = -k_1x_1 = -k_2x_2$.

$$x = x_1 + x_2 = -\frac{F}{k_1} - \frac{F}{k_2} = -F\left(\frac{1}{k_1} + \frac{1}{k_2}\right) = -\frac{F}{k_{eff}} \quad\rightarrow\quad \frac{1}{k_{eff}} = \frac{1}{k_1} + \frac{1}{k_2}$$

$$T = 2\pi\sqrt{\frac{m}{k_{eff}}} = \boxed{2\pi\sqrt{m\left(\frac{1}{k_1} + \frac{1}{k_2}\right)}}$$

(*b*) The block will be in equilibrium when it is stationary, and so the net force at that location is zero. Then, if the block is displaced a distance x to the right in the diagram, then spring # 1 will exert an additional force of $F_1 = -k_1x$, in the opposite direction to x. Likewise, spring # 2 will exert an additional force $F_2 = -k_2x$, in the same direction as F_1. Thus the net force on the displaced block is $F = F_1 + F_2 = -k_1x - k_2x = -(k_1 + k_2)x$. The effective spring constant is thus $k = k_1 + k_2$, and the period is given by

$$T = 2\pi\sqrt{\frac{m}{k}} = \boxed{2\pi\sqrt{\frac{m}{k_1 + k_2}}}$$

31. The spring constant is found from the ratio of applied force to displacement.

$$k = \frac{F}{x} = \frac{95.0\,\text{N}}{0.175\,\text{m}} = 542.9\,\text{N/m}$$

Assuming that there are no dissipative forces acting on the ball, the elastic potential energy in the loaded position will become kinetic energy of the ball.

$$E_i = E_f \quad\rightarrow\quad \tfrac{1}{2}kx_{max}^2 = \tfrac{1}{2}mv_{max}^2 \quad\rightarrow\quad v_{max} = x_{max}\sqrt{\frac{k}{m}} = (0.175\,\text{m})\sqrt{\frac{542.9\,\text{N/m}}{0.160\,\text{kg}}} = \boxed{10.2\,\text{m/s}}$$

37. We assume that the collision of the bullet and block is so quick that there is no significant motion of the large mass or spring during the collision. Linear momentum is conserved in this collision. The speed that the combination has right after the collision is the maximum speed of the oscillating system. Then, the kinetic energy that the combination has right after the collision is stored in the spring when it is fully compressed, at the amplitude of its motion.

$$P_{before} = P_{after} \quad\rightarrow\quad mv_0 = (m + M)v_{max} \quad\rightarrow\quad v_{max} = \frac{m}{m + M}v_0$$

$$\tfrac{1}{2}(m + M)v_{max}^2 = \tfrac{1}{2}kA^2 \quad\rightarrow\quad \tfrac{1}{2}(m + M)\left(\frac{m}{m + M}v_0\right)^2 = \tfrac{1}{2}kA^2 \quad\rightarrow$$

$$v_0 = \frac{A}{m}\sqrt{k(m + M)} = \frac{(9.460 \times 10^{-2}\,\text{m})}{(7.870 \times 10^{-3}\,\text{kg})}\sqrt{(142.7\,\text{N/m})(7.870 \times 10^{-3}\,\text{kg} + 4.648\,\text{kg})} = \boxed{309.8\,\text{m/s}}$$

43. We consider this a simple pendulum. Since the motion starts at the amplitude position at $t = 0$, we may describe it by a cosine function with no phase angle, $\theta = \theta_{max}\cos\omega t$. The angular velocity can be written as a function of the length, $\theta = \theta_{max}\cos\left(\sqrt{\dfrac{g}{l}}\,t\right)$.

(*a*) $\theta(t = 0.35\,\text{s}) = 13°\cos\left(\sqrt{\dfrac{9.80\,\text{m/s}^2}{0.30\,\text{m}}}\,(0.35\,\text{s})\right) = \boxed{-5.4°}$

(b) $\theta(t = 3.45\,\text{s}) = 13° \cos\left(\sqrt{\dfrac{9.80\,\text{m/s}^2}{0.30\,\text{m}}}\,(3.45\,\text{s})\right) = \boxed{8.4°}$

(c) $\theta(t = 6.00\,\text{s}) = 13° \cos\left(\sqrt{\dfrac{9.80\,\text{m/s}^2}{0.30\,\text{m}}}\,(6.00\,\text{s})\right) = \boxed{-13°}$

49. The balance wheel of the watch is a torsion pendulum, described by $\tau = -K\theta$. A specific torque and angular displacement are given, and so the torsion constant can be determined. The angular frequency is given by $\omega = \sqrt{\dfrac{K}{I}}$. Use these relationships to find the mass.

$$\tau = -K\theta \quad \rightarrow \quad K = \left|\dfrac{\theta}{\tau}\right| = \dfrac{1.1\times 10^{-5}\,\text{m·N}}{\pi/4\,\text{rad}}$$

$$\omega = 2\pi f = \sqrt{\dfrac{K}{I}} = \sqrt{\dfrac{K}{mr^2}} \quad \rightarrow$$

$$m = \dfrac{K}{4\pi^2 f^2 r^2} = \dfrac{\dfrac{1.1\times 10^{-5}\,\text{m·N}}{\pi/4\,\text{rad}}}{4\pi^2 (3.10\,\text{Hz})^2 (0.95\times 10^{-2}\,\text{m})^2} = 4.1\times 10^{-4}\,\text{kg} = \boxed{0.41\,\text{g}}$$

55. This is a physical pendulum. Use the parallel-axis theorem to find the moment of inertia about the pin at point A, and then use Eq. 14-14 to find the period.

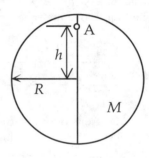

$$I_{\text{pin}} = I_{\text{CM}} + Mh^2 = \tfrac{1}{2}MR^2 + Mh^2 = M\left(\tfrac{1}{2}R^2 + h^2\right)$$

$$T = 2\pi\sqrt{\dfrac{I}{Mgh}} = 2\pi\sqrt{\dfrac{M\left(\tfrac{1}{2}R^2 + h^2\right)}{Mgh}} = 2\pi\sqrt{\dfrac{\left(\tfrac{1}{2}R^2 + h^2\right)}{gh}}$$

$$= 2\pi\sqrt{\dfrac{\tfrac{1}{2}(0.200\,\text{m})^2 + (0.180\,\text{m})^2}{(9.80\,\text{m/s}^2)(0.180\,\text{m})}} = \boxed{1.08\,\text{s}}$$

61. (a) For the "lightly damped" harmonic oscillator, we have $b^2 \ll 4mk \ \rightarrow \ b^2/4m^2 \ll k/m \ \rightarrow \ \omega' \approx \omega_0$. We also assume that the object starts to move from maximum displacement, and so $x = A_0 e^{-bt/2m} \cos\omega't$ and

$$v = \dfrac{dx}{dt} = -\dfrac{b}{2m}A_0 e^{-bt/2m}\cos\omega't - \omega'A_0 e^{-bt/2m}\sin\omega't \approx -\omega_0 A_0 e^{-bt/2m}\sin\omega't.$$

$$E = \tfrac{1}{2}kx^2 + \tfrac{1}{2}mv^2 = \tfrac{1}{2}kA_0^2 e^{-\frac{bt}{m}}\cos^2\omega't + \tfrac{1}{2}m\omega_0^2 A_0^2 e^{-\frac{bt}{m}}\sin^2\omega't$$

$$= \tfrac{1}{2}kA_0^2 e^{-\frac{bt}{m}}\cos^2\omega't + \tfrac{1}{2}kA_0^2 e^{-\frac{bt}{m}}\sin^2\omega't = \tfrac{1}{2}kA_0^2 e^{-\frac{bt}{m}} = \boxed{E_0 e^{-\frac{bt}{m}}}$$

(b) The fractional loss of energy during one period is as follows. Note that we use the approximation that $b/2m \ll \omega_0 = 2\pi/T \ \rightarrow \ bT/m \ll 4\pi \ \rightarrow \ bT/m \ll 1$.

$$\Delta E = E(t) - E(t+T) = E_0 e^{-\frac{bt}{m}} - E_0 e^{-\frac{b(t+T)}{m}} = E_0 e^{-\frac{bt}{m}}\left(1 - e^{-\frac{bT}{m}}\right) \quad \rightarrow$$

$$\dfrac{\Delta E}{E} = \dfrac{E_0 e^{-\frac{bt}{m}}\left(1 - e^{-\frac{bT}{m}}\right)}{E_0 e^{-\frac{bt}{m}}} = 1 - e^{-\frac{bT}{m}} \approx 1 - \left(1 - \dfrac{bT}{m}\right) = \dfrac{bT}{m} = \dfrac{b2\pi}{m\omega_0} = \boxed{\dfrac{2\pi}{Q}}$$

67. Apply the resonance condition, $\omega = \omega_0$, to Eq. 14-23, along with the given condition of $A_0 = 23.7\, F_0/m$. Note that for this condition to be true, the value of 23.7 must have units of s^2.

$$A_0 = \frac{F_0}{m\sqrt{\left(\omega^2 - \omega_0^2\right)^2 + b^2\omega^2/m^2}} \rightarrow$$

$$A_0\left(\omega = \omega_0\right) = \frac{F_0}{m\sqrt{b^2\omega_0^2/m^2}} = \frac{F_0}{m\dfrac{b\omega_0}{m}} = \frac{F_0}{m\dfrac{b\omega_0^2}{m\omega_0}} = \frac{F_0}{m\dfrac{\omega_0^2}{Q}} = \frac{F_0}{m}\frac{Q}{\omega_0^2} = 23.7\frac{F_0}{m} \rightarrow$$

$$Q = 23.7\left(\omega_0^2\right) = 23.7\left(4\pi^2 f^2\right) = \left(23.7\,s^2\right)4\pi^2\left(382\,\text{Hz}\right)^2 = \boxed{1.37 \times 10^8}$$

73. The frequency of a simple pendulum is given by $f = \dfrac{1}{2\pi}\sqrt{\dfrac{g}{L}}$. The pendulum is accelerating vertically which is equivalent to increasing (or decreasing) the acceleration due to gravity by the acceleration of the pendulum.

(a) $f_{new} = \dfrac{1}{2\pi}\sqrt{\dfrac{g+a}{L}} = \dfrac{1}{2\pi}\sqrt{\dfrac{1.50g}{L}} = \sqrt{1.50}\,\dfrac{1}{2\pi}\sqrt{\dfrac{g}{L}} = \sqrt{1.50}\,f = \boxed{1.22\,f}$

(b) $f_{new} = \dfrac{1}{2\pi}\sqrt{\dfrac{g+a}{L}} = \dfrac{1}{2\pi}\sqrt{\dfrac{0.5g}{L}} = \sqrt{0.5}\,\dfrac{1}{2\pi}\sqrt{\dfrac{g}{L}} = \sqrt{0.5}\,f = \boxed{0.71\,f}$

79. The relationship between the velocity and the position of a SHO is given by Eq. 14-11b. Set that expression equal to half the maximum speed, and solve for the displacement.

$$v = \pm v_{max}\sqrt{1 - x^2/A^2} = \tfrac{1}{2}v_{max} \rightarrow \pm\sqrt{1 - x^2/A^2} = \tfrac{1}{2} \rightarrow 1 - x^2/A^2 = \tfrac{1}{4} \rightarrow x^2/A^2 = \tfrac{3}{4} \rightarrow$$

$$\boxed{x = \pm\sqrt{3}A/2 \approx \pm 0.866\,A}$$

85. (a) The relationship between the velocity and the position of a SHO is given by Eq. 14-11b. Set that expression equal to half the maximum speed, and solve for the displacement.

$$v = \pm v_{max}\sqrt{1 - x^2/x_0^2} = \tfrac{1}{2}v_{max} \rightarrow \pm\sqrt{1 - x^2/x_0^2} = \tfrac{1}{2} \rightarrow 1 - x^2/x_0^2 = \tfrac{1}{4} \rightarrow$$

$$x^2/x_0^2 = \tfrac{3}{4} \rightarrow \boxed{x = \pm\sqrt{3}x_0/2 \approx \pm 0.866\,x_0}$$

(b) Since $F = -kx = ma$ for an object attached to a spring, the acceleration is proportional to the displacement (although in the opposite direction), as $a = -xk/m$. Thus the acceleration will have half its maximum value where the displacement has half its maximum value, at $\boxed{\pm\tfrac{1}{2}x_0}$.

91. We must make several assumptions. Consider a static displacement of the trampoline, by someone sitting on the trampoline mat. The upward elastic force of the trampoline must equal the downward force of gravity. We estimate that a 75-kg person will depress the trampoline about 25 cm at its midpoint.

$$kx = mg \rightarrow k = \frac{mg}{x} = \frac{\left(75\,\text{kg}\right)\left(9.80\,\text{m/s}^2\right)}{0.25\,\text{m}} = 2940\,\text{N/m} \approx \boxed{3000\,\text{N/m}}$$

Chapter 15: Wave Motion

Chapter Overview and Objectives

In this chapter, the phenomena associated with mechanical waves are introduced. The properties of waves are described and important aspects of pulse and sinusoidal waves are discussed.

After completing this chapter you should:
- Know what a wave is.
- Know the difference between longitudinal and transverse waves.
- Know the parameters that are necessary to define a sinusoidal wave and the relationship between them.
- Know how wave speed depends on the properties of the material carrying the wave.
- Know the linear wave equation and the nature of solutions of that equation.
- Know how mechanical waves carry energy.
- Know what the wave phenomena of reflection, transmission, refraction, diffraction, and resonance are.
- Know how to calculate the resonant frequencies of a string anchored at both ends.

Summary of Equations

Relationship between period and frequency of a sinusoidal wave:

$$T = \frac{1}{f}$$
(Section 15-1)

Relationship between wave speed, frequency, wavelength, and period:

$$v = \lambda f = \frac{\lambda}{T}$$
(Section 15-1)

General dependence of the wave speed on the properties of materials:

$$v = \sqrt{\frac{elastic\ force\ factor}{inertial\ factor}}$$
(Section 15-2)

Dependence of the transverse wave speed on a string on the tension and the linear density of string:

$$v = \sqrt{\frac{F_\mathrm{T}}{\mu}}$$
(Section 15-2)

Dependence of the longitudinal acoustic wave speed in a solid on the elastic modulus and the density of material:

$$v = \sqrt{\frac{E}{\rho}}$$
(Section 15-2)

Dependence of the longitudinal acoustic wave speed in a fluid on the bulk modulus and the density of material:

$$v = \sqrt{\frac{B}{\rho}}$$
(Section 15-2)

Energy transported by a transverse sinusoidal wave through a cross-sectional area S during a time t:

$$E = 2\pi^2 \rho S v t f^2 A^2 \qquad \text{(Section 15-3)}$$

Average power transported by a transverse sinusoidal wave through a cross-sectional area S:

$$\overline{P} = \frac{E}{t} = 2\pi^2 \rho S v f^2 A^2 \qquad \text{(Section 15-3)}$$

Intensity of a transverse sinusoidal wave:
$$I = \frac{\overline{P}}{S} = 2\pi^2 \rho v f^2 A^2 \qquad \text{(Section 15-3)}$$

Dependence of the intensity of a spherical wave on the radius of the wavefront:

$$\frac{I_2}{I_1} = \frac{r_1^2}{r_2^2} \qquad \text{(Section 15-3)}$$

Dependence of the amplitude of a spherical wave on the radius of the wavefront:

$$\frac{A_2}{A_1} = \frac{r_1}{r_2} \qquad \text{(Section 15-3)}$$

Definition of the wave number:
$$k = \frac{2\pi}{\lambda} \qquad \text{(Section 15-4)}$$

Linear wave equation:
$$\frac{d^2 D}{dt^2} = \frac{1}{v^2}\frac{d^2 D}{dx^2} \qquad \text{(Section 15-5)}$$

Resonant frequencies of a string fixed at both ends: $f_n = \dfrac{nv}{2L}$, where $n = 1,2,3, \ldots$ (Section 15-9)

Relation between the direction of incident and refracted waves and wave speeds:

$$\frac{\sin\theta_2}{\sin\theta_1} = \frac{v_2}{v_1} \qquad \text{(Section 15-10)}$$

Chapter Summary

Section 15-1. Characteristics of Wave Motion

Waves can be divided into two categories:
- **Pulse waves**: waves that last for a limited time.
- **Continuous waves**: waves that last for extended lengths of time.

A special category of continuous waves is the **periodic wave**. A periodic wave has a time dependence that is repetitive. A sinusoidal wave is a particular type of periodic wave in which the motion of the wave at a particular point in space undergoes simple harmonic motion.

A sinusoidal wave can be characterized by quantities similar to those used to describe simple harmonic motion. The **amplitude** A of the wave is the maximum displacement from equilibrium of the material carrying the wave. The **period** T of the wave is the amount of time for one cycle of the sinusoidal motion to take place at a given point in space. The

frequency f of the wave is the number of cycles of sinusoidal motion that take place in one unit of time. The period T and frequency f are related by

$$T = \frac{1}{f}$$

Because the wave moves through space with a constant velocity, there is also a spatial sinusoidal dependence of the displacement of the material at a given time. The spatial extent of one sinusoidal cycle of the wave is called the **wavelength** λ. The **propagation velocity** v of the wave is related to the wavelength and the period or frequency in the following manner:

$$v = \lambda f = \frac{\lambda}{T}$$

Section 15-2. Types of Waves: Transverse and Longitudinal

We can classify waves by the direction of the displacement of the material relative to the direction of propagation of the wave:

- **Longitudinal wave**: a wave for which the direction of the displacement of the material is parallel to the direction of propagation of the wave.
- **Transverse wave**: a wave for which the direction of the displacement of the material is perpendicular to the direction of propagation of the wave.

The propagation velocity of waves is determined by properties of the material in which the waves travel. The general form of the wave velocity for mechanical waves is

$$v = \sqrt{\frac{\text{elastic force factor}}{\text{inertial factor}}}$$

The propagation velocity of waves on string with a mass per unit length μ and stretched with a tension F_T is

$$v = \sqrt{\frac{F_T}{\mu}}$$

The propagation velocity of longitudinal acoustic waves in a solid with density ρ and elastic modulus E is

$$v = \sqrt{\frac{E}{\rho}}$$

The propagation velocity of longitudinal acoustic waves in a gas with density ρ and bulk modulus B is

$$v = \sqrt{\frac{B}{\rho}}$$

Example 15-2-A. Using the speed of sound to determine the bulk modulus of materials. Sound is carried by waves through a medium. The velocity of sound in air is about $v_{air} = 340$ m/s. The density of air is $\rho_{air} = 1.29$ kg/m^3. What is the change in pressure of the air if its volume is reduced by 1%?

Approach: The change in pressure is related to the change in volume. The relative change of these quantities is determined by the bulk modulus B (see Section 12.4). Since the problem does not provide us with information about the bulk modules B, we need to use the relation between the bulk modulus and the speed of sound in order to calculate B.

Solution: The bulk modulus B of the air can be obtained from the speed of sound v_{air} and the density ρ_{air}:

$$v_{air} = \sqrt{\frac{B}{\rho_{air}}} \quad \rightarrow \quad B = \rho_{air} v_{air}^{\;2} = (1.29)(343)^2 = 1.52 \times 10^5 \text{ kg/m s}^2$$

The change in air pressure is related to the fractional change in volume:

$$\Delta P = -B\frac{\Delta V}{V} = -\left(1.52\times10^{5}\right)\left(-0.01\right) = 1.52\times10^{3}\ \text{N/m}^{2}$$

Example 15-2-B: Using wave properties to determine mass densities. A string is stretched to a tension $F_T = 247$ N. Waves traveling on the string with a wavelength $\lambda = 26.4$ cm have a frequency $f = 489$ Hz. What is the linear density of the string?

Approach: A measurement of the propagation velocity of a wave in order to determine the mass density of the medium through which the wave is propagating is an important technique with many applications. Oil companies for example use this technique to determine the densities of underground regions in their quest for oil, using the fact that the mass density of oil is very different from the mass density of rock. The mass density of the string can be obtained on the basis of the propagation velocity, calculated from the wavelength and frequency, and the tension in the string.

Solution: The propagation speed of the wave is equal to

$$v = \lambda f = \left(0.264\right)\left(489\right) = 65.7\ \text{m/s}$$

Since the propagation speed v is related to the tension F_T and the linear density μ we can now calculate μ:

$$v = \sqrt{\frac{F_T}{\mu}} \quad \rightarrow \quad \mu = \frac{F_T}{v^2} = \frac{247}{\left(65.7\right)^2} = 0.0572\ \text{kg/m}$$

Section 15-3. Energy Transported by Waves

A wave traveling along a string transports energy along its direction of propagation. The Figure on the right shows an example of a wave traveling along a string from left to right. When the wave passes position x, it results in a vertical motion of the segment of the string located at that position. The rate of work done on a segment of the string is equal to the downward velocity of the string multiplied by the downward component of the tension in the string. For a wave traveling to the right, the left part of the string in general 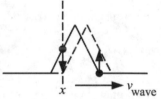 will do positive work on the right part of the string. The right part of the string is almost always doing negative work on the left part of the string as the right-moving wave passes a given point on the string. We obtain similar results for waves propagating in a straight line through a three-dimensional medium.

A sinusoidal wave propagating through a medium of density ρ transports an energy E past a certain point. The energy E transported across a cross-sectional area S during a time interval t is equal to

$$E = 2\pi^2 \rho S v t f^2 A^2$$

where f is the frequency, A is the amplitude, and v is the propagation velocity. The average power transported across the area S during this time interval is

$$\overline{P} = \frac{E}{t} = \frac{2\pi^2 \rho S v t f^2 A^2}{t} = 2\pi^2 \rho S v f^2 A^2$$

The **intensity** of the wave is defined as the average power transmitted per unit area, perpendicular to the direction of propagation:

$$I = \frac{\overline{P}}{S} = 2\pi^2 \rho v f^2 A^2$$

For spherically shaped wavefronts, the area of the wavefront grows as the square of the radius of the wavefront.

Since the energy in the wave is conserved, the intensity of the wave is inversely proportional to the square of the radius of the wavefront:

$$\frac{I_2}{I_1} = \frac{r_1^2}{r_2^2}$$

where I_1 and I_2 are the intensities at radii r_1 and r_2, respectively. Knowing that the intensity of a wave is proportional to the square of the amplitude of the wave, we can use this to conclude that the ratio of amplitudes is equal to

$$\frac{A_2}{A_1} = \frac{r_1}{r_2}$$

Example 15-3-A. Amplitude of sound waves. The intensity of sound waves that is barely audible is $I_{min} = 1 \times 10^{-12}$ W/m^2. What is the displacement amplitude of these sound waves when the speed of sound in air is $v_{air} = 343$ m/s and the frequency of the sound waves is $f = 1000$ Hz? The density of air is $\rho = 1.29$ kg/m^3.

Approach: The expression of the intensity discussed in this section allows us to calculate the intensity of a sinusoidal sound wave propagating through a medium if we know the density of the medium, the propagation velocity, and the frequency and amplitude of the wave. In this problem we know the intensity of the sound wave and can use the remaining information to determine the amplitude of the sound wave.

Solution: Using the information provided we can calculate the amplitude of the sound wave:

$$I_{min} = 2\pi^2 \rho_{air} v_{air} f^2 A^2 \quad \rightarrow \quad A = \sqrt{\frac{I_{min}}{2\pi^2 \rho_{air} v_{air} f^2}} = \frac{1}{\pi f}\sqrt{\frac{I_{min}}{2\rho_{air} v_{air}}} = \frac{1}{\pi(1000)}\sqrt{\frac{1\times10^{-12}}{2(1.29)(343)}} = 1.07\times10^{-11} \text{ m}$$

Note that diameters of atoms are in the order of 10^{-10} m. The amplitude of the sound wave is thus smaller than the typical dimension of the atoms that make up the medium through which the wave propagates.

Section 15-4. Mathematical Representation of a Traveling Wave

A wave can be represented by a wavefunction. A wavefunction is a function of position and time with a value equal to the **displacement** of the wave. In general, we write such a function as $D(x,t)$. A sinusoidal wave, moving in the positive x direction, can be represented by the function

$$D(x,t) = A\sin(kx - \omega t + \phi)$$

where A is the amplitude, k is the **wave number**, ω is the angular frequency, and ϕ is the phase angle. The wave number is related to the wavelength λ:

$$k = \frac{2\pi}{\lambda}$$

A sinusoidal wave, moving in the negative x direction, can be represented by the function

$$D(x,t) = A\sin(kx + \omega t + \phi)$$

Alternate expressions for the wavefunction are

$$D(x,t) = A\sin\left(\frac{2\pi}{\lambda}x - \frac{2\pi}{T}t + \phi\right) = A\sin\left(\frac{2\pi}{\lambda}(x - vt) + \phi\right)$$

where T is the period and v is the propagation velocity of the wave.

Example 15-4-A: Identifying wave properties. A wave is described by the function

$$D(x,t) = 0.02 \sin\left[100(x + 0.05t)\right].$$

What are the amplitude, wavelength, wave number, frequency, period, and speed of the wave? In which direction is the wave traveling?

Approach: In order to determine the values of the various wave properties, we have to compare the functional description of the wavefunction with the representations described in this Section. For example, we recognize that the wavefunction corresponds to a wave traveling in the negative x direction.

Solution: The maximum displacement of the wave is its amplitude. For the wavefunction specified in this problem, the maximum amplitude is 0.02 m and this is the amplitude of the wave. The coefficient of the position x is the wave number k. The wavelength λ can be obtained from the wave number k:

$$\lambda = \frac{2\pi}{k} = \frac{2\pi}{100} = 0.063 \text{ m}$$

The coefficient of time t is the propagation speed and we thus conclude that $v = 0.05$ m/s. We can calculate the frequency from the wavelength and the wave speed:

$$f = \frac{v}{\lambda} = \frac{0.05}{0.063} = 0.79 \text{ Hz}$$

The period of the wave can be obtained from the frequency:

$$T = \frac{1}{f} = \frac{1}{0.79} = 1.3 \text{ s}$$

Finally, the propagation direction of the wave is in the negative x direction because the x and t terms in $D(x, t)$ have the same sign.

Section 15-5. The Wave Equation

Applying Newton's second law to infinitesimal mass elements along a string allows us to derive the equation of motion of the string when it is carrying a wave. The equation that describes the transverse displacement of the string is

$$\frac{d^2 D}{dt^2} = \frac{F_T}{\mu}\frac{d^2 D}{dx^2} = \frac{1}{v^2}\frac{d^2 D}{dx^2}$$

where D is the transverse displacement, F_T is the tension, μ is the mass per unit length, and v is the propagation speed. This equation is called the one-dimensional **linear wave equation**.

Example 15-5-A: Finding solutions of the linear wave equation. Show that $D(x,t) = A \cos(Bx) \sin(Ct)$ is a solution to the wave equation and determine the speed of the waves in this medium.

Approach: To show that $D(x, t)$ is a solution of the linear wave equation we need to calculate d^2D/dt^2 and d^2D/dx^2 and show that the linear wave equation is satisfied.

Solution: The position and time derivatives of the wavefunction are:

$$\frac{d}{dt}D(x,t) = AC\cos(Bx)\cos(Ct) \quad \Rightarrow \quad \frac{d^2}{dt^2}D(x,t) = -AC^2\cos(Bx)\sin(Ct) = -C^2 D(x,t)$$

$$\frac{d}{dx}D(x,t) = -AB\sin(Bx)\sin(Ct) \quad \Rightarrow \quad \frac{d^2}{dx^2}D(x,t) = -AB^2\cos(Bx)\sin(Ct) = -B^2 D(x,t)$$

Substituting these expressions in the linear wave equation we obtain

$$-C^2 D(x,t) = \frac{1}{v^2}\left(-B^2 D(x,t)\right) \quad \rightarrow \quad \frac{B^2}{C^2} = v^2$$

This shows that $D(x,t)$ is a solution to the wave equation if $B/C = \pm v$.

Section 15-6. The Principle of Superposition

Because the linear wave equation only has terms with the first power of the wavefunction or its derivatives, any linear combination of solutions is also a solution to the wave equation. If $f(x,t)$ and $g(x,t)$ are solutions to the linear wave equation, then $A\,f(x,t) + B\,g(x,t)$ is also a solution to the linear wave equation for any A and B. This is called the **principle of superposition**.

Example 15-6-A. Proof the principle of superposition. Confirm that the principle of superposition holds for the linear wave equation using the wavefunctions $f(x,t)$ and $g(x,t)$.

Approach: Use the fact that $f(x,t)$ and $g(x,t)$ satisfy the linear wave equation to show that $A\,f(x,t) + B\,g(x,t)$ also satisfies the linear wave equation.

Solution: Since $f(x,t)$ and $g(x,t)$ are solutions to the linear wave equation, they must satisfy the following relations:

$$\frac{d^2 f(x,t)}{dt^2} = \frac{1}{v^2}\frac{d^2 f(x,t)}{dx^2} \quad \text{and} \quad \frac{d^2 g(x,t)}{dt^2} = \frac{1}{v^2}\frac{d^2 g(x,t)}{dx^2}$$

Now consider the function $A\,f(x,t) + B\,g(x,t)$. The second derivative of this function with respect to time is

$$\frac{d^2}{dt^2}\left[Af(x,t)+Bg(x,t)\right] = A\frac{d^2 f(x,t)}{dt^2} + B\frac{d^2 g(x,t)}{dt^2} = \frac{A}{v^2}\frac{d^2 f(x,t)}{dx^2} + \frac{B}{v^2}\frac{d^2 g(x,t)}{dx^2} = \frac{1}{v^2}\frac{d^2}{dx^2}\left[Af(x,t)+Bg(x,t)\right]$$

This shows the linear superposition of $f(x,t)$ and $g(x,t)$ is also a solution to the wave equation.

Section 15-7. Reflection and Transmission

When waves reach a boundary or discontinuity in the medium in which they are propagating, two new waves will be created:
- A **reflected wave**: a wave that travels back into the initial medium.
- A **transmitted wave**: a wave that travels into the medium past the boundary.

In one dimension, the reflected wave will travel back into the medium with a velocity equal in magnitude but opposite in direction to the incident velocity. The amplitude and phase of the reflected wave may be different than amplitude and phase of the initial incident wave. In two or three dimensions, the reflected wave has only the component of its velocity perpendicular to the boundary reversed. The other components of the velocity remain the same. This is summarized in the **law of reflection**:

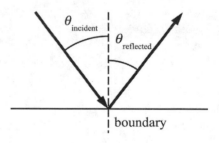

 The angle of incidence is equal to the angle of reflection.

Section 15-8. Interference

When multiple waves are traveling through the same medium, the displacement of the medium at a given point in space is the sum of the displacements associated with the individual waves at that point in space. Since the displacement associated with the individual waves can be positive or negative, the net displacement can be zero, even if the displacements of the individual waves are non-zero. If the displacement is less than the displacement that would result from the individual waves, the condition is called **destructive interference**; the waves are said to be **out of phase**. If the resulting wave displacement is greater than the displacement that would result from the individual waves, the condition is called **constructive interference**; the waves are said to be **in phase**. Interference may be a misleading word in the sense that the waves *do not* interfere with the motion of each other, only in the displacement of the medium in which they travel.

Section 15-9. Standing Waves; Resonance

When two sinusoidal waves of identical frequency and amplitude travel in opposite directions through a medium, the displacement of the media can be described by a **standing wave**. The name "standing wave" results from the fact that under these conditions, even though there are waves traveling in two directions, no motion along the direction of wave travel is apparent.

For a system with fixed dimensions and perfectly reflective boundaries, standing waves are formed only for particular frequencies. These frequencies are called the **natural frequencies** or **resonant frequencies** of the system.

The lowest resonant frequency of a system is called the **fundamental frequency**. The higher resonant frequencies are called **overtones**. Any frequency that is an integer multiple of the fundamental frequency is called a **harmonic** of the fundamental frequency. In some systems, the overtones are harmonics of the fundamental frequency; in other systems they are not harmonics of the fundamental frequency.

For waves on a string that is tied down at both ends, the resonant frequencies are given by the expression

$$f_n = \frac{nv}{2L}, \quad \text{where} \quad n = 1, 2, 3, \ldots$$

where L is the length of the string and v is the propagation speed of the waves. The frequency associated with $n = 1$ is the fundamental frequency.

Example 15-9-A. Changing the fundamental frequency of a string. A string on a guitar has a fundamental frequency of $f_0 = 467$ Hz when its tension is $T_0 = 204$ N. Assuming its linear density and length remain constant when tuning, what should the tension in the string be so that the fundamental frequency of the string is $f_1 = 440$ Hz?

Approach: The wave velocity depends on the linear mass density and the tension in the string. An increase in the tension results in an increase in the wave velocity and thus in an increase of the fundamental frequency. A decrease of the fundamental frequency requires a decrease of the tension in the string.

Solution: The fundamental frequency of a string is equal to

$$f_0 = \frac{v}{2L} = \frac{\sqrt{T_0/\mu}}{2L}$$

The final frequency is f_1 and is related to the final tension T_1:

$$f_1 = \frac{v}{2L} = \frac{\sqrt{T_1/\mu}}{2L}$$

From the ratio of frequencies we can determine the required tension in the string:

$$\frac{f_1}{f_0} = \frac{\left(\dfrac{\sqrt{T_1/\mu}}{2L}\right)}{\left(\dfrac{\sqrt{T_0/\mu}}{2L}\right)} = \sqrt{\frac{T_1}{T_0}} \quad \rightarrow \quad T_1 = T_0\left(\frac{f_1}{f_0}\right)^2 = 204\left(\frac{440}{467}\right)^2 = 181 \text{ N}$$

Section 15-10. Refraction

When waves cross a boundary between two materials, the differences in the propagation velocities in the two materials changes the direction of propagation of the waves. The direction of propagation can be specified by providing the angle of incidence and the angle of refraction. These angles are measured with respect to the normal of the boundary (see Figure).

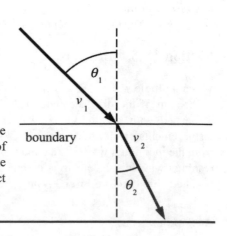

The angles and propagation speeds in the two materials are related in the following way:

$$\frac{\sin\theta_2}{\sin\theta_1} = \frac{v_2}{v_1}$$

Example 15-10-A: Refraction of sound waves. Sound waves in the water of a swimming pool strike the wall of the swimming pool at an angle of $\theta_{in} = 12°$ and are partially transmitted into the wall of the swimming pool where they travel at an angle of $\theta_{out} = 33°$. The speed of sound waves in the water is $v_{water} = 1450$ m/s. What is the speed of the sound waves in the wall of the pool?

Approach: We know how the propagation speeds and the angles of incidence and refraction are related. The problem provides us with information about the angles and the propagation speed in one of the materials (the water). Using this information we can determine the propagation speed in the other material (the wall).

Solution: The angle of incidence and the angle of refraction are related to the velocities in the following manner

$$\frac{v_{wall}}{v_{water}} = \frac{\sin\theta_{wall}}{\sin\theta_{water}} \rightarrow v_{wall} = \left(\frac{\sin\theta_{wall}}{\sin\theta_{water}}\right)v_{water} = \left(\frac{\sin 33°}{\sin 12°}\right)(1450) = 3800 \text{ m/s}$$

Section 15-11 Diffraction

Waves do not cast geometrically sharp shadows. Waves tend to bend around objects placed in their path by a phenomenon called **diffraction**. Diffraction is important when the wavelength of the waves is comparable or larger than the size of the object. The angle of diffraction, which is a measure of the angular spread of waves of wavelength λ after passing an object of size L, is

$$\theta \approx \frac{\lambda}{L}$$

Practice Quiz

1. You watch a wave on string travel from point A to point B. What has physically moved from point A to point B?
    a) Some of the mass of the string
    b) Energy
    c) Air surrounding the string
    d) Nothing

2. A picture of a string is shown at the right. What can you conclude about any transverse waves traveling on the string?
    a) There is no possibility of any waves traveling on the string at this time.
    b) The waves traveling on the string are destructively interfering everywhere and always.
    c) The displacements of a left-traveling and a right-traveling wave might be momentarily canceling each other.
    d) The string is not in motion.

3. Which of the following functions describes a wave moving with a speed v in the positive x direction and keeping a constant shape?
    a) $f(x/v + t)$
    b) $g(x^2 - vt)$
    c) $h(x - vt)$
    d) $i(x - v^2t)$

4. A string has a fundamental frequency of 324 Hz. Which of the following frequencies is not one of the harmonics of the string?
   a) 648 Hz
   b) 972 Hz
   c) 162 Hz
   d) 324,000,000 Hz

5. Harmonic waves travel along a string with a wavelength λ and with an amplitude such that the average power carried by the waves is P_0. What is the average power carried by waves with the same amplitude, but with double the wavelength (2λ)?
   a) $2P_0$
   b) $\frac{1}{2}P_0$
   c) $4P_0$
   d) $\frac{1}{4}P_0$

6. Waves are incident on a boundary at an angle of 42° to the normal. What is the angle of the reflected wave with respect to the normal?
   a) 42°
   b) 21°
   c) 132°
   d) Need to know wave speeds in the two media to answer question.

7. The diagram to the right shows a wave, incident on a boundary between two media, and a refracted wave. The angles of incidence and refraction are θ_1 and θ_2, respectively. The propagation speeds in the two media are v_1 and v_2. Based on the information provided in the Figure, what do you conclude about the speed of the waves in the two media on either side of the boundary?

   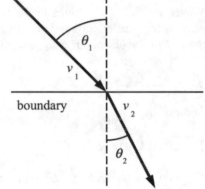

   a) The speed of the waves in the upper medium is greater than the speed of the waves in the lower medium.
   b) The speed of the waves in the upper medium is the same as the speed of the waves in the lower medium.
   c) The speed of the waves in the upper medium is less than the speed of the waves in the lower medium.
   d) No conclusion can be reached about the wave speeds from the drawing.

8. Which of the following actions would raise the fundamental resonant frequency of a string?
   a) Lowering the tension in the string.
   b) Shortening the string.
   c) Making the string more massive.
   d) Increasing the linear density of the string.

9. If you are sitting in a small boat on a lake, you observe that the short wavelength ripples on the surface of the water do not appear on the side of the boat opposite the direction the ripples are traveling on the surface. You do, however, observe the long wavelength swells appearing on the both sides of the boat. This can be explained by
   a) Reflection
   b) Refraction
   c) Resonance
   d) Diffraction

10. A violin player tunes her instrument when it is cold. As the violin warms up, the tension in the strings drop, but the strings get slightly longer. What happens to the fundamental resonant frequencies of the strings?
    a) All fundamental resonant frequencies increase.
    b) All fundamental resonant frequencies decrease.
    c) Some fundamental resonant frequencies decrease, others increase.
    d) It depends on which of the two changes, length or tension, is larger.

11. Determine the elastic modulus for a solid material in which longitudinal waves travel at a speed of 4.2 km/s. The density of the material is 5.32 g/cm³.

12. Spherical waves have an intensity of 3.24 W/m² at a distance of 13.4 m from the source. How much energy does the source emit in the form of waves during a period of 32.5 s?

13. The displacement of a wave is represented by the function

$$D(x,t) = 0.032 \sin(120\ x + 6.8\ t).$$

Determine the amplitude, wavelength, frequency, period, and velocity of the wave.

14. A string has a tension $T = 436$ N, a length $L = 68.4$ cm, and a mass of $m = 13.3$ g. Determine the fundamental frequency of the string.

15. A wave traveling in the first material reaches the boundary between the two materials traveling in the direction shown in the diagram. It is transmitted past the boundary into the second material and leaves the boundary in the direction shown. If the speed of the wave in the first material is 16.4 m/s, what is the speed of the wave in the second material?

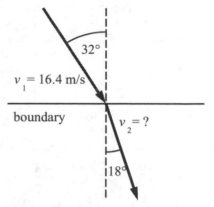

Responses to Select End-of-Chapter Questions

1.   Yes. A simple periodic wave travels through a medium, which must be in contact with or connected to the source for the wave to be generated. If the medium changes, the wave speed and wavelength can change but the frequency remains constant.

7.   The speed of sound is defined as $v = \sqrt{\dfrac{B}{\rho}}$, where B is the bulk modulus and ρ is the density of the material. The bulk modulus of most solids is at least 10^6 times as great as the bulk modulus of air. This difference overcomes the larger density of most solids, and accounts for the greater speed of sound in most solids than in air.

13. Yes, total energy is always conserved. The particles in the medium, which are set into motion by the wave, have both kinetic and potential energy. At the instant in which two waves interfere destructively, the displacement of the medium may be zero, but the particles of the medium will have velocity, and therefore kinetic energy.

19. Waves exhibit diffraction. If a barrier is placed between the energy source and the energy receiver, and energy is still received, it is a good indication that the energy is being carried by waves. If placement of the barrier stops the energy transfer, it may be because the energy is being transferred by particles or that the energy is being transferred by waves with wavelengths smaller than the barrier.

Solutions to Select End-of-Chapter Problems

1.   The wave speed is given by $v = \lambda f$. The period is 3.0 seconds, and the wavelength is 8.0 m.

$$v = \lambda f = \lambda/T = (8.0\,\mathrm{m})/(3.0\,\mathrm{s}) = \boxed{2.7\,\mathrm{m/s}}$$

7.   For a cord under tension, we have from Eq. 15-2 that $v = \sqrt{(F_T/\mu)}$. The speed is also the displacement divided by the elapsed time, $v = \Delta x/\Delta t$. The displacement is the length of the cord.

$$v = \sqrt{\frac{F_T}{\mu}} = \frac{\Delta x}{\Delta t} \ \rightarrow\ F_T = \mu \frac{L^2}{(\Delta t)^2} = \frac{m}{L}\frac{L^2}{(\Delta t)^2} = \frac{mL}{(\Delta t)^2} = \frac{(0.40\,\mathrm{kg})(7.8\,\mathrm{m})}{(0.85\,\mathrm{s})^2} = \boxed{4.3\,\mathrm{N}}$$

13. The speed of the waves on the cord can be found from Eq. 15-2, $v = \sqrt{\dfrac{F_T}{m/l}}$. The distance between the children is the wave speed times the elapsed time.

$$\Delta x = v\Delta t = \Delta t\sqrt{\frac{F_T}{m/\Delta x}} \quad \rightarrow \quad \Delta x = \left(\Delta t\right)^2\frac{F_T}{m} = \left(0.50\,\text{s}\right)^2\frac{35\,\text{N}}{0.50\,\text{kg}} = \boxed{18\,\text{m}}$$

19. (a) The power transmitted by the wave is assumed to be the same as the output of the oscillator. That power is given by Eq. 15-6. The wave speed is given by Eq. 15-2. Note that the mass per unit length can be expressed as the volume mass density times the cross-sectional area.

$$\bar{P} = 2\pi^2\rho S v f^2 A^2 = 2\pi^2\rho S\sqrt{\frac{F_T}{m/l}}\,vf^2 A^2 = 2\pi^2\rho S\sqrt{\frac{F_T}{\rho S}}\,f^2 A^2 = 2\pi^2 f^2 A^2\sqrt{S\rho F_T}$$

$$= 2\pi^2\left(60.0\,\text{Hz}\right)^2\left(0.0050\,\text{m}\right)^2\sqrt{\pi\left(5.0\times10^{-3}\,\text{m}\right)^2\left(7800\,\text{kg/m}^3\right)\left(7.5\,\text{N}\right)} = \boxed{0.38\,\text{W}}$$

(b) The frequency and amplitude are both squared in the equation. Thus as the power is constant, and the frequency doubles, the amplitude must be halved, and so be $\boxed{0.25\,\text{cm}}$.

25. The traveling wave is given by $D(x,t) = \left(0.026\,\text{m}\right)\sin\left[\left(45\,\text{m}^{-1}\right)x - \left(1570\,\text{s}^{-1}\right)t + 0.66\right]$.

(a) $v_x = \dfrac{\partial D(x,t)}{\partial t} = -\left(1570\,\text{s}^{-1}\right)\left(0.026\,\text{m}\right)\cos\left[\left(45\,\text{m}^{-1}\right)x - \left(1570\,\text{s}^{-1}\right)t + 0.66\right] \quad\rightarrow$

$\left(v_x\right)_{\text{max}} = \left(1570\,\text{s}^{-1}\right)\left(0.026\,\text{m}\right) = \boxed{41\,\text{m/s}}$

(b) $a_x = \dfrac{\partial^2 D(x,t)}{\partial t^2} = -\left(1570\,\text{s}^{-1}\right)^2\left(0.026\,\text{m}\right)\sin\left[\left(45\,\text{m}^{-1}\right)x - \left(1570\,\text{s}^{-1}\right)t + 0.66\right] \quad\rightarrow$

$\left(a_x\right)_{\text{max}} = \left(1570\,\text{s}^{-1}\right)^2\left(0.026\,\text{m}\right) = \boxed{6.4\times10^4\,\text{m/s}^2}$

(c) $v_x\left(1.00\,\text{m},2.50\,\text{s}\right) = -\left(1570\,\text{s}^{-1}\right)\left(0.026\,\text{m}\right)\cos\left[\left(45\,\text{m}^{-1}\right)\left(1.00\,\text{m}\right) - \left(1570\,\text{s}^{-1}\right)\left(2.50\,\text{s}\right) + 0.66\right] = \boxed{35\,\text{m/s}}$

$a_x\left(1.00\,\text{m},2.50\,\text{s}\right) = -\left(1570\,\text{s}^{-1}\right)^2\left(0.026\,\text{m}\right)\sin\left[\left(45\,\text{m}^{-1}\right)\left(1.00\,\text{m}\right) - \left(1570\,\text{s}^{-1}\right)\left(2.50\,\text{s}\right) + 0.66\right] = \boxed{3.2\times10^4\,\text{m/s}^2}$

31. To be a solution of the wave equation, the function must satisfy Eq. 15-16, $\dfrac{\partial^2 D}{\partial x^2} = \dfrac{1}{v^2}\dfrac{\partial^2 D}{\partial t^2}$.

$$D = A\sin kx\cos\omega t$$

$$\frac{\partial D}{\partial x} = kA\cos kx\cos\omega t \quad ; \quad \frac{\partial^2 D}{\partial x^2} = -k^2 A\sin kx\cos\omega t$$

$$\frac{\partial D}{\partial t} = -\omega A\sin kx\sin\omega t \quad ; \quad \frac{\partial^2 D}{\partial t^2} = -\omega^2 A\sin kx\cos\omega t$$

This gives $\dfrac{\partial^2 D}{\partial x^2} = \dfrac{k^2}{\omega^2}\dfrac{\partial^2 D}{\partial t^2}$, and since $v = \dfrac{\omega}{k}$ from Eq. 15-12, we have $\dfrac{\partial^2 D}{\partial x^2} = \dfrac{1}{v^2}\dfrac{\partial^2 D}{\partial t^2}$.
$\boxed{\text{Yes, the function is a solution.}}$

37. (a) For the wave in the lighter cord, $D(x,t) = \left(0.050\,\text{m}\right)\sin\left[\left(7.5\,\text{m}^{-1}\right)x - \left(12.0\,\text{s}^{-1}\right)t\right]$.

$$\lambda = \frac{2\pi}{k} = \frac{2\pi}{\left(7.5\,\text{m}^{-1}\right)} = \boxed{0.84\,\text{m}}$$

(b) The tension is found from the velocity, using Eq. 15-2.

$$v = \sqrt{\frac{F_T}{\mu}} \rightarrow F_T = \mu v^2 = \mu \frac{\omega^2}{k^2} = (0.10\,\text{kg/m})\frac{(12.0\,\text{s}^{-1})^2}{(7.5\,\text{m}^{-1})^2} = \boxed{0.26\,\text{N}}$$

(c) The tension and the frequency do not change from one section to the other.

$$F_{T1} = F_{T2} \rightarrow = \mu_1 \frac{\omega_1^2}{k_1^2} = \mu_2 \frac{\omega_2^2}{k_2^2} \rightarrow \lambda_2 = \lambda_1 \sqrt{\frac{\mu_1}{\mu_2}} = \frac{2\pi}{k_1}\sqrt{\frac{\mu_1}{\mu_2}} = \frac{2\pi}{(7.5\,\text{m}^{-1})}\sqrt{0.5} = \boxed{0.59\,\text{m}}$$

43. The fundamental frequency of the full string is given by $f_{\text{unfingered}} = v/2l = 441\,\text{Hz}$. If the length is reduced to 2/3 of its current value, and the velocity of waves on the string is not changed, then the new frequency will be as follows.

$$f_{\text{fingered}} = \frac{v}{2\left(\frac{2}{3}l\right)} = \frac{3}{2}\frac{v}{2l} = \left(\frac{3}{2}\right)f_{\text{unfingered}} = \left(\frac{3}{2}\right)(441\,\text{Hz}) = \boxed{662\,\text{Hz}}$$

49. Since $f_n = nf_1$, two successive overtones differ by the fundamental frequency, as shown below.

$$\Delta f = f_{n+1} - f_n = (n+1)f_1 - nf_1 = f_1 = 320\,\text{Hz} - 240\,\text{Hz} = \boxed{80\,\text{Hz}}$$

55. (a) The given wave is $D_1 = 4.2\sin(0.84x - 47t + 2.1)$. To produce a standing wave, we simply need to add a wave of the same characteristics but traveling in the opposite direction. This is the appropriate wave.

$$\boxed{D_2 = 4.2\sin(0.84x + 47t + 2.1)}$$

(b) The standing wave is the sum of the two component waves. We use the trigonometric identity that $\sin\theta_1 + \sin\theta_2 = 2\sin\frac{1}{2}(\theta_1 + \theta_2)\cos\frac{1}{2}(\theta_1 - \theta_2)$.

$$D = D_1 + D_2 = 4.2\sin(0.84x - 47t + 2.1) + 4.2\sin(0.84x + 47t + 2.1)$$
$$= 4.2(2)\left\{\sin\frac{1}{2}\left[(0.84x - 47t + 2.1) + (0.84x + 47t + 2.1)\right]\right\}$$
$$\left\{\cos\frac{1}{2}\left[(0.84x - 47t + 2.1) - (0.84x + 47t + 2.1)\right]\right\}$$
$$= 8.4\sin(0.84x + 2.1)\cos(-47t) = \boxed{8.4\sin(0.84x + 2.1)\cos(47t)}$$

We note that the origin is NOT a node.

61. Any harmonic with a node directly above the pickup will NOT be "picked up" by the pickup. The pickup location is exactly 1/4 of the string length from the end of the string, so a standing wave with a frequency corresponding to 4 (or 8 or 12 etc.) loops will not excite the pickup. So $\boxed{n = 4, 8, \text{and } 12}$ will not excite the pickup.

67. The angle of refraction can be found from the law of refraction, Eq. 15-19. The relative velocities can be found from the relationship given in the problem.

$$\frac{\sin\theta_2}{\sin\theta_1} = \frac{v_2}{v_1} = \frac{331 + 0.60T_2}{331 + 0.60T_1} \rightarrow \sin\theta_2 = \sin33°\frac{331 + 0.60(-15)}{331 + 0.60(25)} = \sin33°\frac{322}{346} = 0.5069$$
$$\theta_2 = \sin^{-1}0.5069 = \boxed{30°}\ (2\text{ sig. fig.})$$

73. The speed of a longitudinal wave in a solid is given by Eq. 15-3, $v = \sqrt{E/\rho}$. Let the density of the less dense material be ρ_1, and the density of the more dense material be ρ_2. The less dense material will have the higher speed, since the speed is inversely proportional to the square root of the density.

$$\frac{v_1}{v_2} = \frac{\sqrt{E/\rho_1}}{\sqrt{E/\rho_2}} = \sqrt{\frac{\rho_2}{\rho_1}} = \sqrt{2.5} \approx \boxed{1.6}$$

79. (a)

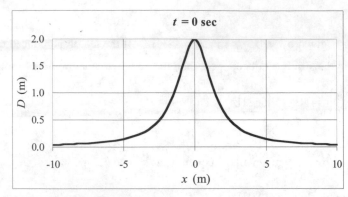

(b) $$D = \frac{4.0\,\text{m}^3}{\left[x - \left(2.4\,\text{m/s}\right)t\right]^2 + 2.0\,\text{m}^2}$$

(c)

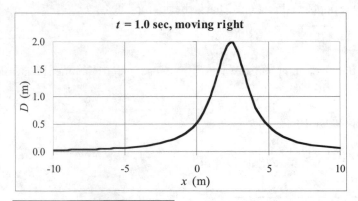

(d) $$D = \frac{4.0\,\text{m}^3}{\left[x + \left(2.4\,\text{m/s}\right)t\right]^2 + 2.0\,\text{m}^2}$$

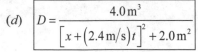

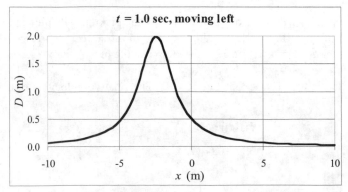

85. For a resonant condition, the free end of the string will be an antinode, and the fixed end of the string will be a node. The minimum distance from a node to an antinode is $\lambda/4$. Other wave patterns that fit the boundary conditions of a node at one end and an antinode at the other end include $3\lambda/4$, $5\lambda/4$, ... See the diagrams. The general

    relationship is $l = (2n - 1)\lambda/4$, $n = 1, 2, 3, \ldots$. Solving for the wavelength gives $\boxed{\lambda = \dfrac{4l}{2n-1} , n = 1, 2, 3, \cdots}$.

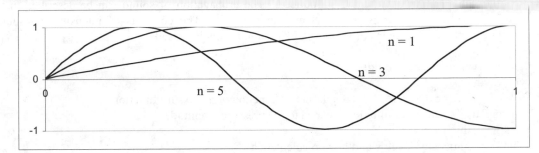

91. Because the radiation is uniform, the same energy must pass through every spherical surface, which has the surface area $4\pi r^2$. Thus the intensity must decrease as $1/r^2$. Since the intensity is proportional to the square of the amplitude, the amplitude will decrease as $1/r$. The radial motion will be sinusoidal, and so we have

$$D = \left(\frac{A}{r}\right)\sin\left(kr - \omega t\right).$$

Chapter 16: Sound

Chapter Overview and Objectives

In this chapter, the phenomenon of sound is introduced. The propagation of sound can be described in terms of the displacement of molecules or in terms of the change in pressure. The concepts of intensity and sound level are introduced and the vibration frequencies of strings and air columns are discussed.

After completing this chapter you should:
- Know what sound is.
- Know that sound can be described by a displacement function or a pressure function.
- Know how the displacement amplitude is related to the pressure amplitude.
- Know what intensity is.
- Know what sound intensity level is and how to calculate it.
- Know what the inverse square law is.
- Know how to determine vibrational frequencies of strings and air columns.
- Know how to determine the beat frequency of two sound sources.
- Know how to calculate frequency changes that are the result of the Doppler effect.
- Know what a shock wave is, how it is produced, and how to calculate the apex angle of a shock wave.

Summary of Equations

Approximate speed of sound in air: $v = (331 + 0.60T)\,\text{m/s}$ (Section 16-1)

Expression for harmonic displacement wave: $D(x,t) = A\sin(kx - \omega t)$ (Section 16-2)

Expression for associated harmonic pressure wave: $\Delta P(x,t) = -(BAk)\cos(kx - \omega t)$ (Section 16-2)

Relationships between pressure and displacement amplitudes:

$$\Delta P_M = BAk$$ (Section 16-2)

$$\Delta P_M = 2\pi \rho v A f$$ (Section 16-2)

Definition of sound intensity level: $\beta = 10\log_{10}\dfrac{I}{I_0}$ (Section 16-3)

Inverse square law: $I \propto \dfrac{1}{r^2}$ (Section 16-3)

Vibration frequencies of a string or air column with identical end conditions:

$$f_n = \frac{nv}{2\ell} \qquad n = 1, 2, 3, 4, \ldots$$ (Section 16-4)

Vibration frequencies of an air column with one open end and one closed end:

$$f_n = \frac{nv}{4\ell} \qquad n = 1,3,5,\ldots \qquad \text{(Section 16-4)}$$

Beat frequency of two sources:
$$f_b = |f_1 - f_2| \qquad \text{(Section 16-6)}$$

Doppler shifted frequency:
$$f' = \left(\frac{v_{snd} \pm v_{obs}}{v_{snd} \mp v_{source}}\right) f \qquad \text{(Section 16-7)}$$

Apex angle of shock wave:
$$\sin\theta = \frac{v_{snd}}{v_{obj}} \qquad \text{(Section 16-8)}$$

Chapter Summary

Section 16-1. Characteristics of Sound

Sound is a mechanical wave that travels through materials. Its velocity in air is a function of temperature:

$$v_{sound} = (331 + 0.60T)\,\text{m/s}$$

where T is the temperature of the air in units of degrees Celsius.

Section 16-2. Mathematical Representation of Longitudinal Waves

Sound needs a medium to propagate. The propagation of sound through a medium can be described mathematically by a function that describes the displacement of the atoms or molecules of the medium as a function of position and time. A harmonic sound wave, with wave number k and angular frequency ω, has a displacement D that is a function of position x and time t:

$$D(x,t) = A\sin(kx - \omega t)$$

where A is the amplitude of the displacement. Sound waves in air are longitudinal waves. The displacement is parallel to the direction of propagation.

Rather than describing the sound wave in terms of the displacement of the molecules of the medium, we can also describe the propagation of sound waves in terms of the change in pressure in the medium. The change in pressure ΔP corresponding to a displacement $D(x, t)$ is given by

$$\Delta P(x,t) = -B\frac{\partial D(x,t)}{\partial x} = -(BAk)\cos(kx - \omega t)$$

where B is the bulk modulus of the medium. We see that the pressure amplitude ΔP_M is related to the displacement amplitude A by

$$\Delta P_M = BAk$$

We can rewrite this expression in terms of the wave speed v and the wave frequency f and obtain the following relation:

$$\Delta P_M = 2\pi\rho vAf$$

where ρ is the density of the medium.

Example 16-2-A. Displacement amplitude. The pressure amplitude of a loud sound wave is 1.00 N/m^2. If this sound wave has a frequency of 1000 Hz, what is the displacement amplitude of this sound wave? Assume that the speed of sound in air is 340 m/s.

Approach: To solve this problem we use the relation between pressure amplitude and displacement amplitude. We know the frequency of the wave, the propagation velocity, and the pressure amplitude, and can determine the displacement amplitude.

Solution: The relationship between pressure and displacement amplitudes can be used to determine the displacement amplitude:

$$\Delta P_M = 2\pi \rho v A f \quad \Rightarrow \quad A = \frac{\Delta P_M}{2\pi \rho v f} = \frac{(1.00)}{2\pi(1.29)(340)(1000)} = 3.6 \times 10^{-7} \text{ m}$$

Section 16-3. Intensity of Sound; Decibels

As with all mechanical waves, sound waves carry energy. The energy picked up by a receiver is proportional to the energy content or **intensity** of the wave and the area of the receiver. The intensity I of the wave is the energy per unit area per unit time:

$$I = \frac{E}{At} = \frac{P}{A}$$

The intensities to which the human ear is sensitive range from about 10^{-12} W/m^2 to about 1 W/m^2. Due to the large range of intensities, the **sound level** β is usually specified in terms of the logarithm of the ratio of the intensity I and the threshold intensity I_0, which is taken to be 10^{-12} W/m^2 and is the minimum intensity that is audible to a good ear. Although the ratio of intensities is unitless, it is common to identify the sound level specified in this manner by using the unit decibels (dB):

$$\beta = 10 \log_{10} \frac{I}{I_0}$$

Sound waves that are free to expand in all directions from their source, have an intensity I that is proportional to the inverse square of the distance r from the source:

$$I \propto \frac{1}{r^2}$$

This is known as the **inverse square law**.

Example 16-3-A. Checking the inverse square law. At a distance of $r_1 = 3.8$ m from a source of sound, you measure the sound intensity level to be $\beta_1 = 79$ dB. At a distance of $r_2 = 42.6$ m from the same source, you measure the sound intensity level to be $\beta_2 = 62$ dB. Does the sound field behave according to the inverse square law?

Approach: If the measured intensities behave according to the inverse square law, the ratios of the intensities at the two distances should be related to the ratio of distances in the following manner:

$$\frac{I_2}{I_1} = \frac{r_1^2}{r_2^2}$$

By calculating the ratio of the intensities at the two distances, we can determine whether or not the observed intensities are consistent with the inverse square law.

Solution: We can use the definition of sound level to determine the intensity:

$$\beta = 10 \log_{10} \frac{I}{I_0} \quad \Rightarrow \quad I = I_0 10^{\frac{\beta}{10}}$$

The ratio of I_2 and I_1 is equal to:

$$\frac{I_2}{I_1} = \frac{I_0 10^{\frac{\beta_2}{10}}}{I_0 10^{\frac{\beta_1}{10}}} = 10^{\frac{\beta_2-\beta_1}{10}} = 10^{\frac{62-79}{10}} = 0.020$$

The ratio of the inverse distances squared is

$$\frac{r_1^2}{r_2^2} = \frac{(3.8)^2}{(42.6)^2} = 0.0080$$

We conclude that

$$\frac{I_2}{I_1} = 0.020 \neq \frac{r_1^2}{r_2^2} = 0.0080$$

It is clear that the sound field is not obeying the inverse square law. The intensity falls off slower than predicted by the inverse square law. This can indicate that the intensity recorded at position 2 is increased due to reflection of sound waves from the surrounding surfaces or that the source produces directional sound.

Section 16-4. Sources of Sound: Vibrating Strings and Air Columns

A string vibrates with a frequency that depends on the mass m, the length ℓ, and the tension F_T. The possible frequencies of vibration are

$$f_n = \frac{nv}{2\ell} \qquad n = 1, 2, 3, 4, \ldots$$

where v is the propagation speed of waves traveling along the string. The frequencies can be expressed in terms of the mass and tension of the string:

$$f_n = \frac{n}{2\ell}\sqrt{\frac{F_T}{m/\ell}} \qquad n = 1, 2, 3, 4, \ldots$$

The frequency f_1 is called the **fundamental frequency** of the string. The other frequencies are called the **overtones**. For a string, the overtones are harmonics of the fundamental frequency. The nth harmonic of the fundamental frequency has a frequency nf_1.

A column of air in a tube with both ends open to the atmosphere supports frequencies of vibration identical to those given for the string:

$$f_n = \frac{nv}{2\ell} \qquad n = 1, 2, 3, 4, \ldots$$

A column of air in a tube with one end closed and the other end open to the atmosphere supports frequencies of vibration given by

$$f_n = \frac{nv}{4\ell} \qquad n = 1, 3, 5, \ldots$$

where v is the speed of sound in the air and ℓ is the length of the air column.

Example 16-4-A. Frequencies on a string. What should the length of a string be that has a 3rd overtone frequency $f = 940$ Hz and a tension $F_T = 240$ N? A piece of the string of length $L = 2.8$ m has a mass $M = 12$ g.

Approach: In order to solve this problem, we have to recognize that M/L in the expression for f_n is the density of the string, which is a constant, independent of the length of the string. Based on the information provided, we can now calculate the length ℓ required to generate the 3rd overtone. Be careful in determining the value of n to use! The fundamental frequency has $n = 1$. The first overtone frequency has $n = 2$. This implies that the third overtone frequency has $n = 4$. We thus need to consider f_4.

Solution: We know that the frequencies of the string are given by the expression

$$f_n = \frac{n}{2\ell}\sqrt{\frac{F_T}{M/L}} \qquad n = 1, 2, 3, 4, \ldots$$

The mass per unit length, M/L, is the same, regardless of the length of the string. We can calculate M/L from the information provided in the problem:

$$M/L = \frac{0.012}{2.8} = 4.29 \times 10^{-3} \text{ kg/m}$$

We can now determine the length ℓ required to generate the 3rd overtone:

$$\ell = \frac{n}{2f_n}\sqrt{\frac{F_T}{M/L}} = \frac{4}{2(940)}\sqrt{\frac{240}{4.29 \times 10^{-3}}} = 0.50 \text{ m}$$

Section 16-5. Quality of Sound, and Noise; Superposition

Most sound waves are not harmonic waves of a single frequency. However, many sound waves are approximately periodic for a finite length of time. Periodic waves can be constructed as a sum of harmonic waves that have frequencies that are multiples of the fundamental frequency of the wave. The fundamental frequency is the frequency that corresponds to the period of the wave. The relative amplitudes of the different components and their relative phases determine the waveform of the resulting wave.

Section 16-6. Interference of Sound Waves; Beats

When waves are traveling through a medium from different sources, the principle of superposition tells us that the waves can interfere with each other. The displacement of the medium at a given time and place is the sum of displacements associated with the individual waves. We will consider two important special cases of interference:

- Two sources located at different positions emit identical sound waves. Constructive interference will occur at those positions whose distance from the first source differs from the distance to the second source by a whole number of wavelengths. Destructive interference will occur at positions whose distance from one source differs from the distance to the other source by an odd number of half wavelengths.
- Two sources located at the same position emit sound waves that have the same amplitude but slightly different frequencies f_1 and f_2. The result of summing the displacements of two such waves can be described as a wave with a frequency equal to the average of the two frequencies, but an amplitude that varies with time. The frequency of the amplitude variation is called the **beat frequency**, f_b. The beat frequency is related to the frequencies of the sources by

$$f_b = |f_1 - f_2|$$

Example 16-6-A. Beat frequency. Two sound waves are reaching an observer. The observer hears a beat frequency f_b = 5.8 Hz. The observer knows that one of the frequencies is f_1 = 984.6 Hz. What is the other frequency?

Approach: The beat frequency is equal to the magnitude of the difference between the two source frequencies. The problem specifies the beat frequency and one of the source frequencies. The second source frequency will differ from the first source frequency by the beat frequency. Note that there will be two possible solutions for the second source frequency.

Solution: The beat frequency is related to the two frequencies of the sound waves by

$$f_b = |f_1 - f_2|$$

This implies either

$$f_b = f_1 - f_2 \qquad \Rightarrow \qquad f_2 = f_1 - f_b = 984.6 \text{ Hz} - 5.8 \text{ Hz} = 978.8 \text{ Hz}$$

or

$$f_b = f_2 - f_1 \quad \Rightarrow \quad f_2 = f_1 + f_b = 984.6\,\text{Hz} + 5.8\,\text{Hz} = 990.4\,\text{Hz}$$

Section 16-7. Doppler Effect

When a source and observer of sound waves are in relative motion to each other, the frequency of sound waves received by the observer is different from the frequency of sound waves emitted by the source. The relationship between these frequencies is given by

$$f' = \left(\frac{v_{snd} \pm v_{obs}}{v_{snd} \mp v_{source}} \right) f$$

where v_{snd} is the speed of the sound waves in the medium, v_{obs} is the speed of the observer relative to the medium, and v_{source} is the speed of the source relative to the medium.

Example 16-7-A. Doppler shift of a car horn. A car is traveling north with a speed of $v_{car} = 24$ m/s. The car is catching up to a truck that is also traveling north with a speed of $v_{truck} = 18$ m/s. A wind is blowing toward the south with a speed of $v_{wind} = 5$ m/s. The car driver honks the horn and emits a sound with a frequency $f = 400$ Hz. What is the frequency of sound observed by the truck driver?

Approach: To calculate the frequency observed by the truck driver, we apply the Doppler shift equation. In this problem we have to recognize that the medium is not at rest with respect to the surface of the Earth.

Solution: All of the velocities in the Doppler shift expression are relative to the medium. It is easy to see that the speed of the car is 29 m/s toward the north, relative to the air, and the velocity of the truck is 23 m/s toward the north, relative to the air. We also need to determine the correct signs to use in the equation. The car, which is the source of the sound, is moving toward the position of the truck. This means that a negative sign should be used in the denominator of the expression. The truck, the observer, is moving away from the position of the source. This means that the negative sign should be used in the numerator of the expression also. Solving for the frequency of the observer:

$$f' = \left(\frac{v_{snd} \pm v_{truck}}{v_{snd} \mp v_{car}} \right) f = \left(\frac{(340) - (23)}{(340) - (29)} \right)(400) = 408\,\text{Hz}$$

Note: even though the distance between the source and the observer is decreasing with increasing time, the observer is still considered to be moving away from the source position.

Section 16-8. Shock Waves and the Sonic Boom

A **shock wave** is generated whenever a sound source is moving relative to the medium at a speed greater than the speed of sound in that medium. The shock wave moves outward from the source with a conical wavefront. The apex angle, θ, of the conical wavefront depends on the ratio of the wave speed to the source speed as

$$\sin\theta = \frac{v_{wave}}{v_{source}}$$

Section 16-9. Applications: Sonar, Ultrasound, and Medical Imaging

Sound waves can be used as an exploratory and diagnostic tool. **Sonar** and **ultrasound imaging** are two of the applications of sound waves of this sort.

Practice Quiz

1. What happens to the wavelength of a particular frequency of sound in air as the temperature rises?
    a) Wavelength remains the same.
    b) Wavelength becomes longer.
    c) Wavelength becomes shorter.
    d) Wavelength is unpredictable.

2. Which perceptual judgment of sound is primarily determined by the intensity of a sound wave?
    a) Pitch
    b) Loudness
    c) Timbre
    d) Tonal quality

3. In which of the following materials is the speed of sound the greatest?
    a) Air
    b) Helium
    c) Water
    d) Iron

4. What type of wave is a sound wave?
    a) A longitudinal wave.
    b) A transverse wave.
    c) A shear wave.
    d) A torsional wave.

5. The string of a guitar has a frequency that is 2% too low. What can be done to correct the frequency of the string?
    a) Increase the tension in the string by 2%.
    b) Increase the tension in the string by 4%.
    c) Decrease the tension in the string by 2%.
    d) Decrease the tension in the string by 1%.

6. Sound B has a sound level that is 10 dB higher than the sound level of sound A. The ratio of the intensity of sound B to the intensity of sound A is 10. Sound C has a sound level that is 20 dB higher than the sound level of sound A. What is the ratio of the intensity of sound C to the intensity of sound A?
    a) 20
    b) 40
    c) 100
    d) 200

7. A column of air that is open at one end and closed at the other has a fundamental resonant frequency of 300 Hz. What is the frequency of the first overtone?
    a) 150 Hz
    b) 300 Hz
    c) 600 Hz
    d) 900 Hz

8. You know that a source that is stationary relative to you emits a frequency of 400 Hz. When it is moving, you hear a frequency of 404 Hz. What can you say about the motion of the source?
    a) The source is not moving.
    b) The source is moving directly toward you.
    c) The source is moving with a component of its velocity toward you.
    d) The source is moving with a component of its velocity away from you.

9. Velocities are relative to a reference frame. What must be the reference frame for the velocities in the Doppler shift equation?
   a) Any frame of reference.
   b) The frame of reference at rest relative to the source of the sound.
   c) The frame of reference at rest relative to the observer of the sound.
   d) The frame of reference at rest relative to the medium carrying the sound.

10. You detect the following frequencies being emitted from an air column: 100 Hz, 300 Hz, 500 Hz, 700 Hz. What can you conclude about the air column?
   a) It is open at one end and closed at the other.
   b) It is closed at both ends.
   c) It is open at both ends.
   d) You cannot exclude any of the options in a, b, and c.

11. Determine the lowest three frequencies of vibration of a string that is 69 cm long, has a mass of 12.5 g, and has a tension of 232 N.

12. The sound field of a source follows the inverse square law. At a distance of 12.4 m from the source, the sound intensity level is 98 dB. What is the sound intensity level at a distance 29.6 m from the source?

13. A person is driving a car toward a large building. When the person honks the horn of the car, which emits a 466 Hz tone, the driver hears a 6 Hz beat frequency. What is the speed of the car toward the building? Assume the speed of sound in air is 341 m/s.

14. The apex angle of a shock wave produced by a source moving through the air is 35.6°. Assume the speed of sound in air is 340 m/s. What is the speed of the sound source relative to the air?

15. A trombone player tunes his instrument when the trombone is at 20°C. By what percentage will the trombone be out of tune when it and the air inside warm up to 32°C?

Responses to Select End-of-Chapter Questions

1. Sound exhibits diffraction, refraction, and interference effects that are characteristic of waves. Sound also requires a medium, a characteristic of mechanical waves.

7. The speed of sound in a medium is equal to $v = \sqrt{\dfrac{B}{\rho}}$, where B is the bulk modulus and ρ is the density of the medium. The bulk moduli of air and hydrogen are very nearly the same. The density of hydrogen is less than the density of air. The reduced density is the main reason why sound travels faster in hydrogen than in air.

13. Standing waves are generated by a wave and its reflection. The two waves have a constant phase relationship with each other. The interference depends only on where you are along the string, on your position in space. Beats are generated by two waves whose frequencies are close but not equal. The two waves have a varying phase relationship, and the interference varies with time rather than position.

19. The child will hear the highest frequency at position C, where her speed toward the whistle is the greatest.

Solutions to Select End-of-Chapter Problems

1. The round trip time for sound is 2.0 seconds, so the time for sound to travel the length of the lake is 1.0 second. Use the time and the speed of sound to determine the length of the lake.

$$d = vt = (343\,\text{m/s})(1.0\text{ s}) = 343\text{ m} \approx \boxed{340\,\text{m}}$$

7. The total time T is the time for the stone to fall (t_{down}) plus the time for the sound to come back to the top of the cliff (t_{up}): $T = t_{up} + t_{down}$. Use constant acceleration relationships for an object dropped from rest that falls a distance h in order to find t_{down}, with down as the positive direction. Use the constant speed of sound to find t_{up} for the sound to travel a distance h.

$$\text{down: } y = y_0 + v_0 t_{down} + \tfrac{1}{2} a t_{down}^2 \rightarrow h = \tfrac{1}{2} g t_{down}^2 \qquad \text{up: } h = v_{snd} t_{up} \rightarrow t_{up} = \frac{h}{v_{snd}}$$

$$h = \tfrac{1}{2} g t_{down}^2 = \tfrac{1}{2} g \left(T - t_{up} \right)^2 = \tfrac{1}{2} g \left(T - \frac{h}{v_{snd}} \right)^2 \rightarrow h^2 - 2 v_{snd} \left(\frac{v_{snd}}{g} + T \right) h + T^2 v_{snd}^2 = 0$$

This is a quadratic equation for the height. This can be solved with the quadratic formula, but be sure to keep several significant digits in the calculations.

$$h^2 - 2 \left(343 \, \text{m/s} \right) \left(\frac{343 \, \text{m/s}}{9.80 \, \text{m/s}^2} + 3.0 \, \text{s} \right) h + \left(3.0 \, \text{s} \right)^2 \left(343 \, \text{m/s} \right)^2 = 0 \rightarrow$$

$$h^2 - \left(26068 \, \text{m} \right) h + 1.0588 \times 10^6 \, \text{m}^2 = 0 \rightarrow h = 26028 \, \text{m}, \, 41 \, \text{m}$$

The larger root is impossible since it takes more than 3.0 sec for the rock to fall that distance, so the correct result is $H = \boxed{41 \, \text{m}}$.

13. The pressure wave is $\Delta P = \left(0.0035 \, \text{Pa} \right) \sin \left[\left(0.38\pi \, \text{m}^{-1} \right) x - \left(1350\pi \, \text{s}^{-1} \right) t \right]$.

(a) $\lambda = \dfrac{2\pi}{k} = \dfrac{2\pi}{0.38\pi \, \text{m}^{-1}} = \boxed{5.3 \, \text{m}}$

(b) $f = \dfrac{\omega}{2\pi} = \dfrac{1350\pi \, \text{s}^{-1}}{2\pi} = \boxed{675 \, \text{Hz}}$

(c) $v = \dfrac{\omega}{k} = \dfrac{1350\pi \, \text{s}^{-1}}{0.38\pi \, \text{m}^{-1}} = 3553 \, \text{m/s} \approx \boxed{3600 \, \text{m/s}}$

(d) Use Eq. 16-5 to find the displacement amplitude.

$$\Delta P_M = 2\pi \rho v A f \rightarrow$$

$$A = \frac{\Delta P_M}{2\pi \rho v f} = \frac{\left(0.0035 \, \text{Pa} \right)}{2\pi \left(2300 \, \text{kg/m}^3 \right) \left(3553 \, \text{m/s} \right) \left(675 \, \text{Hz} \right)} = \boxed{1.0 \times 10^{-13} \, \text{m}}$$

19. The intensity can be found from the decibel value.

$$\beta = 10 \log \frac{I}{I_0} \rightarrow I = 10^{\beta/10} I_0 = 10^{12} \left(10^{-12} \, \text{W/m}^2 \right) = 1.0 \, \text{W/m}^2$$

Consider a square perpendicular to the direction of travel of the sound wave. The intensity is the energy transported by the wave across a unit area perpendicular to the direction of travel, per unit time. So $I = \Delta E / (S\Delta t)$, where S is the area of the square. Since the energy is "moving" with the wave, the "speed" of the energy is v, the wave speed. In a time Δt, a volume equal to $\Delta V = Sv\Delta t$ would contain all of the energy that had been transported across the area S. Combine these relationships to find the energy in the volume.

$$I = \frac{\Delta E}{S\Delta t} \rightarrow \Delta E = IS\Delta t = \frac{I\Delta V}{v} = \frac{\left(1.0 \, \text{W/m}^2 \right) \left(0.010 \, \text{m} \right)^3}{343 \, \text{m/s}} = \boxed{2.9 \times 10^{-9} \, \text{J}}$$

25. The intensity of the sound is defined to be the power per unit area. We assume that the sound spreads out spherically from the loudspeaker.

(a) $I_{250} = \dfrac{250\,\text{W}}{4\pi(3.5\,\text{m})^2} = 1.624\,\text{W/m}^2 \quad \beta_{250} = 10\log\dfrac{I_{250}}{I_0} = 10\log\dfrac{1.624\,\text{W/m}^2}{1.0\times10^{-12}\,\text{W/m}^2} = \boxed{122\,\text{dB}}$

$I_{40} = \dfrac{40\,\text{W}}{4\pi(3.5\,\text{m})^2} = 0.2598\,\text{W/m}^2 \quad \beta_{40} = 10\log\dfrac{I_{40}}{I_0} = 10\log\dfrac{0.2598\,\text{W/m}^2}{1.0\times10^{-12}\,\text{W/m}^2} = \boxed{114\,\text{dB}}$

(b) According to the textbook, for a sound to be perceived as twice as loud as another means that the intensities need to differ by a factor of 10. That is not the case here – they differ only by a factor of 1.624/0.2598 = 6.

$\boxed{\text{The expensive amp will not sound twice as loud as the cheaper one.}}$

31. (a) We find the intensity of the sound from the decibel value, and then calculate the displacement amplitude from Eq. 15-7.

$\beta = 10\log\dfrac{I}{I_0} \;\rightarrow\; I = 10^{\beta/10}I_0 = 10^{12}\left(10^{-12}\,\text{W/m}^2\right) = 1.0\,\text{W/m}^2$

$I = 2\rho v \pi^2 f^2 A^2 \;\rightarrow\;$

$A = \dfrac{1}{\pi f}\sqrt{\dfrac{I}{2\rho v}} = \dfrac{1}{\pi(330\,\text{Hz})}\sqrt{\dfrac{1.0\,\text{W/m}^2}{2(1.29\,\text{kg/m}^3)(343\,\text{m/s})}} = \boxed{3.2\times10^{-5}\,\text{m}}$

(b) The pressure amplitude can be found from Eq. 16-7.

$I = \dfrac{(\Delta P_M)^2}{2v\rho} \;\rightarrow\;$

$\Delta P_M = \sqrt{2v\rho I} = \sqrt{2(343\,\text{m/s})(1.29\,\text{kg/m}^3)(1.0\,\text{W/m}^2)} = \boxed{30\,\text{Pa}\,(2\,\text{sig. fig.})}$

37. For a pipe open at both ends, the fundamental frequency is given by $f_1 = v/(2L)$, and so the length for a given fundamental frequency is $L = v/(2f_1)$.

$L_{20\,\text{Hz}} = \dfrac{343\,\text{m/s}}{2(20\,\text{Hz})} = \boxed{8.6\,\text{m}} \qquad L_{20\,\text{kHz}} = \dfrac{343\,\text{m/s}}{2(20{,}000\,\text{Hz})} = \boxed{8.6\times10^{-3}\,\text{m}}$

43. For a tube open at both ends, all harmonics are allowed, with $f_n = nf_1$. Thus consecutive harmonics differ by the fundamental frequency. The four consecutive harmonics give the following values for the fundamental frequency.

$f_1 = 523\,\text{Hz} - 392\,\text{Hz} = 131\,\text{Hz},\;\; 659\,\text{Hz} - 523\,\text{Hz} = 136\,\text{Hz},\;\; 784\,\text{Hz} - 659\,\text{Hz} = 125\,\text{Hz}$

The average of these is $f_1 = \frac{1}{3}(131\,\text{Hz} + 136\,\text{Hz} + 125\,\text{Hz}) \approx 131\,\text{Hz}$. We use that for the fundamental frequency.

(a) $f_1 = \dfrac{v}{2l} \;\rightarrow\; l = \dfrac{v}{2f_1} = \dfrac{343\,\text{m/s}}{2(131\,\text{Hz})} = \boxed{1.31\,\text{m}}$

Note that the bugle is coiled like a trumpet so that the full length fits in a smaller distance.

(b) $f_n = nf_1 \;\rightarrow\; n_{G4} = \dfrac{f_{G4}}{f_1} = \dfrac{392\,\text{Hz}}{131\,\text{Hz}} = 2.99\;;\; n_{C5} = \dfrac{f_{C5}}{f_1} = \dfrac{523\,\text{Hz}}{131\,\text{Hz}} = 3.99\;;$

$n_{E5} = \dfrac{f_{E5}}{f_1} = \dfrac{659\,\text{Hz}}{131\,\text{Hz}} = 5.03\;;\; n_{G5} = \dfrac{f_{G5}}{f_1} = \dfrac{784\,\text{Hz}}{131\,\text{Hz}} = 5.98$

The harmonics are $\boxed{3,\,4,\,5,\,\text{and}\,6}$.

49. A tube closed at both ends will have standing waves with displacement nodes at each end, and so has the same harmonic structure as a string that is fastened at both ends. Thus the wavelength of the fundamental frequency is twice the length of the hallway, $\lambda_1 = 2l = 16.0$ m.

$$f_1 = \frac{v}{\lambda_1} = \frac{343\,\text{m/s}}{16.0\,\text{m}} = \boxed{21.4\,\text{Hz}} \quad ; \quad f_2 = 2f_1 = \boxed{42.8\,\text{Hz}}$$

55. Since there are 4 beats/s when sounded with the 350 Hz tuning fork, the guitar string must have a frequency of either 346 Hz or 354 Hz. Since there are 9 beats/s when sounded with the 355 Hz tuning fork, the guitar string must have a frequency of either 346 Hz or 364 Hz. The common value is $\boxed{346\,\text{Hz}}$.

61. (*a*) Observer moving towards stationary source.

$$f' = \left(1 + \frac{v_{obs}}{v_{snd}}\right) f = \left(1 + \frac{30.0\,\text{m/s}}{343\,\text{m/s}}\right)(1350\,\text{Hz}) = \boxed{1470\,\text{Hz}}$$

(*b*) Observer moving away from stationary source.

$$f' = \left(1 - \frac{v_{obs}}{v_{snd}}\right) f = \left(1 - \frac{30.0\,\text{m/s}}{343\,\text{m/s}}\right)(1350\,\text{Hz}) = \boxed{1230\,\text{Hz}}$$

67. We assume that the comparison is to be made from the frame of reference of the stationary tuba. The stationary observers would observe a frequency from the moving tuba of

$$f_{obs} = \frac{f_{source}}{\left(1 - \dfrac{v_{source}}{v_{snd}}\right)} = \frac{75\,\text{Hz}}{\left(1 - \dfrac{12.0\,\text{m/s}}{343\,\text{m/s}}\right)} = 78\,\text{Hz} \qquad f_{beat} = 78\,\text{Hz} - 75\,\text{Hz} = \boxed{3\,\text{Hz}}\,.$$

73. (*a*) The Mach number is the ratio of the object's speed to the speed of sound.

$$M = \frac{v_{obs}}{v_{sound}} = \frac{\left(1.5 \times 10^4\,\text{km/hr}\right)\left(\dfrac{1\,\text{m/s}}{3.6\,\text{km/hr}}\right)}{45\,\text{m/s}} = 92.59 \approx \boxed{93}$$

(*b*) Use Eq. 12.5 to find the angle.

$$\theta = \sin^{-1}\frac{v_{snd}}{v_{obj}} = \sin^{-1}\frac{1}{M} = \sin^{-1}\frac{1}{92.59} = \boxed{0.62°}$$

79. Assume that only the fundamental frequency is heard. The fundamental frequency of an open pipe is given by $f = v/(2L)$.

(*a*) $\quad f_{3.0} = \dfrac{v}{2L} = \dfrac{343\,\text{m/s}}{2(3.0\,\text{m})} = \boxed{57\,\text{Hz}} \qquad f_{2.5} = \dfrac{v}{2L} = \dfrac{343\,\text{m/s}}{2(2.5\,\text{m})} = \boxed{69\,\text{Hz}}$

$\quad f_{2.0} = \dfrac{v}{2L} = \dfrac{343\,\text{m/s}}{2(2.0\,\text{m})} = \boxed{86\,\text{Hz}} \qquad f_{1.5} = \dfrac{v}{2L} = \dfrac{343\,\text{m/s}}{2(1.5\,\text{m})} = 114.3\,\text{Hz} \approx \boxed{110\,\text{Hz}}$

$\quad f_{1.0} = \dfrac{v}{2L} = \dfrac{343\,\text{m/s}}{2(1.0\,\text{m})} = 171.5\,\text{Hz} \approx \boxed{170\,\text{Hz}}$

(*b*) On a noisy day, there are a large number of component frequencies to the sounds that are being made – more people walking, more people talking, etc. Thus it is more likely that the frequencies listed above will be a component of the overall sound, and then the resonance will be more prominent to the hearer. If the day is quiet, there might be very little sound at the desired frequencies, and then the tubes will not have any standing waves in them to detect.

85. The gain is given by $\beta = 10\log\dfrac{P_{out}}{P_{in}} = 10\log\dfrac{125\,\text{W}}{1.0\times10^{-9}\,\text{W}} = \boxed{51\,\text{dB}}$

91. The fundamental frequency of a tube closed at one end is given by $f_1 = v/(4l)$. The change in air temperature will change the speed of sound, resulting in two different frequencies.

$$\frac{f_{30.0°C}}{f_{25.0°C}} = \frac{\dfrac{v_{30.0°C}}{4l}}{\dfrac{v_{25.0°C}}{4l}} = \frac{v_{30.0°C}}{v_{25.0°C}} \rightarrow f_{30.0°C} = f_{25.0°C}\left(\frac{v_{30.0°C}}{v_{25.0°C}}\right)$$

$$\Delta f = f_{30.0°C} - f_{25.0°C} = f_{25.0°C}\left(\frac{v_{30.0°C}}{v_{25.0°C}} - 1\right) = (349\,\text{Hz})\left(\frac{331+0.60(30.0)}{331+0.60(25.0)} - 1\right) = \boxed{3\,\text{Hz}}$$

97. (a) Since both speakers are moving towards the observer at the same speed, both frequencies have the same Doppler shift, and the observer hears $\boxed{\text{no beats}}$.
(b) The observer will detect an increased frequency from the speaker moving towards him and a decreased frequency from the speaker moving away. The difference in those two frequencies will be the beat frequency that is heard.

$$f'_{towards} = f\frac{1}{\left(1-\dfrac{v_{train}}{v_{snd}}\right)} \qquad f'_{away} = f\frac{1}{\left(1+\dfrac{v_{train}}{v_{snd}}\right)}$$

$$f'_{towards} - f'_{away} = f\frac{1}{\left(1-\dfrac{v_{train}}{v_{snd}}\right)} - f\frac{1}{\left(1+\dfrac{v_{train}}{v_{snd}}\right)} = f\left[\frac{v_{snd}}{\left(v_{snd}-v_{train}\right)} - \frac{v_{snd}}{\left(v_{snd}+v_{train}\right)}\right]$$

$$(348\,\text{Hz})\left[\frac{343\,\text{m/s}}{\left(343\,\text{m/s}-10.0\,\text{m/s}\right)} - \frac{343\,\text{m/s}}{\left(343\,\text{m/s}+10.0\,\text{m/s}\right)}\right] = \boxed{20\,\text{Hz}}$$

(c) Since both speakers are moving away from the observer at the same speed, both frequencies have the same Doppler shift, and the observer hears $\boxed{\text{no beats}}$.

103. The person will hear a frequency $f'_{towards} = f\left(1+\dfrac{v_{walk}}{v_{snd}}\right)$ from the speaker that they walk towards. The person will hear a frequency $f'_{away} = f\left(1-\dfrac{v_{walk}}{v_{snd}}\right)$ from the speaker that they walk away from. The beat frequency is the difference in those two frequencies.

$$f'_{towards} - f'_{away} = f\left(1+\frac{v_{walk}}{v_{snd}}\right) - f\left(1-\frac{v_{walk}}{v_{snd}}\right) = 2f\frac{v_{walk}}{v_{snd}} = 2(282\,\text{Hz})\frac{1.4\,\text{m/s}}{343\,\text{m/s}} = \boxed{2.3\,\text{Hz}}$$

109.(*a*)

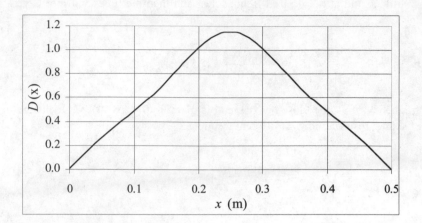

(*b*)

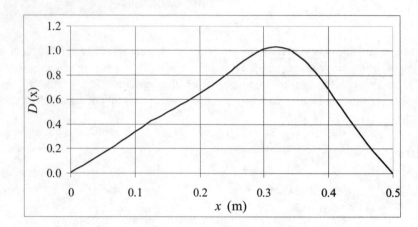

Chapter 17:

Temperature, Thermal Expansion, and the Ideal Gas Law

Chapter Overview and Objectives

In this chapter, the concept of temperature is introduced. Temperature-related phenomena, such as thermal equilibrium and thermal expansion are described. The gas laws are discussed.

After completing this chapter you should:
- Know what the atomic theory of matter is.
- Know the conditions for thermal equilibrium.
- Know what temperature measures.
- Know what the absolute temperature scale is.
- Know how to calculate the amount of thermal expansion of a material.
- Know how to use the gas laws and under what conditions they are valid.

Summary of Equations

Atomic Mass Units:	$1u = 1.66 \times 10^{-27} \text{ kg}$	(Section 17-1)
Conversions between Celsius and Fahrenheit:	$T\left(^{\circ}C\right) = \tfrac{5}{9}\left[T\left(^{\circ}F\right) - 32\right]$	(Section 17-2)
	$T\left(^{\circ}F\right) = \tfrac{9}{5}T\left(^{\circ}C\right) + 32$	(Section 17-2)
Linear Thermal Expansion:	$\ell = \ell_0\left(1 + \alpha\,\Delta T\right)$	(Section 17-4)
Volume Thermal Expansion:	$V = V_0\left(1 + \beta\,\Delta T\right)$	(Section 17-4)
Thermal Stress:	$\dfrac{F}{A} = \alpha\,E\Delta T$	(Section 17-5)
Ideal Gas Law:	$PV = nRT = NkT$	(Section 17-7)

Chapter Summary

Section 17-1. Atomic Theory of Matter

The atomic theory of matter models matter as a collection of **atoms**. A unit of mass that is commonly used for atoms and molecules is the **unified atomic mass** unit (u):

$$1u = 1.66 \times 10^{-27} \text{ kg}$$

In 1905, Albert Einstein estimated the mass of molecules from measurements made on the random motion of dust particles suspended in air. This motion is called **Brownian Motion**.

Example 17-1-A: Calculating the number of atoms in macroscopic samples. The atomic mass of silicon is 28.1 u. How many atoms of silicon are contained in a 17.4-g sample of silicon?

Approach: The atomic mass provides us with information about the mass of a single atom of silicon. By taking the ratio of the mass of our sample of silicon and the mass of a single silicon atom we can determine the number of silicon atoms in the sample.

Solution: The mass of a single silicon atom is

$$m_{Si} = 28.1\,\text{u} = (28.1)\left(1.66 \times 10^{-27}\right) = 4.665 \times 10^{-26}\,\text{kg}$$

The number of atoms in the 17.4-g sample of silicon is

$$N = \frac{m_{sample}}{m_{Si}} = \frac{17.4 \times 10^{-3}}{4.665 \times 10^{-26}} = 3.73 \times 10^{23}$$

Section 17-2. Temperature and Thermometers

Thermometers are used to measure temperature. All thermometers rely on the temperature dependence of some property of a material, such as length, density, and electrical resistivity, to measure temperature.

There are two temperature scales in use in everyday life. These are the **Fahrenheit** scale (°F) and the **Celsius** scale (°C). These temperature scales are related in the following manner:

$$T\left(^\circ\text{C}\right) = \tfrac{5}{9}\left[T\left(^\circ\text{F}\right) - 32\right]$$

$$T\left(^\circ\text{F}\right) = \tfrac{9}{5}T\left(^\circ\text{C}\right) + 32$$

Section 17-3. Thermal Equilibrium and the Zeroth Law of Thermodynamics

Two systems at the same temperature will not exchange energy through the process of heat transfer when in thermal contact. These systems are said to be in **thermal equilibrium**. If two systems with different temperatures are in thermal contact, heat will flow from the system with the higher temperature to the system with the lower temperature until both systems have the same temperature.

The **zeroth law of thermodynamics** states

if two systems are in thermal equilibrium with a third system, then they are in thermal equilibrium with each other.

The zeroth law implies that assigning a temperature to a system is meaningful. Temperature is the only property that determines if two systems are in thermal equilibrium.

Section 17-4. Thermal Expansion

Most materials expand when their temperature is increased and contract when their temperature is decreased. The amount of the expansion or contraction depends on the type of material. The change in the linear dimension $\Delta \ell$ of a material that undergoes a temperature change ΔT can be approximated by the following linear relation:

$$\Delta \ell = \alpha\, \ell_0\, \Delta T$$

where ℓ_0 is the original linear dimension of the material and α is the **coefficient of linear expansion**. The values of α for various materials are given in Table 17-1 of the textbook. Typical values of α for solids have a magnitude of about $10^{-5}/^\circ\text{C}$. The length ℓ at the new temperature can be written as:

$$\ell = \ell_0 + \Delta \ell = \ell_0 \left(1 + \alpha\, \Delta T\right)$$

Similar relationships hold for the volume change ΔV of a material with temperature change ΔT:

$$\Delta V = V_o \beta \Delta T \quad \rightarrow \quad V = V_0 + \Delta V = V_0 \left(1 + \beta \Delta T \right)$$

where the initial volume of the material is V_0 and β is the **coefficient of volume expansion**. The coefficient of volume expansion is typically three times the coefficient of linear expansion. The increase in volume with increasing temperature implies that the potential energy of atoms or molecules as a function of separation distance is asymmetrical. Water in the temperature range of 0°C to 4°C behaves differently than most other materials. In that temperature range, water decreases in volume with increasing temperature.

Example 17-4-A: Dimensional changes with temperature. A copper ring of diameter $d = 3.47$ cm fits snugly around an iron rod of the same diameter at a temperature $T = 20$°C. To remove the ring easily from this rod, it is necessary for the ring to have a diameter that is $\Delta d = 0.05$ mm larger than that of the rod. Determine the temperature the rod and ring must be heated or cooled to so that the ring slides easily off of the rod.

Approach: The thermal expansion coefficients for copper and iron are different ($\alpha_{Cu} = 17 \times 10^{-6}$ °C^{-1} and $\alpha_{Fe} = 12 \times 10^{-6}$ °C^{-1}). When the temperature of the system is increased, the copper will expand more than the iron. Using the equation of linear expansion we can determine by how much the temperature must be increased in order to change the dimension of the copper ring by 0.05 mm more than the dimension of the iron rod.

Solution: Consider a change in the temperature ΔT. The change in the diameters of the copper ring and the iron rod are equal to

$$\Delta \ell_{ring} = \alpha_{Cu} \ell_0 \Delta T \qquad \Delta \ell_{rod} = \alpha_{Fe} \ell_0 \Delta T$$

The difference in the change in diameters of the two objects is equal to

$$\Delta \ell_{ring} - \Delta \ell_{rod} = \Delta d = \left(\alpha_{Cu} - \alpha_{Fe} \right) \ell_0 \Delta T$$

The required change in the temperature can now be calculated:

$$\Delta T = \frac{\Delta d}{\left(\alpha_{Cu} - \alpha_{Fe} \right) \ell_0} = \frac{0.05 \times 10^{-3}}{\left(17 \times 10^{-6} - 12 \times 10^{-6} \right) \left(3.47 \times 10^{-2} \right)} = 288\text{°C}$$

The final temperature of the system must thus be equal to 288°C + 20°C = 308°C.

Example 17-4-B. Calibration of a thermometer. A bulb type thermometer holds a volume $V_0 = 0.132$ cm^3 of ethyl alcohol. The bore of the capillary tube in which the level of the alcohol rises with temperature has a diameter of $d = 0.061$ cm. Assuming the thermal expansion of the quartz bulb and capillary are negligible, what distance should be marked as one degree Celsius on the thermometer?

Approach: This problem is an example of volume expansion. Since the bulb and the capillary are assumed not to expand with increasing temperature, the change in the height of the ethyl alcohol in the capillary is a direct result of the change of its volume with increasing temperature. We can determine the change in the volume of the ethyl alcohol and use this information to determine the change in the height of the alcohol in the capillary.

Solution: The change in the volume of the liquid for a one-degree change in temperature is given by:

$$\Delta V = \beta V_o \Delta T$$

The increase of the volume of the alcohol increases the height of the alcohol in the capillary by h. This increase in height is equal to

$$h = \frac{\Delta V}{A} = \frac{\Delta V}{\left(\frac{1}{4} \pi d^2 \right)} = \frac{4 \left(\beta V_o \Delta T \right)}{\left(\pi d^2 \right)} = \frac{4 \left(1100 \times 10^{-6} \right) \left(0.132 \times 10^{-6} \right) (1)}{\pi \left(0.061 \times 10^{-2} \right)^2} = 5.0 \times 10^{-4} \text{ m} = 0.05 \text{ cm}$$

A one-degree temperature difference results in a change in height of the alcohol column by 0.05 cm.

Section 17-5. Thermal Stresses

If materials are not allowed to freely expand, the strain of thermal expansion will induce a stress called **thermal stress**. As an example, consider a piece of material with elastic constant E that is constrained from increasing its length. If the material is not constrained, the result of an increase in its temperature by ΔT will change its length by

$$\Delta \ell = \alpha \, \ell_0 \, \Delta T$$

When both ends are fixed, there must be an elastic strain of equal magnitude but opposite in direction. The required elastic strain must thus be equal to

$$\frac{F}{A} = E \frac{\Delta \ell}{\ell_0} = E \alpha \, \Delta T$$

Example 17-5-A: Compressive forces on steel beams. A steel beam of length $\ell = 10.6$ m with a cross-sectional area $A = 0.183$ m^2 is wedged between two bedrock outcroppings when it is at a temperature of $T_1 = 14.6°C$. On a hot day, the temperature of the steel beam rises to $T_2 = 46.2°C$. What is the compressive force on the beam if the spacing between the bedrock outcroppings does not change?

Approach: The compressive force must result in a compression that has the same magnitude as the thermal expansion, but with an opposite sign. Since we know the thermal expansion we can determine the required compression force.

Solution: If the beam were free to expand, the change in length of the beam would be

$$\Delta \ell = \alpha \ell_0 \Delta T$$

Because the ends of beam are not allowed to move, an elastic compression equal to this thermal expansion must exist in the beam. The required elastic compression is related to the stress in the beam by Hooke's law:

$$\frac{F}{A} = E \frac{\Delta \ell}{\ell_0}$$

E is the Young's modulus, which for steel is equal to 200×10^9 N/m^2 (see Table 12.1 in the textbook.) The cross-sectional area A of the beam is known, so we can determine the compressive force F:

$$F = EA \frac{\Delta \ell}{\ell_0} = EA \frac{\alpha \ell_0 \Delta T}{\ell_0} = EA\alpha\Delta T = \left(200 \times 10^9\right)\left(0.183\right)\left(12 \times 10^{-6}\right)\left(46.2 - 14.6\right) = 1.39 \times 10^7 \text{ N}$$

Section 17-6. The Gas Laws and Absolute Temperature

The equation of state of a system in equilibrium provides a relation between the volume, pressure, temperature, and the number of atoms/molecules that are part of the system. Gases have relatively simple equations of state, particularly when the volume of the molecules is small compared to the volume of the gas and the temperature of the gas is much greater than the temperature at which the gas would condense.

For a fixed quantity of an ideal gas, the volume of the gas is inversely proportional to the pressure of the gas at fixed temperature:

$$PV = \text{Constant}$$

This relationship is known as **Boyle's law**.

For a fixed quantity of an ideal gas, the volume of the gas is directly proportional to the absolute temperature of the gas when it is held at a fixed pressure:

$$V = \text{Constant} \times T$$

This relationship is known as **Charles's law**.

The temperature T in Charles's law is measured on the **absolute temperature scale**. A plot of the volume of a gas, maintained at a moderate pressure, as function of temperature (measured on for example the Celsius scale) shows a linear relation between volume and temperature. However, the volume is not proportional to temperature (e.g., when $T =$

$0, V \neq 0$). Extrapolating the observed linear relation to $V = 0$, it is determined that $T = 0$ on the absolute temperature scale corresponds to $-273.15°C$, independent of the type of gas. The absolute temperature scale is defined such that V is proportional to T. The unit of the absolute temperature scale is the **Kelvin** (K). The relation between the absolute temperature scale and the Celsius scale is as follows:

$$T(K) = T(°C) + 273.15$$

A third relationship between gas quantities relates the pressure of a gas to its absolute temperature. **Gay-Lussac's law** states that the pressure of a gas is directly proportional to the absolute temperature of the gas when the volume is held constant:

$$P = \text{Constant} \times T$$

Section 17-7. The Ideal Gas Law

The **ideal gas law** relates pressure, volume, temperature, and the number of moles of a gas to each other:

$$PV = nRT$$

where P is the absolute pressure, V is the volume, n is the number of moles, and T is the absolute temperature. The constant R is called the **universal gas constant** and has a value of 8.315 J/(mol K). The relationship is also called the **equation of state** of an ideal gas.

Note: A **mole** of a substance is an amount that contains as many molecules as there are carbon atoms in 12.00 g of carbon.

Section 17-8. Problem Solving with the Ideal Gas Law

When we apply the ideal gas law to solve problems, we must make sure to use the absolute pressure and temperature. **Standard temperature and pressure** (STP) refer to a temperature of 273 K and a pressure of 1.013×10^5 N/m². The volume of 1 mole of an ideal gas at STP is 22.4 liters or 2.24×10^{-2} m³.

Example 17-8-A. Floating in a balloon. A hot air balloon has a volume of $V = 134$ m³ when inflated and the payload has negligible volume. The balloon needs to lift a total mass $M = 23.8$ kg. The temperature of the ambient air is $T_{air} = 26°C$. What does the temperature of the air inside the balloon need to be to generate enough buoyant force to let the balloon and its payload float in the air? Assume that the mass of one mole of air at 26°C is 0.029 kg.

Approach: The magnitude of the buoyant force must be equal to the weight of the balloon and its payload and the weight of the hot air contained inside the balloon. The buoyant force is determined by the weight of the air that is displaced and thus depends on the volume of the balloon and the density of the outside air. The weight of the hot air inside the balloon can be determined by using the ideal gas law, assuming that the pressure inside and outside the balloon is the same (and equal to 1 atm). The weight of the hot air inside the balloon will decrease with increasing temperature, and the required buoyant force to float will thus decrease.

Solution: The buoyant force of the balloon can be obtained by calculating the mass of the air that is displaced. This number can be obtained on the basis of the volume of the balloon and the temperature of the ambient air:

$$n_{air\ displaced} = \frac{P_{air} V_{balloon}}{RT_{air}}$$

The buoyant force is equal to the weight of the air that is displaced.

$$F_{buoyant} = \rho n_{air\ displaced} g = \rho \frac{P_{air} V_{balloon}}{RT_{air}} g$$

where ρ is the mass of one mole of air (0.029 kg). The weight of the air inside the balloon can be calculated using a similar expression, except that the temperature used is the temperature of the air inside the balloon:

$$W_{air\ inside} = \rho \frac{P_{air} V_{balloon}}{RT_{balloon}} g$$

In order for the balloon to float, the buoyant force due to the displaced air must be equal in magnitude to the weight of the air inside the balloon and the weight of the balloon and its payload:

$$F_{buoyant} = \rho \frac{P_{air} V_{balloon}}{RT_{air}} g = \rho \frac{P_{air} V_{balloon}}{RT_{balloon}} g + Mg$$

This equation can be rewritten as

$$\rho \frac{P_{air} V_{balloon}}{R} \left(\frac{1}{T_{air}} - \frac{1}{T_{balloon}} \right) = M$$

The only unknown in this equation is the temperature of the air inside the balloon, $T_{balloon}$, which can be expressed in terms of the information provided:

$$T_{balloon} = \frac{1}{\left(\dfrac{1}{T_{air}} - \dfrac{MR}{\rho P_{air} V_{balloon}} \right)} = \frac{1}{\left(\dfrac{1}{(26+273)} - \dfrac{(23.8)(8.315)}{(0.029)(1.01\times 10^5)(134)} \right)} = 352 \text{ K} = 79^\circ \text{C}$$

To carry out this calculation we have to be sure to express the temperature in units of Kelvin and the pressure in units of N/m^2. The calculated temperature will be in units of Kelvin.

Example 17-8-B. Lifting off in a balloon. To what temperature must we heat the air inside the balloon discussed in Example 17-8-A to establish a vertical acceleration of $a = 0.5$ m/s^2?

Approach: In order to have an upward vertical acceleration, there must be a net force acting in this direction. The buoyant force must thus exceed the weight of the balloon, its payload, and the air inside it. The procedure followed to solve this problem is similar to the procedure used in Example 17-8-A.

Solution: In Example 17-8-A we have calculated the mass of the air inside the balloon. The net force acting on the balloon must be equal to the product of the total mass of the balloon (including the mass of the air inside the balloon and the mass of payload) and the required acceleration. This requirement can be expressed as

$$F_{net} = F_{buoyant} - \left(\rho \frac{P_{air} V_{balloon}}{RT_{balloon}} + M \right) g = \left(\rho \frac{P_{air} V_{balloon}}{RT_{balloon}} + M \right) a$$

Using the expression for the buoyant force obtained in example 17-8-A we can rewrite this equation as

$$F_{buoyant} = \rho \frac{P_{air} V_{balloon}}{RT_{air}} g = \left(\rho \frac{P_{air} V_{balloon}}{RT_{balloon}} + M \right) (g+a)$$

The only unknown in this equation is the temperature of the air inside the balloon, $T_{balloon}$, which can be expressed in terms of the information provided:

$$T_{balloon} = \frac{1}{\left(\dfrac{1}{T_{air}} \dfrac{g}{(g+a)} - \dfrac{MR}{\rho P_{air} V_{balloon}} \right)} = \frac{1}{\left(\dfrac{1}{(26+273)} \dfrac{(9.80)}{(10.30)} - \dfrac{(23.8)(8.315)}{(0.029)(1.01\times 10^5)(134)} \right)} = 373 \text{ K} = 100^\circ \text{C}$$

In order to ensure that we did not make a mistake in our derivation, we can double check that if we set the acceleration a to zero, we obtain the answer to Example 17-8-A.

Example 17-8-C: Applications of the ideal gas law. A container of volume $V = 0.180$ m^3 contains liquid nitrogen of mass $M = 1.22$ kg and nitrogen gas in the remainder of the volume. The gas is maintained at atmospheric pressure and a temperature of $T_0 = 78$ K. The container is sealed and the temperature rises to $T_1 = 300$ K. At this temperature, all of the liquid has become a gas. What is the pressure inside the container now? The density of liquid nitrogen is $\rho_{liquid} = 810$ kg/m^3 and the molecular weight of nitrogen gas is $\rho_{mol} = 0.028$ kg/mol.

Approach: The problem provides us with information about the volume and temperature of the final gas. In order to determine the pressure in the container we need to determine the number of moles of the gas in the container. This information can be obtained by using the information provided on the initial conditions of the system to determine the number of moles in the liquid and in the gas.

Solution: The number of moles in the liquid can be determined by dividing the mass of liquid by the mass per mole:

$$n_{liquid} = \frac{M}{\rho_{mol}} = \frac{1.22}{0.028} = 43.6 \text{ moles}$$

The ideal gas law can be used to determine the number of moles of nitrogen gas under the initial conditions. The initial volume of the gas in the container is equal to the volume of the container minus the volume of the liquid:

$$V_{gas,initial} = V_{container} - V_{liquid} = V_{container} - \frac{M}{\rho_{liquid}}$$

Using the ideal gas law, we can now determine the number of moles of the gas:

$$n_{gas} = \frac{PV}{RT} = \frac{P_{initial}\left(V - \dfrac{M}{\rho_{liquid}}\right)}{RT_0} = \frac{\left(1.01\times10^5\right)\left((0.180) - \dfrac{(1.22)}{(810)}\right)}{(8.315)(78)} = 27.8 \text{ moles}$$

The total number of moles of gas at 300 K will be the sum of the moles in the initial liquid and gas:

$$n = n_{liquid} + n_{gas} = 43.6 + 27.8 = 71.4$$

We can now use the ideal gas law to determine the final pressure:

$$P = \frac{nRT_1}{V} = \frac{(71.4)(8.315)(300)}{0.180} = 9.9\times10^5 \text{ N/m}^2$$

Section 17-9. Ideal Gas Law in Terms of Molecules: Avogadro's Number

The number of molecules in a mole of substance is called **Avogadro's number** (N_A):

$$N_A = 6.022 \times 10^{23}$$

The ideal gas law can be written in the form

$$PV = NkT$$

where N is the number of molecules of a substance and $k = R/N_A$ and is called the **Boltzmann's constant**:

$$k = \frac{R}{N_A} = \frac{8.315}{6.022\times10^{23}} = 1.381\times10^{-23} \text{ J/K}$$

Example 17-9-A: Calculating the number of air molecules. At atmospheric pressure and at a temperature of 20°C, how many air molecules are there in a room of dimensions 20 m × 15 m × 3 m?

Approach: Since we know the pressure, the temperature, and the volume of the room, we can use the ideal gas law to determine the number of molecules.

Solution: By rewriting the ideal gas law we can calculate the number of air molecules N:

$$N = \frac{PV}{kT} = \frac{\left(1.01\times10^5\right)\left(20\times15\times3\right)}{\left(1.38\times10^{-23}\right)\left(20+273\right)} = 2.2\times10^{28}$$

We need to express the pressure in units of N/m^2 and the temperature in Kelvin.

Section 17-10. Ideal Gas Temperature Scale – a Standard

In order to create a standard of temperature, one particular method of measuring temperature must be agreed upon and used to define a standard, and all other methods must be calibrated against that standard by placing those thermometers in equilibrium with the standard thermometer.

The absolute temperature T, in Kelvin, is defined using a **constant volume gas thermometer**:

$$T = 273.16\left(\frac{P}{P_{tp}}\right)$$

where P is the absolute pressure of the gas, and P_{tp} is the absolute pressure of the gas at the **triple point** of water (that temperature at which ice, liquid water, and water vapor are in thermal equilibrium). No real gas behaves exactly as an ideal gas, but as the pressure of the gas decreases at temperatures far from the liquefaction temperature, the behavior approaches ideal gas behavior. The **ideal gas temperature scale** is defined as the limit of T when the pressure of the real gas approaches zero pressure:

$$T = 273.16\lim_{P\to 0}\left(\frac{P}{P_{tp}}\right)$$

Different type of gases will have the same limit and the **ideal gas temperature scale** has the characteristics of universality and repeatability that makes it a good choice as a temperature standard.

Practice Quiz

1. The random motion of dust particles suspended in a gas or liquid is
    a) Brownian motion.
    b) Impossible.
    c) The same regardless of the temperature.
    d) Stopped at a temperature of 0 °C.

2. When are two systems in thermal equilibrium?
    a) When they have the same thermal expansion.
    b) When one system has the negative of the thermal stress of the other system.
    c) When the two systems have the same pressure.
    d) When the two systems have the same temperature.

3. How does the coefficient of linear expansion in units of $(°F)^{-1}$ compare to the coefficient of linear expansion in units of $(°C)^{-1}$?
    a) They are the same.
    b) The coefficient of linear expansion in $(°F)^{-1}$ is 5/9 the coefficient of linear expansion in $(°C)^{-1}$.
    c) The coefficient of linear expansion in $(°F)^{-1}$ is 9/5 the coefficient of linear expansion in $(°C)^{-1}$.
    d) The coefficient of linear expansion in $(°F)^{-1}$ is 5/9 the coefficient of linear expansion in $(°C)^{-1} + 32$.

4. Can thermal stress ever be tensile rather than compressive?
    a) No; objects always expand on heating and it takes a compressive stress to prevent the expansion.
    b) Yes, but only if the material has a negative coefficient of expansion.
    c) Yes, when objects are cooled to a lower temperature.
    d) No, because it is impossible to cool an object below absolute zero temperature.

5. What is meant by the anomalous behavior of water in regards to thermal expansion?
    a) Water has a zero coefficient of linear thermal expansion at all temperatures.
    b) Water has a negative coefficient of linear thermal expansion at all temperatures.
    c) Water has a negative coefficient of thermal expansion at temperatures below 4 °C.
    d) Water has an unusually large coefficient of thermal expansion.

6. For most materials, how is the coefficient of volume expansion related to the coefficient of linear expansion?
    a) The coefficient of volume expansion is equal to the coefficient of linear expansion.
    b) The coefficient of volume expansion is one-third the coefficient of linear expansion.
    c) The coefficient of volume expansion is three times the coefficient of linear expansion.
    d) The coefficient of volume expansion is twice the coefficient of linear expansion.

7. When applying the ideal gas law to a system that changes temperature, the temperature can be measured on which temperature scale?
    a) Fahrenheit.
    b) Celsius.
    c) Kelvin.
    d) Any scale.

8. Which of the gas laws relates the volume of gas to its pressure at a fixed temperature and fixed amount of gas?
    a) Boyle's law.
    b) Charles's law.
    c) Gay-Lussac's law.
    d) Zeroth law of thermodynamics.

9. Which of the gas laws relates the temperature of gas to its pressure at a fixed volume and fixed amount of gas?
    a) Boyle's law.
    b) Charles's law.
    c) Gay-Lussac's law.
    d) Zeroth law of thermodynamics.

10. What is special about the ideal gas temperature scale?
    a) It is the best temperature scale to use for finding the temperature of gases.
    b) It defines the temperature scale independently from the properties of a particular gas.
    c) It is the most practical scale for determining the fixed points of the scale at home.
    d) It uses the nicest set of numbers for working with everyday temperatures.

11. The floors of a building are to be made level. As the building is being built, the 12-foot-long posts that support the second floor on the south side of the building are exposed to the sun and warm to 120 °F. The posts on the north side remain at the 70°F ambient temperature. The second floor is leveled under these conditions. How far out of level will the floor be when the building all cools down to 60 °F at night? Do you think builders bother to consider this when constructing a building?

12. Determine the temperature on the Celsius scale at which the Kelvin temperature scale and the Fahrenheit temperature scale have the same value.

13. A 134-m-long concrete sidewalk is made without expansion joints when the temperature is 45 °F. The temperature of the concrete rises to 110° F on a sunny day. What is the thermal stress in the sidewalk? Assume the ends of the sidewalk are fixed in position.

14. A container of nitrogen gas contains 128 moles of nitrogen at a pressure of 2.4×10^5 N/m^2 at a temperature of 290 K. As nitrogen is added to the container, the pressure rises to 6.9×10^5 N/m^2 and the temperature rises to 354 K. How many moles of nitrogen were added to the container?

15. The unit of pressure **torr** is often used to specify the pressure in a vacuum system. One torr is equal to the pressure that supports a one-mm-high column of mercury. Determine the number of molecules per cubic centimeter in a vacuum of 10^{-10} torr at a temperature of 300 K.

Responses to Select End-of-Chapter Questions

1. 1 kg of aluminum will have more atoms. Aluminum has an atomic mass less than iron. Since each Al atom is less massive than each Fe atom, there will be more Al atoms than Fe atoms in 1 kg.

7. Aluminum. Al has a larger coefficient of linear expansion than Fe, and so will expand more than Fe when heated and will be on the outside of the curve.

13. When the cold thermometer is placed in the hot water, the glass part of the thermometer will expand first, as heat is transferred to it first. This will cause the mercury level in the thermometer to decrease. As heat is transferred to the mercury inside the thermometer, the mercury will expand at a rate greater than the glass, and the level of mercury in the thermometer will rise.

19. Helium. If we take the atomic mass of 6.7×10^{-27} kg and divide by the conversion factor from kg to u (atomic mass units), which is 1.66×10^{-27} kg/u, we get 4.03 u. This corresponds to the atomic mass of helium.

Solutions to Select End-of-Chapter Problems

1. The number of atoms in a pure substance can be found by dividing the mass of the substance by the mass of a single atom. Take the atomic masses of gold and silver from the periodic table.

$$\frac{N_{Au}}{N_{Ag}} = \frac{\dfrac{2.15\times10^{-2}\,\text{kg}}{(196.96655)(1.66\times10^{-27}\,\text{kg/atom})}}{\dfrac{2.15\times10^{-2}\,\text{kg}}{(107.8682)(1.66\times10^{-27}\,\text{kg/atom})}} = \frac{107.8682}{196.96655} = 0.548 \quad\rightarrow\quad \boxed{N_{Au} = 0.548 N_{Ag}}$$

Because a gold atom is heavier than a silver atom, there are fewer gold atoms in the given mass.

7. Take the 300-m height to be the height in January. Then the increase in the height of the tower is given by Eq. 17-1a.

$$\Delta l = \alpha l_0 \Delta T = (12\times10^{-6}/\text{C}°)(300\text{ m})(25°\text{C} - 2°\text{C}) = \boxed{0.08\,\text{m}}$$

13. The rivet must be cooled so that its diameter becomes the same as the diameter of the hole.

$$\Delta l = \alpha l_0 \Delta T \rightarrow l - l_0 = \alpha L_0 (T - T_0) \rightarrow T = T_0 + \frac{l - l_0}{\alpha l_0} = 20 + \frac{1.870 - 1.872}{(12\times10^{-6})(1.872)} = \boxed{-69°\text{C}}$$

The temperature of "dry ice" is about -80°C, so this process will be successful.

19. We model the vessel as having a constant cross-sectional area A. Then a volume V_0 of fluid will occupy a length l_0 of the tube, given that $V_0 = Al_0$. Likewise, $V = Al$.

$$\Delta V = V - V_0 = Al - Al_0 = A\Delta l \text{ and } \Delta V = \beta V_0 \Delta T = \beta Al_0 \Delta T.$$

Equate the two expressions for ΔV, and get

$$A\Delta l = \beta Al_0 \Delta T \rightarrow \Delta l = \beta l_0 \Delta T$$

But $\Delta l = \alpha l_0 \Delta T$, so we see that under the conditions of the problem, $\boxed{\alpha = \beta}$.

25. The thermal stress must compensate for the thermal expansion. E is Young's modulus for the aluminum.

$$\text{Stress} = F/A = \alpha E \Delta T = (25\times10^{-6}/\text{C}°)(70\times10^9\text{ N/m}^2)(35°\text{C} - 18°\text{C}) = \boxed{3.0\times10^7\text{ N/m}^2}$$

31. Assume the gas is ideal. Since the amount of gas is constant, the value of PV/T is constant.

$$\frac{P_1 V_1}{T_1} = \frac{P_2 V_2}{T_2} \rightarrow$$

$$V_2 = V_1 \frac{P_1}{P_2} \frac{T_2}{T_1} = (3.80 \, m^3)\left(\frac{1.00 \, atm}{3.20 \, atm}\right)\frac{(273+38.0)\,K}{273 \, K} = \boxed{1.35 \, m^3}$$

37. (a) Assume the nitrogen is an ideal gas. The number of moles of nitrogen is found from the atomic weight, and then the ideal gas law is used to calculate the volume of the gas.

$$n = (28.5 \, kg)\frac{1 \, mole \, N_2}{28.01 \times 10^{-3} \, kg} = 1017 \, mol$$

$$PV = nRT \rightarrow V = \frac{nRT}{P} = \frac{(1017 \, mol)(8.315 \, J/mol\cdot K)(273 \, K)}{1.013 \times 10^5 \, Pa} = 22.79 \, m^3$$

$$\approx \boxed{22.8 \, m^3}$$

(b) Hold the volume and temperature constant, and again use the ideal gas law.

$$n = (28.5 \, kg + 25.0 \, kg)\frac{1 \, mole \, N_2}{28.01 \times 10^{-3} \, kg} = 1910 \, mol$$

$$PV = nRT \rightarrow$$

$$P = \frac{nRT}{V} = \frac{(1910 \, mol)(8.315 \, J/mol\cdot K)(273 \, K)}{22.79 \, m^3} = \boxed{1.90 \times 10^5 \, Pa = 1.88 \, atm}$$

43. Assume the oxygen is an ideal gas. Since the amount of gas is constant, the value of PV/T is constant.

$$\frac{P_1 V_1}{T_1} = \frac{P_2 V_2}{T_2} \rightarrow P_2 = P_1 \frac{V_1}{V_2}\frac{T_2}{T_1} = (2.45 \, atm)\left(\frac{61.5 \, L}{48.8 \, L}\right)\frac{(273+56.0)\,K}{(273+18.0)\,K} = \boxed{3.49 \, atm}$$

49. We calculate the density of water vapor, with a molecular mass of 18.0 grams per mole, from the ideal gas law.

$$PV = nRT \rightarrow \frac{n}{V} = \frac{P}{RT} \rightarrow$$

$$\rho = \frac{m}{V} = \frac{Mn}{V} = \frac{MP}{RT} = \frac{(0.0180 \, kg/mol)(1.013 \times 10^5 \, Pa)}{(8.315 \, J/mol\cdot K)(373 \, K)} = \boxed{0.588 \, m^3}$$

The density from Table 13-1 is 0.598 m³. Because this gas is very "near" a phase change state (water can also exist as a liquid at this temperature and pressure), we would not expect it to act like an ideal gas. It is reasonable to expect that the molecules will have other interactions besides purely elastic collisions. That is evidenced by the fact that steam can form droplets, indicating an attractive force between the molecules.

55. Assume the gas is ideal at those low pressures, and use the ideal gas law.

$$PV = NkT \rightarrow \frac{N}{V} = \frac{P}{kT} = \frac{1 \times 10^{-12} \, N/m^2}{(1.38 \times 10^{-23} \, J/K)(273 \, K)} = \left(3 \times 10^8 \frac{molecules}{m^3}\right)\left(\frac{10^{-6} \, m^3}{1 \, cm^3}\right)$$

$$= \boxed{300 \frac{molecules}{cm^3}}$$

61. Since the volume is constant, the temperature of the gas is proportional to the pressure of the gas. First we calculate the two temperatures of the different amounts of gas.

$$T_1 = (273.16\,\text{K})\frac{P_{1\,\text{melt}}}{P_{1\,\text{tp}}} = (273.16\,\text{K})\frac{218}{286} = 208.21\,\text{K}$$

$$T_2 = (273.16\,\text{K})\frac{P_{2\,\text{melt}}}{P_{2\,\text{tp}}} = (273.16\,\text{K})\frac{128}{163} = 214.51\,\text{K}$$

Assume that there is a linear relationship between the melting-point temperature and the triple-point pressure, as shown in Fig. 17-17. The actual melting point is the *y*-intercept of that linear relationship. We use Excel to find that *y*-intercept. The graph is shown below.

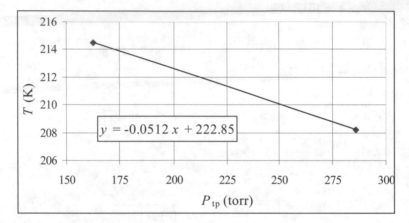

We see the melting temperature is $\boxed{223\,\text{K}}$.

67. Assume that the air in the lungs is an ideal gas, that the amount of gas is constant, and that the temperature is constant. The ideal gas law then says that the value of PV is constant. The pressure a distance h below the surface of a fluid is given by Eq. 13-6b, $P = P_0 + \rho g h$, where P_0 is atmospheric pressure and ρ is the density of the fluid. We assume that the diver is in sea water.

$$(PV)_{\text{surface}} = (PV)_{\text{submerged}} \quad \rightarrow \quad V_{\text{surface}} = V_{\text{submerged}}\frac{P_{\text{submerged}}}{P_{\text{surface}}} = V_{\text{submerged}}\frac{P_{\text{atm}} + \rho g h}{P_{\text{atm}}}$$

$$= (5.5\,\text{L})\frac{1.01\times10^5\,\text{Pa} + (1025\,\text{kg/m}^3)(9.80\,\text{m/s}^2)(8.0\,\text{m})}{1.01\times10^5\,\text{Pa}} = \boxed{9.9\,\text{L}}$$

This is obviously very dangerous, to have the lungs attempt to inflate to twice their volume. Thus it is $\boxed{\text{not advisable}}$ to quickly rise to the surface.

73. Since the pressure is force per unit area, if the pressure is multiplied by the surface area of the Earth, the force of the air is found. If we assume that the force of the air is due to its weight, then the mass of the air can be found. The number of molecules can then be found using the molecular mass of air (calculated in problem 71) and Avogadro's number.

$$P = \frac{F}{A} \quad \rightarrow \quad F = PA \quad \rightarrow \quad Mg = P4\pi R_{\text{Earth}}^2 \quad \rightarrow$$

$$M = \frac{4\pi R_{\text{Earth}}^2 P}{g} = \frac{4\pi(6.38\times10^6\,\text{m})^2(1.01\times10^5\,\text{Pa})}{9.80\,\text{m/s}^2} = 5.27\times10^{18}\,\text{kg}$$

$$N = 5.27\times10^{18}\,\text{kg}\left(\frac{1\,\text{mole}}{29\times10^{-3}\,\text{kg}}\right)\left(\frac{6.02\times10^{23}\,\text{molecules}}{1\,\text{mole}}\right) = \boxed{1.1\times10^{44}\,\text{molecules}}$$

79. (*a*) Assume that a mass M of gasoline with volume V_0 at 0°C is under consideration, and so its density is $\rho_0 = M/V_0$. At a temperature of 35°C, the same mass has a volume $V = V_0 (1+\beta\Delta T)$.

$$\rho = \frac{M}{V} = \frac{M}{V_0(1+\beta\Delta T)} = \frac{\rho_0}{1+\beta\Delta T} = \frac{0.68\times10^3\,\text{kg/m}^3}{1+\left(950\times10^{-6}/\text{C}°\right)\left(35\text{C}°\right)} = 0.6581\times10^3\,\text{kg/m}^3$$

$$\approx \boxed{660\,\text{kg/m}^3}$$

(*b*) Calculate the percentage change in the density.

$$\%\text{ change} = \frac{\left(0.6581-0.68\right)\times10^3\,\text{kg/m}^3}{0.68\times10^3\,\text{kg/m}^3\,V}\times100 = \boxed{-3\%}$$

85. We assume ideal gas behavior for the air in the lungs, and a constant temperature for the air in the lungs. When underwater, we assume the relaxed lung of the diver is at the same pressure as the surrounding water, which is given by Eq. 13-6b, $P = P_0 + \rho g h$. In order for air to flow through the snorkel from the atmospheric air above the water's surface (assume to be at atmospheric pressure), the diver must reduce the pressure in his lungs to atmospheric pressure or below, by increasing the volume of the lungs. We assume that the diver is in sea water.

$$P_{\text{relaxed}}V_{\text{relaxed}} = P_{\text{inhaling}}V_{\text{inhaling}} \rightarrow \frac{V_{\text{inhaling}}}{V_{\text{relaxed}}} = \frac{P_{\text{relaxed}}}{P_{\text{inhaling}}} = \frac{P_{\text{underwater}}}{P_{\text{atmospheric}}} = \frac{P_0+\rho g h}{P_0} = 1 + \frac{\rho g h}{P_0}$$

$$\frac{\Delta V}{V_{\text{relaxed}}} = \frac{V_{\text{inhaling}}-V_{\text{relaxed}}}{V_{\text{relaxed}}} = \frac{V_{\text{inhaling}}}{V_{\text{relaxed}}}-1 = \frac{\rho g h}{P_0} = \frac{\left(1025\,\text{kg/m}^3\right)\left(9.80\,\text{m/s}^2\right)\left(0.30\text{m}\right)}{1.013\times10^5\,\text{Pa}} = \boxed{0.030}$$

This is a 3% increase.

Chapter 18: Kinetic Theory of Gases

Chapter Overview and Objectives

In this chapter, the kinetic theory of gases is introduced. You will see that the ideal gas law is consistent with a model that describes the gas as a collection of independently moving particles that only interact via collisions. The concepts of saturated vapor pressure, relative humidity, and partial pressure of gases are presented. Non-ideal equations of state for gases are described and the concepts of the mean free path and diffusion are introduced.

After completing this chapter you should:
- Understand the microscopic model of an ideal gas as a collection of particles.
- Know how the mean kinetic energy of the particles in a gas is related to the absolute temperature.
- Know how the speeds of the molecules of an ideal gas are distributed.
- Know how the apply Maxwell's distribution of molecular speeds.
- Know the difference in behavior of real gases and ideal gases.
- Understand the concepts of saturated vapor pressure, partial pressure, and relative humidity.
- Be familiar with the Clausius and the van der Waals equations of state of a gas.
- Understand the concepts of mean free path and diffusion.

Summary of Equations

Mean kinetic energy of particles in an ideal gas:
$$\bar{K} = \left[\frac{1}{2} m \left(v^2 \right) \right]_{av} = \frac{3}{2} kT$$
(Section 18-1)

Root-mean-square speed of particles in an ideal gas:
$$v_{RMS} = \sqrt{\frac{3kT}{m}}$$
(Section 18-1)

Maxwell distribution of speeds in an ideal gas:
$$f(v) = 4\pi N \left(\frac{m}{2\pi kT} \right)^{\frac{3}{2}} v^2 e^{-\frac{1}{2} \frac{mv^2}{kT}}$$
(Section 18-2)

Definition of relative humidity:
$$\text{Relative humidity} = \frac{\text{Partial pressure } H_2O}{\text{Saturated Vapor pressure of } H_2O} \times 100\%$$
(Section 18-4)

Clausius equation of state:
$$P(V - nb) = nRT$$
(Section 18-5)

Van der Waals equation of state:
$$\left(P + \frac{a}{(V/n)^2} \right) \left(\frac{V}{n} - b \right) = RT$$
(Section 18-5)

Mean free path in a gas:
$$l_M = \frac{1}{4\pi\sqrt{2} r^2 (N/V)}$$
(Section 18-6)

Diffusion equation or Fick's law:
$$J = DA \frac{dC}{dx}$$
(Section 18-7)

Chapter Summary

Section 18-1. The Ideal Gas Law and the Molecular Interpretation of Temperature

An ideal gas is a model of the gas phase of matter. The main properties of the ideal gas model are:
- The particles of the gas have a negligible volume compared to the volume of the container
- The number of particles in the gas is large.
- Particles of the gas move independently and interact only during collisions.
- All collisions with walls of containers and other gas particles are elastic collisions.

The average kinetic energy $\bar{K}$ of the gas molecules is proportional to the absolute temperature T of the gas:

$$\bar{K} = \left[\frac{1}{2} m \left(v^2 \right) \right]_{av} = \frac{3}{2} kT$$

where m is the mass of the gas molecules, v is their speed, and k is the Boltzmann's constant. The root-mean-square speed, v_{RMS}, of the gas particles is

$$v_{RMS} = \sqrt{\frac{3kT}{m}}$$

Root-mean-square (RMS) means the square root of the mean (average) of the square of the quantity. **Root-mean-square speed** means the square root of the average of the square of the speed of each particle. This is different from the mean speed $\bar{v}$, which is the average of the speeds of each particle.

Example 18-1-A: RMS speeds of dust particles. Suppose you have a "gas" of spherical dust particles with each dust particle having a radius $r = 0.1$ μm and a density of $\rho = 1800$ kg/m^3. What is the RMS speed of these particles at a temperature 300 K?

Approach: To calculate the rms speed of the dust particles we need to know their temperature and their mass. The mass of the dust particles can be calculated from the size and the density of the dust particles.

Solution: The mass of each dust particle is:

$$m = \rho V = \rho \left(\frac{4}{3} \pi r^3 \right) = (1800) \frac{4}{3} \pi \left(0.1 \times 10^{-6} \right)^3 = 7.5 \times 10^{-18} \, \text{kg}$$

The RMS speed is then:

$$v_{RMS} = \sqrt{\frac{3kT}{m}} = \sqrt{\frac{3 \left(1.38 \times 10^{-23} \right) (300)}{\left(7.5 \times 10^{-18} \right)}} = 0.041 \, \text{m/s} = 4.1 \, \text{cm/s}$$

Compare this result with the result of Example 18-2 in the textbook where the RMS speed for air molecules is calculated. The RMS velocity of the dust particles (0.041 m/s) is very small compared to RMS velocity of the air molecules (480 m/s).

Section 18-2. Distribution of Molecular Speeds

The speeds of particles in an ideal gas are not all the same but are distributed according to a distribution called the **Maxwell distribution of speeds**. Consider an ideal gas containing N particles with temperature T. The number of particles in this gas with a velocity between v and $v + dv$ is $f(v)dv$ where $f(v)$ is the Maxwell distribution of speeds:

$$f(v) = 4\pi N \left(\frac{m}{2\pi kT} \right)^{\frac{3}{2}} v^2 e^{-\frac{1}{2} \frac{mv^2}{kT}}$$

In this expression, m is the mass of the gas particles, k is Boltzmann's constant, T is the absolute temperature, and N is the number of particles in the gas.

Example 18-2-A: Applications of the Maxwell distribution. For a fixed speed v, determine the ratio of the number of molecules at temperature T_2 to the number of molecules at temperature T_1 within the interval v to $v + dv$.

Approach: The number of particles with a velocity between v and $v + dv$ is equal to $f(v)dv$. The Maxwell distribution depends on the temperature and the number of particles with a velocity between v and $v + dv$ will thus depend on temperature. The ratio of the number of particles at the two temperatures is just the ratio of the Maxwell distributions at these two temperatures.

Solution: The ratio of the number of molecules within the given velocity interval will be the ratio of the Maxwell distribution at those two temperatures:

$$\frac{f_{T_2}(v)}{f_{T_1}(v)} = \frac{4\pi N\left(\dfrac{m}{2\pi kT_2}\right)^{\frac{3}{2}} v^2 e^{-\frac{1}{2}\frac{mv^2}{kT_2}}}{4\pi N\left(\dfrac{m}{2\pi kT_1}\right)^{\frac{3}{2}} v^2 e^{-\frac{1}{2}\frac{mv^2}{kT_1}}} = \left(\frac{T_1}{T_2}\right)^{\frac{3}{2}} e^{-\frac{1}{2}\frac{mv^2}{k}\left(\frac{1}{T_2}-\frac{1}{T_1}\right)}$$

Section 18-3. Real Gases and Changes of Phase

At high temperatures and low densities, real gases have a behavior that is well described by the ideal gas model. However, when the temperature becomes low or the density becomes high, the behavior of real gases can depart significantly from that predicted by the ideal gas model. At high densities, the ideal gas model fails since the volume of the gas molecules is not insignificant. At these densities, the probability of collisions increases which causes the pressure in a real gas to be greater than that predicted by the ideal gas law. At low temperatures, the average kinetic energy of the gas particles becomes comparable to the size of the potential energy associated with the interaction between the gas molecules, and the gas particles no longer move independently.
At very low temperatures or very high densities, the molecular interaction is so large that the molecules tend to stick together to form liquids or solids. There is a maximum temperature at which a liquid can exist. This temperature is called the **critical temperature**. The **critical point** of a material is determined by the critical density and pressure at the critical temperature.
A **phase diagram** is a plot of the pressure versus the temperature that shows the equilibrium phases for a given pressure and temperature. The **triple point** of a material is the point on the pressure–temperature plot of the material at which the three phases of matter can coexist in equilibrium.

Section 18-4. Vapor Pressure and Humidity

The molecules in a liquid can escape from the surface if their kinetic energy exceeds the potential energy that binds them to the remaining liquid. At any finite temperature, the speeds of the molecules have a distribution that ensures at any given instant that there is a finite probability that a molecule has enough kinetic energy to escape from the surface of the liquid. The escape of molecules from the surface of a liquid is called **evaporation**.
The molecules that escape from the surface of the liquid undergo collisions with the gas molecules above the liquid. There is some probability that the molecules will return to the liquid. This process is called **condensation**. The rate of condensation increases with an increased density of molecules in the gas above the liquid. In a closed container, with both liquid and vapor molecules, the density of vapor molecules will eventually reach an equilibrium such that the condensation rate will equal the evaporation rate. The pressure in the gas when the system is in equilibrium is called the **saturated vapor pressure**. The saturated vapor pressure for a given liquid depends on the temperature. When the saturated pressure equals the external pressure, **boiling** will occur.
In the ideal gas model, the total pressure in a gas is the sum of the pressure for each of the constituents of the gas. For example, consider a gas containing two different types of molecules A and B. If only the A molecules were present, the pressure would be P_A. If only the B molecules were present, the pressure would be P_B. Each of these pressures is called the **partial pressure** of the corresponding molecule. The pressure in the gas composed of both types of molecules will be the sum of all the partial pressures. For the gas that contains molecules A and B, the gas pressure P is equal to

$$P = P_A + P_B.$$

The **relative humidity** is the ratio of the partial pressure of water in the air to the saturated vapor pressure at the air temperature:

$$\text{relative humidity} = \frac{\text{partial pressure } H_2O}{\text{saturated vapor pressure of } H_2O} \times 100\%$$

It is possible for the air to have a partial pressure of water vapor greater than the saturated vapor pressure. In this condition, the air is **supersaturated** with water vapor.

The **dew point** is the temperature at which the existing partial pressure of water in the air would be equal to the saturated vapor pressure of water. When the temperature drops below the dew point, the water vapor will condense and fog will form.

Example 18-4-A. Calculating the relative humidity. If 0.500 kg of water evaporates in a room of dry air at a temperature of 20°C, what will the relative humidity be? The room is 3.00 m × 4.00 m × 2.5 m. Treat the water vapor as an ideal gas.

Approach: Since the air initially is dry air, we assume that no water vapor is present. Based on the information provided, we can determine the number of moles of water molecules in the air. Since we can treat the water as an ideal gas, we can use this information to calculate the partial pressure due to the water molecules. Table 18.2 in the textbook provides us with information about the saturated vapor pressure at 20°C and this information can be used to calculate the relative humidity.

Solution: The number of moles of water that will be in the air after it evaporates is equal to

$$n_{H_2O} = \frac{M}{m} = \frac{0.500\,\text{kg}}{0.018\,\text{kg}} = 27.8 \text{ moles}$$

Using the ideal gas law we can determine the partial pressure of the water vapor:

$$P_{H_2O} = \frac{n_{H_2O}RT}{V} = \frac{(27.8)(8.31)(293.15)}{(3.00)(4.00)(2.5)} = 2.26 \times 10^3 \text{ N/m}^2$$

The relative humidity can now be calculated using this partial pressure and the saturated vapor pressure of water at this temperature:

$$\text{relative humidity} = \frac{\text{partial pressure } H_2O}{\text{saturated vapor pressure of } H_2O} \times 100\% = \frac{2.26 \times 10^3}{2.33 \times 10^3} \times 100\% = 97\%$$

About one pint of water is enough to almost saturate the air of a small room with water vapor at 20°C!

Example 18-4-B. Calculating the dew point. By how much do we have to lower the room temperature in the previous problem for the relative humidity to become 100%. This temperature is the dew point.

Approach: When we lower the temperature, the saturated vapor pressure will decrease. We can interpolate the change in the vapor pressure using the data provided in Table 18-2. However, the partial pressure also decreases when we decrease the temperature. We have to take both changes into consideration in order to determine the required change in temperature.

Solution: Interpolating the vapor pressures in Table 18-2 between 20°C and 15°C we obtain:

$$\text{saturated vapor pressure of } H_2O = 2.33 \times 10^3 + 124(T - 293.15)$$

The partial pressure of the water vapor depends on temperature and is equal to

$$P_{H_2O} = \frac{n_{H_2O}RT}{V} = \frac{(27.8)(8.31)T}{(3.00)(4.00)(2.5)} = 7.70T$$

We now have to determine the temperature at which the relative humidity becomes 100%. This requires that

$$\text{relative humidity} = \frac{\text{partial pressure } H_2O}{\text{saturated vapor pressure of } H_2O} \times 100\% = \frac{7.70T}{2.33 \times 10^3 + 124(T - 293.15)} \times 100\% = 100\%$$

We can rewrite this requirement as

$$\left\{2.33 \times 10^3 + 124(T - 293.15)\right\} = -3.40 \times 10^4 + 124T = 7.70T \quad \rightarrow \quad T = \frac{3.40 \times 10^4}{124 - 7.70} = 292.52 \text{ K}$$

Section 18-5. Van der Waals Equation of State

Shortcomings of the ideal gas model in describing real gases can be addressed by applying various corrections:
- **Volume correction.** If the molecules of the gas have a volume that is not negligible, a correction must be made to the ideal gas law by reducing the actual volume by the volume of the gas molecules:

$$P(V - nb) = nRT$$

  where nb is the total volume of all the molecules in the gas. This equation is called the **Clausius equation of state**. If the volume of the molecules nb is negligible compared to the volume of the gas V then the Clausius equation of state reduces to the ideal gas law. When V becomes equal to the volume of the gas molecules, we see that the pressure P will climb without limit. That is what would be expected for a gas made of rigid molecules of finite size.
- **Attractive force correction.** A correction must also be made to correct for the attractive force between the gas molecules. The **van der Waals equation of state** includes a correction to account for these attractive forces:

$$\left(P + \frac{a}{(V/n)^2}\right)\left(\frac{V}{n} - b\right) = RT$$

  In this equation, a is a constant, proportional to the magnitude of the attractive interaction between the gas molecules. When a becomes small, the van der Waals equation of state approaches the Clausius equation of state.

Example 18-5-A. Working with the van der Waals equation of state. A fit of the van der Waals equation of state for a particular gas results in constants $a = 2.6 \times 10^{-2}$ Nm^4/mol^2 and $b = 6.3 \times 10^{-5}$ m^3/mol. At what temperature are the pressures given by the van der Waals equation of state and the ideal gas law identical if the molar density of the gas is 76 moles/m^3?

Approach: In this problem we consider a volume V of a certain gas and try to determine the temperature T at which the pressure consistent with the ideal gas law is the same as the pressure that is consistent with the van der Waals equation of state. For each equation of state we have to determine the pressure for a specific temperature T. We will have two equations with two unknown, and this set of equations can be solved.

Solution: We will denote the pressure from the van der Waals equation of state as P_{vdw} and the pressure from the ideal gas law as P_i. The van der Waals equation of state and the ideal gas law for a given temperature T are:

$$\left(P_{vdw} + \frac{a}{\left(\frac{V}{n}\right)^2}\right)\left(\frac{V}{n} - b\right) = RT \quad \text{and} \quad P_i V = nRT$$

Each equation of state can be used to express the pressure in terms of the temperature. We obtain:

$$P_{vdw} = \frac{RT}{\left(\frac{V}{n} - b\right)} - \frac{a}{\left(\frac{V}{n}\right)^2} \quad \text{and} \quad P_i = \frac{nRT}{V}$$

Setting the two pressures equal we obtain:

$$\frac{RT}{\left(\dfrac{V}{n}-b\right)} - \frac{a}{\left(\dfrac{V}{n}\right)^2} = \frac{nRT}{V}$$

This equation has only one unknown, the temperature T, which can now be determined:

$$T = \frac{a}{R\left(\dfrac{V}{n}\right)^2} \frac{1}{\dfrac{1}{\left(\dfrac{V}{n}-b\right)} - \dfrac{n}{V}} = \frac{\left(2.6 \times 10^{-2}\right)}{(8.31)\left(\dfrac{1}{76}\right)^2} \frac{1}{\dfrac{1}{\left(\dfrac{1}{76}-6.3\times10^{-5}\right)} - 76} = 49\,\text{K}$$

Section 18-6. Mean Free Path

Because of the finite size of the molecules in a real gas, the particles undergo collisions with each other. A quantity that is useful to interpret certain macroscopic phenomena in terms of the microscopic properties of the gas molecules is the **mean free path**. The mean free path is the average distance a gas molecule travels between collisions with other gas molecules. It is easy to see that the mean free path depends on the size of the gas molecules and the average spacing between them. For spherical gas molecules, with radius r, that have a distribution of speeds given by the Maxwell distribution, the mean free path, ℓ_M, is given by

$$\ell_M = \frac{1}{4\pi\sqrt{2}\,r^2\left(\dfrac{N}{V}\right)}$$

where N is the number of molecules in the gas, and V is the volume of the gas.

Example 18-6-A: Changing pressure to adjust the mean free path. Many processes are carried out in a vacuum. In a good vacuum, the number of molecules per unit volume is very small, and the mean free path of these molecules is very large. These molecules can move from one part of the system to another with a small probability of colliding with other gas molecules. By adjusting the pressure of the gas, we can adjust the mean free path of the molecules. What must the pressure be in a system that needs a mean free path for air molecules of 10.0 cm at a temperature 300 K? Use a diameter of 3×10^{-10} m for air molecules.

Approach: The mean free path of the molecules can be fine-tuned by adjusting the N/V ratio of the gas. Using the ideal gas law, we can determine the pressure that corresponds to this N/V ratio.

Solution: The mean free path depends on the N/V ratio of the molecules. In this problem we are told what the required mean free path is, and we can thus determine what the required N/V ratio is:

$$\frac{N}{V} = \frac{1}{4\pi\sqrt{2}\,r^2\ell_M}$$

Assuming that the gas behaves like an ideal gas, the ratio N/V is also related to the pressure and temperature of the gas:

$$\frac{N}{V} = \frac{P}{kT}$$

Combining the last two expressions, we can relate the mean free path to the gas pressure:

$$\frac{1}{4\pi\sqrt{2}\,r^2\ell_M} = \frac{P}{kT} \quad \rightarrow \quad P = \frac{kT}{4\pi\sqrt{2}\,r^2\ell_M} = \frac{\left(1.38\times10^{-23}\right)(300)}{4\pi\sqrt{2}\left(3\times10^{-10}\right)^2(0.10)} = 0.026\,\text{N/m}^2$$

This is less than one-millionth of atmospheric pressure, but is easily obtainable with vacuum systems used for the processes that require this size of mean free path.

Section 18-7. Diffusion

In order to describe the motion of molecules over distances that are large compared to the mean free path, statistical techniques must be employed. A good example of a statistical process is **diffusion**. It is observed that if a difference in concentration of a certain species of gas molecules exists, the average motion of these molecules will form regions of high concentration to regions of low concentration. This motion will continue until all regions have the same concentration.

The rate of diffusion J of a given species of molecules across a surface of area A is related to the gradient of the concentration of that species in the direction perpendicular to the area A. The relation between the diffusion rate and the concentration gradient is called the **diffusion equation** or **Fick's law**. The diffusion rate is proportional to the surface area A through which the molecules are diffusing and the concentration gradient dC/dx:

$$J = DA\frac{dC}{dx}$$

In this equation, D is called the **diffusion constant**. The diffusion constant depends on the species that is diffusing, on the material they are diffusing through, and on the temperature and density.

Example 18-7-A. Diffusion of air. Consider a room of volume V connected to the outside air by a narrow passage with a cross-sectional area A and length L. In the room, there are perfume molecules uniformly distributed with a concentration C_0. Estimate the concentration of the perfume molecules in the room as a function of time. Assume the dimensions of the room are small compared to the length of the passage. The diffusion constant of the perfume molecules is D.

Approach: We will assume that the density of perfume molecules in the outside air is 0. Since the volume of the outside air is large compared to the volume of the room, the density of perfume molecules in the outside air will remain approximately 0. The concentration of perfume molecules in the room, $C(t)$, will be a function of time; the gradient of the concentration will be equal to $C(t)/L$. Using Fick's law we can obtain a relation between the number of perfume molecules left in the room and the rate of diffusion.

Solution: The rate of diffusion of perfume molecules out of the room is given by Fick's law:

$$J = DA\frac{dC}{dx} = DA\frac{C(t)}{L}$$

The concentration of perfume molecules depends on the number of perfume molecules, N, in the room:

$$C(t) = \frac{N(t)}{V}$$

Due to diffusion, the number of perfume molecules in the room will change. The rate of change is related to the diffusion rate:

$$\frac{dN}{dt} = -J = -\frac{DAC(t)}{L}$$

The rate of change dN/dt can also be expressed in terms of the rate of change of the concentration $C(t)$:

$$N(t) = VC(t) \quad \rightarrow \quad \frac{dN}{dt} = V\frac{dC}{dt}$$

Combining the two expressions for dN/dt we obtain

$$-\frac{DAC(t)}{L} = V\frac{dC}{dt} \quad \rightarrow \quad \frac{dC}{C} = -\frac{DA}{VL}dt$$

Both sides of this equation can be integrated:

$$\int_{C(0)}^{C(t)} \frac{dC(t)}{C(t)} = -\frac{DA}{L}\int_0^t dt \Rightarrow \ln C(t) - \ln C(0) = -\frac{DA}{L}t$$

Taking the exponential function of each side of the equation gives

$$C(t) = C(0)e^{-\frac{DA}{L}t}$$

We need to realize that our assumptions that the concentration in the room is uniform and the concentration outside is 0 are in general not exactly true. The solution obtained here is thus an approximate solution.

Practice Quiz

1. Consider the three following gas models: the Clausius equation of state, the van der Waals equation of state, and the equation of state of an ideal gas. Which statement about pressure predictions of the models is definitely true for a given density and temperature? Assume that the constants a and b of the Clausius and van der Waals equations of state are positive.
    a) The ideal gas pressure is greater than the Clausius pressure.
    b) The van der Waals pressure is greater than the Clausius pressure.
    c) The Clausius pressure is greater than the ideal gas pressure.
    d) The van der Waals pressure is greater than the ideal gas pressure.

2. Why is one of the assumptions of the kinetic theory of the ideal gas model that there are a large number of molecules?
    a) Many of the quantities derived require averages over many molecules.
    b) If you lose some molecules by leakage, it won't matter as much.
    c) To ensure that the pressure will be large.
    d) To ensure that there are molecules traveling in every direction.

3. You place a 1-m-tall beaker, filled with water, on a hot plate. A 10-cm beaker, filled with water, is placed on the same hot plate. Of course, it takes longer for the water in the 1-m-tall beaker to boil because there is more water in the tall beaker to heat. We observe that when the water is boiling, the temperature at the bottom of the tall beaker is higher than the temperature at the bottom of the short beaker when the water is boiling. Why?
    a) The top of the tall beaker is at a lower atmospheric pressure.
    b) There is more water in the tall beaker.
    c) The pressure is higher at the bottom of the tall beaker.
    d) The volume of the large beaker is greater.

4. Why is a humidifier necessary in many heated homes during the winter?
    a) The pressure of the air in the house drops as it is heated and the partial pressure of the added water is increased to help the indoor pressure match the outdoor pressure.
    b) As the air is heated, the average speed of the molecules in the air increases. Water molecules move slower than other air molecules and help to slow the other molecules down.
    c) Water molecules are less massive than most air molecules and therefore move faster than the other air molecules and therefore make it easier to warm up the air.
    d) As the air is heated, its relative humidity drops. To keep the relative humidity high enough for comfort, water is added to the air.

5. By what factor does the RMS speed of an ideal gas change when the absolute temperature of the gas is doubled?
    a) 2
    b) 4
    c) ½
    d) √2

6. Why is the diffusion constant for hydrogen molecules larger than the diffusion constant for oxygen molecules?
    a) Hydrogen is more reactive than oxygen.
    b) Hydrogen molecules are less massive and smaller than oxygen molecules.
    c) Hydrogen molecules have fewer protons than oxygen molecules.
    d) Hydrogen molecules are larger than oxygen molecules.

7. Which of the following changes increases the mean free path of the molecules in a gas that is maintained at a constant volume (assume it remains above the critical temperature)?
    a) Increasing the number of molecules per unit volume.
    b) Increasing the temperature of the gas.
    c) Decreasing the temperature of the gas.
    d) Decreasing the radius of the molecules.

8. What condition exists when the partial pressure of a vapor exceeds the saturated vapor pressure of the vapor?
    a) Saturation
    b) Evaporation
    c) Boiling
    d) Supersaturation

9. Why does it take longer to cook food in boiling water at the top of a mountain than at sea level?
    a) The water's saturated vapor pressure reaches the atmospheric pressure at a lower temperature at the top of a mountain than at sea level.
    b) The food has more potential energy stored in it at the top of the mountain.
    c) It is usually colder on top of a mountain.
    d) The relative humidity is low at the top of a mountain.

10. A small drop of dye is placed in the center of a beaker of water that has a diameter D. It takes a time T for the dye molecules to diffuse to the edge of the beaker. If the measurement is repeated with a beaker that has a diameter $2D$, how much time does it take the dye molecules to diffuse to the edge of the beaker?
    a) T
    b) $2T$
    c) $3T$
    d) $4T$

11. At what temperature would the RMS speed of hydrogen molecules be 1.00 cm/s?

12. Using the Maxwell distribution of speeds, calculate the expected number of gas particles that have a kinetic energy between E and $E + dE$ in a gas of N particles at temperature T.

13. Determine the total mass of water in a cubic meter of air with a relative humidity of 85% at a temperature 25° C.

14. What is the number of molecules per cubic meter in a gas with a mean free path of 10 cm? Assume the effective diameter of the molecules is 3.0×10^{-10} m.

15. Determine constants a and b of the van der Waals equation of state for 4 moles of a gas that has a pressure 1.000×10^5 N/m^2 at a temperature of 300.0 K and volume 0.0900 m^3 and a pressure 0.600×10^5 N/m^2 at a temperature of 240.0 K and the same volume.

Responses to Select End-of-Chapter Questions

1. One of the fundamental assumptions for the derivation of the ideal gas law is that the average separation of the gas molecules is much greater than the diameter of the molecules. This assumption eliminates the need to consider the different sizes of the molecules.

7. For an absolute vacuum, no. But for most "vacuums," there are still a few molecules in the containers, and the temperature can be determined from the (very low) pressure.

13. Evaporation. Only molecules in a liquid that are traveling fast enough will be able to escape the surface of the liquid and evaporate.

19. The freezing point of water decreases slightly with higher pressure. The wire exerts a large pressure on the ice (due to the weights hung at each end). The ice under the wire will melt, allowing the wire to move lower into the block. Once the wire has passed a given position, the water now above the wire will have only atmospheric pressure on it and will refreeze. This process allows the wire to pass all the way through the block and yet leave a solid block of ice.

25. Liquid CO_2 can exist at temperatures between $-56.6°C$ and $31°C$ and pressures between 5.11 atm and 73 atm. (See Figure 18-6.) CO_2 can exist as a liquid at normal room temperature, if the pressure is between 56 and 73 atm.

Solutions to Select End-of-Chapter Problems

1. (a) The average translational kinetic energy of a gas molecule is $\frac{3}{2}kT$.

    $$KE_{avg} = \tfrac{3}{2}kT = \tfrac{3}{2}\left(1.38\times10^{-23}\,J/K\right)\left(273\,K\right) = \boxed{5.65\times10^{-21}\,J}$$

    (b) The total translational kinetic energy is the average kinetic energy per molecule, times the number of molecules.

    $$KE_{total} = N\left(KE_{avg}\right) = \left(1.0\,mol\right)\left(\frac{6.02\times10^{23}\,molecules}{1}\right)\tfrac{3}{2}\left(1.38\times10^{-23}\,J/K\right)\left(298\,K\right)$$

    $$= \boxed{3700\,J}$$

7. The mean (average) speed is as follows.

    $$v_{avg} = \frac{6.0+2.0+4.0+6.0+0.0+4.0+1.0+8.0+5.0+3.0+7.0+8.0}{12} = \frac{54.0}{12} = \boxed{4.5}\ .$$

    The rms speed is the square root of the mean (average) of the squares of the speeds.

    $$v_{rms} = \sqrt{\frac{6.0^2+2.0^2+4.0^2+6.0^2+0.0^2+4.0^2+1.0^2+8.0^2+5.0^2+3.0^2+7.0^2+8.0^2}{12}}$$

    $$= \sqrt{\frac{320}{12}} = \boxed{5.2}$$

13. From Eq. 18-5, we have $v_{rms} = \sqrt{\dfrac{3kT}{m}}$

    (a) $\dfrac{dv_{rms}}{dT} = \dfrac{d}{dT}\left(\dfrac{3kT}{m}\right)^{1/2} = \dfrac{1}{2}\left(\dfrac{3k}{m}\right)^{1/2}\dfrac{1}{T^{1/2}} = \dfrac{1}{2}\left(\dfrac{3kT}{m}\right)^{1/2}\dfrac{1}{T} = \boxed{\dfrac{1}{2}\dfrac{v_{rms}}{T}}$

    $\Delta v_{rms} \approx \dfrac{dv_{rms}}{dT}\Delta T = \dfrac{1}{2}\dfrac{v_{rms}}{T}\Delta T \rightarrow \dfrac{\Delta v_{rms}}{v_{rms}} \approx \boxed{\dfrac{1}{2}\dfrac{\Delta T}{T}}$

    (b) The temperature must be calculated in Kelvin for the formula to be applicable. We calculate the percent change relative to the winter temperature.

    $\dfrac{\Delta v_{rms}}{v_{rms}} \approx \dfrac{1}{2}\dfrac{\Delta T}{T} = \dfrac{1}{2}\left(\dfrac{30\,K}{268\,K}\right) = 0.056 = \boxed{5.6\%}$

19. In the Maxwell distribution, Eq. 18-6, we see that the mass and temperature always occur as a ratio. Thus if the mass has been doubled, $\boxed{\text{doubling the temperature}}$ will keep the velocity distribution constant.

25. (*a*) From Fig. 18-5, water is $\boxed{\text{vapor}}$ when the pressure is 0.01 atm and the temperature is 90°C.

    (*b*) From Fig. 18-5, water is $\boxed{\text{solid}}$ when the pressure is 0.01 atm and the temperature is -20°C.

31. At the boiling temperature, the air pressure equals the saturated vapor pressure. The pressure of 0.75 atm is equal to 7.60×10^4 Pa. From Table 18-2, the temperature is between 90°C and 100°C. Since there is no entry for 7.60×10^4 Pa, the temperature can be estimated by a linear interpolation. Between 90°C and 100°C, the temperature change per Pa is as follows:

$$\frac{(100-90)\,^\circ\text{C}}{(10.1-7.01)\times 10^4\,\text{Pa}} = 3.236 \times 10^{-4}\,^\circ\text{C/Pa}.$$

Thus the temperature corresponding to 7.60×10^4 Pa is

$$90^\circ\text{C} + \left[(7.60-7.01)\times 10^4\,\text{Pa}\right]\left(3.236 \times 10^{-4}\,^\circ\text{C/Pa}\right) = 91.9^\circ\text{C} \approx \boxed{92^\circ\text{C}}.$$

37. At 30.0°C, the saturated vapor pressure as found in Table 18-2 is 4240 Pa. We can find the partial pressure of the water vapor by using the equation given immediately before Example 18-6.

$$\text{Rel. Hum.} = \frac{P_{\text{partial}}}{P_{\text{saturated}}} \times 100 \quad \rightarrow \quad P_{\text{partial}} = P_{\text{saturated}} \frac{\text{Rel. Hum.}}{100} = (4240\,\text{Pa})(0.45) = 1908\,\text{Pa}$$

The dew point is that temperature at which 1908 Pa is the saturated vapor pressure. From Table 18-2, we see that will be between 15°C and 20°C.

$$T_{1920} = T_{1710} + \frac{1908\,\text{Pa} - 1710\,\text{Pa}}{2330\,\text{Pa} - 1710\,\text{Pa}}(5^\circ\text{C}) = \boxed{16.6^\circ\text{C}}$$

43. (*a*) The van der Waals equation of state is given by Eq. 18-9, $P = \dfrac{RT}{(V/n)-b} - \dfrac{a}{(V/n)^2}$. At the critical point, both the first and second derivatives of P with respect to V are 0. Use those conditions to find the critical volume, and then evaluate the critical temperature and critical pressure.

$$P = \frac{RT}{(V/n)-b} - \frac{a}{(V/n)^2} = \frac{nRT}{V-nb} - \frac{an^2}{V^2}$$

$$\frac{dP}{dV} = -\frac{nRT}{(V-nb)^2} + \frac{2an^2}{V^3} \quad ; \quad \frac{dP}{dV} = 0 \quad \rightarrow \quad T_{\text{crit}} = \frac{2an(V_{\text{crit}}-nb)^2}{RV_{\text{crit}}^3}$$

$$\frac{d^2P}{dV^2} = \frac{2nRT}{(V-nb)^3} - \frac{6an^2}{V^4} \quad ; \quad \frac{d^2P}{dV^2} = 0 \quad \rightarrow \quad T_{\text{crit}} = \frac{3an(V_{\text{crit}}-nb)^3}{RV_{\text{crit}}^4}$$

Set the two expressions for the critical temperature equal to each other, and solve for the critical volume. Then use that expression to find the critical temperature, and finally the critical pressure.

$$\frac{2an(V_{\text{crit}}-nb)^2}{RV_{\text{crit}}^3} = \frac{3an(V_{\text{crit}}-nb)^3}{RV_{\text{crit}}^4} \quad \rightarrow \quad V_{\text{crit}} = 3nb$$

$$T_{\text{crit}} = \frac{2an(V_{\text{crit}}-nb)^2}{RV_{\text{crit}}^3} = \frac{2an(3nb-nb)^2}{R(3nb)^3} = \boxed{\frac{8a}{27bR}}$$

$$P_{\text{crit}} = \frac{nRT_{\text{crit}}}{V_{\text{crit}}-nb} - \frac{an^2}{V_{\text{crit}}^2} = \boxed{\frac{a}{27b^2}}$$

(b) To evaluate the constants, use the ratios $\dfrac{T_{crit}^2}{P_{crit}}$ and $\dfrac{T_{crit}}{P_{crit}}$.

$$\frac{T_{crit}^2}{P_{crit}} = \frac{\left(\dfrac{8a}{27bR}\right)^2}{\dfrac{a}{27b^2}} = \frac{64a}{27R^2} \quad \rightarrow$$

$$a = \frac{27R^2T_{crit}^2}{64P_{crit}} = \frac{27\left(8.314\dfrac{\text{J}}{\text{mol}\bullet\text{K}}\right)^2(304\,\text{K})^2}{64(72.8)(1.013\times10^5\,\text{Pa})} = \boxed{0.365\,\frac{\text{N}\bullet\text{m}^4}{\text{mol}^2}}$$

$$\frac{T_{crit}}{P_{crit}} = \frac{\dfrac{8a}{27bR}}{\dfrac{a}{27b^2}} = \frac{8b}{R} \quad \rightarrow \quad b = \frac{RT_{crit}}{8P_{crit}} = \frac{\left(8.314\dfrac{\text{J}}{\text{mol}\bullet\text{K}}\right)(304\,\text{K})}{8(72.8)(1.013\times10^5\,\text{Pa})} = \boxed{4.28\times10^{-5}\,\text{m}^3/\text{mol}}$$

49. (a) If the average speed of a molecule is $\overline{v}$, then the average time between collisions (seconds per collision) is the mean free path divided by the average speed. The reciprocal of that average time (collisions per second) is the frequency of collisions. Use Eq. 18-10b for the mean free path. The typical size of an air molecule is given in problem 45, which can be used for the size of the nitrogen molecule.

$$\Delta t_{avg} = \frac{l_M}{\overline{v}} \quad \rightarrow \quad f = \frac{1}{\Delta t_{avg}} = \frac{\overline{v}}{l_M} = \boxed{4\sqrt{2}\pi r^2 \overline{v}\frac{N}{V}}$$

(b) From Eq. 18-7b, $\overline{v} = \sqrt{\dfrac{8kT}{\pi m}}$, and from the ideal gas law, $PV = NkT \rightarrow \dfrac{N}{V} = \dfrac{P}{kT}$. Use these relationships to calculate the collision frequency.

$$f = 4\sqrt{2}\pi r^2 \overline{v}\frac{N}{V} = 4\sqrt{2}\pi r^2 \sqrt{\frac{8kT}{\pi m}}\frac{P}{kT} = 16Pr^2\sqrt{\frac{\pi}{mkT}}$$

$$= 16(0.010)(1.013\times10^5\,\text{Pa})(1.5\times10^{-10}\,\text{m})^2 \times$$

$$\sqrt{\frac{\pi}{28(1.66\times10^{-27}\,\text{kg})(1.38\times10^{-23}\,\text{J/K})(293\,\text{K})}} = \boxed{4.7\times10^7\,\text{collisions/s}}$$

55. (a) Use the ideal gas law to find the concentration of the oxygen. We assume that the air pressure is 1.00 atm, and so the pressure caused by the oxygen is 0.21 atm.

$$PV = nRT \quad \rightarrow$$

$$\frac{n}{V} = \frac{P}{RT} = \frac{(0.21\,\text{atm})(1.013\times10^5\,\text{Pa/atm})}{(8.315\,\text{J/mol}\bullet\text{K})(293\,\text{K})} = 8.732\,\text{mol/m}^3 \approx \boxed{8.7\,\text{mol/m}^3}$$

(b) Use Equation 18-11 to calculate the diffusion rate.

$$J = DA\frac{C_1 - C_2}{\Delta x} = (1\times10^{-5}\,\text{m}^2/\text{s})(2\times10^{-9}\,\text{m}^2)\left(\frac{8.732\,\text{mol/m}^3 - 4.366\,\text{mol/m}^3}{2\times10^{-3}\,\text{m}}\right)$$

$$= 4.366\times10^{-11}\,\text{mol/s} \approx \boxed{4\times10^{-11}\,\text{mol/s}}$$

(*c*) From Example 18-9, we have an expression for the time to diffuse a given distance.

$$t = \frac{\bar{C}}{\Delta C}\frac{(\Delta x)^2}{D} = \frac{\frac{1}{2}\left(8.732\,\text{mol/m}^3 + 4.366\,\text{mol/m}^3\right)}{\left(8.732\,\text{mol/m}^3 - 4.366\,\text{mol/m}^3\right)}\frac{\left(2\times10^{-3}\,\text{m}\right)^2}{1\times10^{-5}\,\text{m}^2/\text{s}} = \boxed{0.6\,\text{s}}$$

61. Calculate the volume per molecule from the ideal gas law, and assume the molecular volume is spherical.

$$PV = NkT \;\rightarrow\; \frac{V}{N} = \frac{kT}{P} = \tfrac{4}{3}\pi r^3 \;\rightarrow$$

$$r_{\text{Volume}} = \left(\frac{3kT}{4\pi P}\right)^{1/3} = \left(\frac{3\left(1.38\times10^{-23}\,\text{J/K}\right)\left(273\,\text{K}\right)}{4\pi\left(1.01\times10^5\,\text{Pa}\right)}\right)^{1/3} = 2.07\times10^{-8}\,\text{m}$$

The intermolecular distance would be twice this "radius," so about 4×10^{-8} m. This is about 130 times larger than the molecular diameter: $\dfrac{d_{\text{Volume}}}{d_{\text{molecule}}} \approx \dfrac{4\times10^{-8}\,\text{m}}{3\times10^{-10}\,\text{m}} = 133$. So if we say the molecular diameter is 4 cm, then the intermolecular distance would 133 times that, or about $\boxed{5\text{ meters}}$.

67. The temperature can be found from the rms speed by Eq. 18-5, $v_{\text{rms}} = \sqrt{3kT/m}$. The molecular mass of nitrogen molecules is 28.

$$v_{\text{rms}} = \sqrt{3kT/m} \;\rightarrow$$

$$T = \frac{mv_{\text{rms}}^2}{3k} = \frac{(28)\left(1.66\times10^{-27}\,\text{kg}\right)\left[\left(4.2\times10^4\,\text{km/h}\right)\left(\dfrac{1\,\text{m/s}}{3.6\,\text{km/h}}\right)\right]^2}{3\left(1.38\times10^{-23}\,\text{J/K}\right)} = \boxed{1.5\times10^5\,\text{K}}$$

73. Assume that the water is an ideal gas, and that the temperature is constant. From Table 13-3, saturated vapor pressure at 90°C is 7.01×10^4 Pa, and so to have a relative humidity of 10%, the vapor pressure will be 7.01×10^3 Pa. Use the ideal gas law to calculate the amount of water.

$$PV = nRT \;\rightarrow$$

$$n = \frac{PV}{RT} = \frac{\left(7.01\times10^3\,\text{Pa}\right)\left(8.5\,\text{m}^3\right)}{\left(8.314\,\text{J/mol·K}\right)\left(273+90\right)\text{K}} = 19.74\,\text{moles}\left(\frac{18\times10^{-3}\,\text{kg}}{1\,\text{mole}}\right) = \boxed{0.36\,\text{kg}}$$

Chapter 19: Heat and the First Law of Thermodynamics

Chapter Overview and Objectives

In this chapter, heat is defined as the flow of energy between bodies at different temperatures and the first law of thermodynamics is presented. Specific heat and the latent heat of phase change are introduced. Examples of calculations of the work done by a system in general are given, and several specific thermodynamic processes are discussed in detail. Adiabatic processes in ideal gases and the mechanisms of heat transfer are described.

After completing this chapter you should:
- Know what heat is.
- Know the first law of thermodynamics.
- Know the relationship between heat, specific heat, and change of temperature.
- Know what Calories, calories, and kilocalories are and how they are related to the Joule.
- Know how the internal energy of a monatomic ideal gas is related to temperature.
- Know how to calculate the work done by a system during a process.
- Know the definitions of isothermal, adiabatic, isochoric, and isobaric processes.
- Know what is meant by equipartition of energy.
- Know what the three mechanisms of heat transfer are and how conductivity and radiation rates depend on temperature.

Summary of Equations

Mechanical equivalent of heat:	$4.186\,\text{J} = 1\,\text{calorie (cal)}$	(Section 19-1)
	$4186\,\text{J} = 1\,\text{kilocalorie (kcal)}$	(Section 19-1)
Internal energy of ideal monatomic gas:	$E_{int} = \dfrac{3}{2}nRT$	(Section 19-2)
Relationship between heat and temperature change:	$Q = mc\Delta T$	(Section 19-3)
Latent heat of phase change:	$Q = mL$	(Section 19-5)
First law of thermodynamics:	$\Delta E_{int} = Q - W$	(Section 19-6)
Differential form of the first law of thermodynamics:	$dE_{int} = dQ - dW$	(Section 19-6)
Work done by a system during a process:	$W = \displaystyle\int_{V_i}^{V_f} P\,dV$	(Section 19-7)
Work done during an isothermal process:	$W_{\substack{isothermal \\ ideal\ gas}} = nRT \ln \dfrac{V_b}{V_a}$	(Section 19-7)
Work done during an isobaric process:	$W_{isobaric} = P\left(V_B - V_A\right)$	(Section 19-7)
Heat added during an adiabatic process:	$Q_{adiabatic} = 0$	(Section 19-7)
Work done during an isovolumetric process:	$W_{isovolumetric} = 0$	(Section 19-7)

Molar specific heat at constant volume:

$$Q = nC_V \Delta T$$

(Section 19-8)

Molar specific heat at constant pressure:

$$Q = nC_p \Delta T$$

(Section 19-8)

Relationship between C_P and C_V:

$$C_P - C_V = R$$

(Section 19-8)

Definition of the adiabatic gas constant:

$$\gamma = \frac{C_P}{C_V}$$

(Section 19-9)

Relationship of pressure and volume during an adiabatic process:

$$PV^\gamma = \text{constant}$$

(Section 19-9)

Rate of heat conduction:

$$\frac{\Delta Q}{\Delta t} = kA \frac{T_1 - T_2}{\ell}$$

(Section 19-10)

$$\frac{dQ}{dt} = -kA \frac{dT}{dx}$$

(Section 19-10)

Stefan-Boltzmann equation:

$$\frac{dQ}{dt} = e\sigma AT^4$$

(Section 19-10)

Net rate of radiant heat flow from a body at a temperature T_1 in an environment at a temperature T_2:

$$\frac{\Delta Q}{\Delta t} = e\sigma A\left(T_1^4 - T_2^4\right)$$

(Section 19-10)

Chapter Summary

Section 19-1. Heat as Energy Transfer

Heat is a measure of the spontaneous flow of energy from one material to another driven by a temperature difference. Heat flows from materials at a higher temperature to materials at a lower temperature. The unit of heat is the Joule, but it is often specified in units of **calories** (cal) or **kilocalories** (kcal). A calorie is the amount of heat needed to raise the temperature of one gram of water from 14.5°C to 15.5°C. A kilocalorie is 1000 calories and is sometimes written as **Calorie**. Care must be taken to notice whether the heat unit is calorie or Calorie; it is often difficult to determine because handwritten uppercase C's are similar to lowercase c's.

It can be determined experimentally that the **mechanical equivalent of heat** is 4.186 J. This amount of mechanical work will transfer the same amount of energy to an object as 1 calorie of heat:

$$4.186\,\text{J} = 1\,\text{calorie}$$

or, in terms of kilocalories:

$$4186\,\text{J} = 1\,\text{kilocalorie}$$

Section 19-2. Internal Energy

The **internal energy** is the kinetic and potential energy associated with the random motion of atoms and molecules. Sometimes the internal energy is also called **thermal energy**.

The internal energy, E_{int}, of a monatomic ideal gas is given by:

$$E_{int} = \frac{3}{2} nRT$$

where n is the number of moles of the gas, R is the gas constant, and T is the absolute temperature. For a polyatomic ideal gas, the internal energy at a given temperature is higher than the internal energy of a monatomic ideal gas because vibrational and rotational energies are present, in addition to the translational kinetic energy.

The internal energy of liquids and solids is difficult to determine because of the large contribution of potential energy due to the interatomic forces, which make contributions to the internal energy that are of the same order of magnitude as the kinetic energy associated with the translational motion.

Example 19-2-A. Adding heat to a gas. A monatomic ideal gas is maintained at a pressure $P_1 = 1.00 \times 10^5$ N/m^2 and occupies a volume $V_1 = 1.65$ m^3. Energy is added to the gas, resulting in a change of its volume and pressure. Its new pressure is $P_2 = 2.00 \times 10^5$ N/m^2 and its new volume is $V_2 = 2.44$ m^3. How much energy was added to the gas?

Approach: The internal energy of the gas only depends on its temperature. Using the ideal gas law we can express the internal energy of the gas in terms of its pressure and volume.

Solution: We know that the internal energy E_{int} of the gas is related to the temperature:

$$E_{int} = \frac{3}{2} nRT$$

Using the ideal gas law, we can rewrite this expression as

$$E_{int} = \frac{3}{2} nRT = \frac{3}{2} PV$$

The initial internal energy is

$$E_{int,1} = \frac{3}{2} P_1 V_1$$

The final internal energy is

$$E_{int,2} = \frac{3}{2} P_2 V_2$$

The change in the internal energy of the gas is equal to

$$\Delta E_{int} = E_{int,2} - E_{int,1} = \frac{3}{2} \left(P_2 V_2 - P_1 V_1 \right) =$$

$$= \frac{3}{2} \left[\left(2.00 \times 10^5 \right) \left(2.44 \right) - \left(1.00 \times 10^5 \right) \left(1.65 \right) \right] = 4.85 \times 10^5 \text{ J}$$

We thus conclude that 4.85×10^5 J of energy was added to the gas.

Section 19-3. Specific Heat

At most temperatures, the amount of heat added to an object is approximately proportional to the change in its temperature. The amount of heat required to achieve a particular temperature change is also proportional to the mass of the object. We can write this relationship as

$$Q = mc\Delta T$$

where Q is the amount of heat added to the object, m is its mass, c is called the **specific heat** of the object (which depends on the type of material and may be a function of temperature and pressure), and ΔT is the change in the temperature of the object.

Example 19-3-A. Heating water. A typical water heater for a home holds 40 gallons of water. How much heat is required to raise the temperature of the water from 55°F to 105°F?

Approach: This problem can be solved using the specific heat of water (which is 4186 J/kg/°C). We need to make sure that the units being used are consistent. If we express the specific heat in units of J/kg/°C we need to express the temperature difference in units of °C and the mass of the water in units of kg.

Solution: The volume of the water is 40 gallons which is (40×0.00378) m^3 = 0.151 m^3. The mass of the water is

$$m = \rho V = (1000)(0.151) = 151 \, \text{kg}$$

The temperature difference of the water of 50°F must be converted to °C:

$$\Delta T_C = \frac{5}{9} \Delta T_F = \frac{5}{9}(50) = 27.8°C$$

The heat that must be added to the water is

$$Q = mc\Delta T = (151)(4186)(27.8) = 1.8 \times 10^7 \, \text{J}$$

Section 19-4. Calorimetry - Solving Problems

Consider two systems brought in thermal contact, which allows heat to flow from one system to another. Systems in thermal contact proceed toward a state of thermal equilibrium (equal temperature) with each other. The equilibrium temperature can be calculated if we treat the two systems as isolated, that is, the two systems are not in thermal contact with any other system and no mechanical work is done on the system. Conservation of energy implies that whatever heat is lost by one system must be gained by the other system:

$$Q_1 + Q_2 = 0$$

where Q_1 is the heat added to the first system and Q_2 is the heat added to the second system. Note: Q is positive if heat is added to the system; when heat is lost, Q is negative.

Example 19-4-A. Applications of conservation of energy. A solid of mass m_s = 0.349 kg and temperature T_s = 55°C is added to water of mass m_w = 0.894 kg and temperature T_w = 22°C. The final temperature of the mixture is T_f = 28°C. What is the specific heat of the solid? Assume no heat is exchanged with the environment.

Approach: This is an example of an important class of problems where changes in temperature are being used to determine the specific heat of a material. In this problem, we can determine the heat required to raise the water temperature by 6°C and use this information to determine the heat capacity of the solid, which provides this energy by dropping its temperature by 27°C. The heat capacity of the water is 1000 cal/kg/°C.

Solution: Since no work is done on the components of the system, and no heat is exchanged with the environment, the heat added to the water plus the heat added to the solid must be 0 J:

$$Q_w + Q_s = 0$$

where the w subscripts refer to the water and the s subscripts refer to the solid. Expressing the heat in terms of the temperature change, we can rewrite the previous equation as

$$m_w c_w \Delta T_w + m_s c_s \Delta T_s = m_w c_w (T_f - T_w) + m_s c_s (T_f - T_s) = 0$$

The only unknown in this equation is the specific heat of the solid (c_S) which can now be determined.

$$c_s = \frac{-m_w c_w (T_f - T_w)}{m_s (T_f - T_s)} = -\frac{(0.894)(1000)(28 - 22)}{(0.349)(28 - 55)} = 570 \, \text{cal/kg/C}°$$

Note: since a temperature difference of 1 K is equal to a temperature difference of 1°C, we do not need to convert the temperatures from Celsius to Kelvin in the ratio of temperature differences.

Section 19-5. Latent Heat

When a material changes from a solid to a liquid, from a liquid to a solid, from a gas to a liquid, or from liquid to a gas, it undergoes a **phase change**. All phase changes are accompanied by some type of discontinuity in a property of the material. Many of the phase changes, but not all, have a discontinuity in the internal energy as a function of temperature. When a material undergoes a phase change, energy must be added to or removed from one phase to create the other phase, before the temperature can change. This energy difference per unit mass is called the **latent heat** of the phase change. We can write the amount of heat needed to change the phase of a material of mass m as

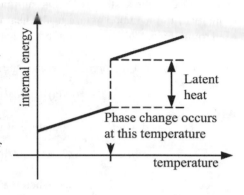

$$Q = mL$$

L is the latent heat of phase change. The **heat of fusion** is the latent heat required to melt a solid. The **heat of vaporization** is the latent heat required to vaporize a liquid. When two systems are in thermal contact and proceeding toward equilibrium, latent heats must be accounted for if any phase change is encountered as the temperatures head toward the equilibrium temperature.

Example 19-5-A. Melting ice. How much ice at -12°C must be added to 0.566 kg of water at 20° C to cool the water down to 5° C?

Approach: In this problem we know how much heat we need to extract from the water to reduce its temperature by 15°C. If we assume that the system reaches a state of equilibrium, all ice will melt and the resulting liquid will increase its temperature from 0°C to 5°C. Assuming that the mass of the ice is m_{ice}, we can calculate the heat released during this transition. Note: the heat capacity of ice is 0.5 kcal/kg/°C, the heat of fusion for water is 80 kcal/kg/°C, and the heat capacity of water is 1 kcal/kg/°C.

Solution: To calculate the total heat removed from the ice of mass m_{ice} during this process we need to examine the following three steps:
- The heat added to the ice when it changes its temperature from -12°C to 0°C is equal to

$$Q_1 = m_{ice}c_{ice}\Delta T = m_{ice}(0.5)(12) = 6m_{ice}$$

- The heat required to melt the ice at 0°C is equal to

$$Q_2 = m_{ice}L_{fusion} = m_{ice}(80) = 80m_{ice}$$

- The heat required to change the temperature of the melted ice from 0°C to 5°C is equal to

$$Q_3 = m_{ice}c_{water}\Delta T = m_{ice}(1)(5) = 5m_{ice}$$

The total heat added to the ice is the sum of these three components:

$$Q_{ice} = Q_1 + Q_2 + Q_3 = 91m_{ice}$$

The heat added to the water when it is cooled from 20°C to 5°C is equal to

$$Q_{water} = mc\Delta T = (0.566)(1.0)(5-20) = -8.5\,\text{kcal}$$

Since the total heat added to the mixture of ice and water is zero, we can calculate the mass of the ice:

$$Q = Q_{ice} + Q_{water} = 91m_{ice} - 8.5 = 0 \quad \Rightarrow$$

$$m_{ice} = \frac{8.5}{91} = 0.093 \text{ kg}$$

Example 19-5-B. Ice and water in equilibrium. A 0.500 kg block of ice at -10°C is added to 0.500 kg of water at 20°C. How much ice will there be when the water and ice reach an equilibrium temperature?

Approach: The problem indicates that the water and ice will be in equilibrium. This can only happen when the final temperature of the system is 0°C, assuming that the system is at atmospheric pressure. By comparing the heat required to increase the ice temperature to the heat released when the water is cooled down, we can determine how much heat is used to melt ice, and thus the amount of ice that melts.

Solution: The heat required to increase the temperature of the ice to 0° C will be

$$Q_{ice} = m_{ice}c_{ice}\Delta T = (0.500)(0.50)(0-(-10)) = 2.5\,\text{kcal}$$

The heat released when the temperature of the water is decreased to 0° C will be

$$Q_{water} = m_{water}c_{water}\Delta T = (0.500)(1.00)(0-20) = -10.0\,\text{kcal}$$

We see that more heat is released when the temperature of the water is reduced than is required to heat the ice. The extra heat is used to melt some of the ice: Q_{melt} = 7.5 kcal. The amount of ice melted by this much heat is determined from the latent heat of fusion of water:

$$Q_{melt} = m_{melted}L_{fusion} \quad \Rightarrow \quad m_{melted} = \frac{Q_{melt}}{L_{fusion}} = \frac{7.5}{80} = 0.094\,\text{kg}$$

This will leave 0.500 kg - 0.094 kg = 0.406 kg of ice when the water and ice reaches equilibrium.

Section 19-6. The First Law of Thermodynamics

The **first law of thermodynamics** is a statement of conservation of energy, applied to the internal energy of a system. A system can exchange energy with its environment, either by an exchange of heat Q or by the work W being done by or on the system. Q is taken to be positive when heat is added to the system; W is positive when work is done by the system (e.g., the expansion of a gas). The change in internal energy, ΔE_{int}, of a system is related to the work done by the system and the heat added to the system

$$\Delta E_{int} = Q - W$$

This statement is the first law of thermodynamics. It can also be written in differential form:

$$dE_{int} = dQ - dW$$

Example 19-6-A. Applying the first law of thermodynamics. While the temperature of a gas remains constant, 536 J of heat are added to the gas. How much work is done by the gas?

Approach: Since the temperature of the system is constant, there will be no change in the internal energy of the system. The first law of thermodynamics thus tells us that $Q = W$.

Solution: Since $Q = W$ we conclude that the work done by the gas is 536 J:

$$W = Q - \Delta E_{int} = 536 - 0 = 536\,\text{J}$$

Section 19-7. Applying the First Law of Thermodynamics; Calculating the Work

It is very useful to consider processes that occur in the system by following the process on a pressure–volume diagram. The work done by the system when the system changes its volume from V_i to V_f is

$$W = \int_{V_i}^{V_f} P \; dV$$

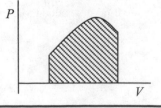

where P is the volume-dependent pressure of the system. The work done by the system is equal to the area under the PV curve that describes the process under consideration.

We will now consider some specific examples of important processes in an ideal gas:

- **Isothermal Processes.** An isothermal process is a process that occurs while the temperature of the system remains constant. During an isothermal process the product of the pressure and volume remains constant. The work done during an isothermal process is equal to:

$$W_{isothermal} = \int_{V_A}^{V_B} P\,dV = \int_{V_A}^{V_B} \frac{nRT}{V}\,dV = nRT\ln\frac{V_B}{V_A}$$

Because the internal energy E_{int} depends on the temperature, it remains constant during an isothermal process. Using the first law of thermodynamics we conclude that

$$Q = \Delta E_{int} + W = W$$

- **Adiabatic processes.** An adiabatic process is a process in which no heat enters or leaves the system: $Q = 0$. The first law of thermodynamics tells us that the work done by the system is equal to the opposite of the change in the internal energy of the system:

$$W_{adiabatic} = -\Delta E_{int}$$

- **Isovolumetric processes.** Isovolumetric processes are processes in which the volume of the system remains constant. If the volume remains constant, no work is done by the system:

$$W_{isovolumetric} = 0$$

- **Isobaric processes.** Isobaric processes are processes in which the pressure remains constant. It is relatively simple to calculate the work done during an isobaric process:

$$W_{isobaric} = \int_{V_A}^{V_B} P\,dV = P\left(V_B - V_A\right)$$

Example 19-7-A. Work done by a hand pump. A hand pump has a volume of $V_a = 740$ cm^3 when the handle is pulled back and the chamber is filled with an ideal gas at atmospheric pressure $P_a = 1.01 \times 10^5$ N/m^2 and temperature $T_a = 300$ K. The handle is pushed down and the air is compressed to a volume of $V_b = 148$ cm^3. The compression is done slowly so that the process is isothermal. How much work is done in compressing the gas, and what is the final pressure of the gas?

Approach: The work done by the gas depends on its initial and final volume and on its temperature. In order to calculate the final pressure we need to know the number of moles of gas. This information can be obtained from the initial conditions. Note: we need to make sure that all our variables are specified in the proper units: volume in m^3, pressure in N/m^2, and temperature in K.

Solution: Since the process is isothermal, the work done is equal to

$$W = nRT\ln\left\{\frac{V_b}{V_a}\right\}$$

In order to calculate the work, we need the value of n. The number of moles of gas can be calculated using the ideal gas law:

$$P_a V_a = nRT \quad\Rightarrow\quad n = \frac{P_a V_a}{RT}$$

Using this expression of n we can rewrite our expression of the work done by the gas as

$$W = P_a V_a \ln\left\{\frac{V_b}{V_a}\right\} = \left(1.01\times10^5\right)\left(7.40\times10^{-4}\right)\ln\left\{\frac{1.48\times10^{-4}}{7.40\times10^{-4}}\right\} = -120 \text{ J}$$

This is the work done by the gas; the work done by the outside force to compress the gas is +120 J.

Since the process is isothermal, nRT is constant, and the product of the pressure and volume is constant. This conclusion can be used to determine the final pressure P_b:

$$P_b = \frac{P_a V_a}{V_b} = \frac{\left(1.01 \times 10^5\right)\left(740\right)}{148} = 5.05 \times 10^5 \text{ N/m}^2$$

Section 19-8. Molar Specific Heats for Gases, and the Equipartition of Energy

The molar specific heat of a substance is the heat that must be added to one mol to raise its temperature by 1°C. The **molar specific heat at constant volume** C_V applies to processes for which the volume remains constant:

$$Q = nC_V \Delta T \qquad\qquad \text{Constant volume}$$

The **molar specific heat at constant pressure** C_P applies to processes in which the pressure remains constant:

$$Q = nC_P \Delta T \qquad\qquad \text{Constant pressure}$$

There is a simple relationship between C_P and C_V:

$$C_P - C_V = R$$

where R is the gas constant. For a monatomic ideal gas, $C_V = {}^3/_2 R$ and $C_P = {}^5/_2 R$.

In general, each **degree of freedom** of a system in thermal equilibrium will have an average energy ${}^1/_2 kT$. This is called the **principle of equipartition of energy**. A measurement of C_V allows us to probe the number of degrees of freedom.

Section 19-9. Adiabatic Expansion of a Gas

During an adiabatic expansion of an ideal gas, the pressure and volume of the gas are related by

$$PV^\gamma = \text{constant}$$

where γ is the ratio of the specific heat at constant pressure to the specific heat at constant volume: $\gamma = C_P/C_V$.

Example 19-9-A. Work done by a hand pump - Part 2. A hand pump has a volume of $V_a = 740 \text{ cm}^3$ when the handle is pulled back and the chamber is filled with an ideal gas at atmospheric pressure $P_a = 1.01 \times 10^5 \text{ N/m}^2$, and a temperature of $T_a = 300$ K. The handle is pushed down and the air is compressed to a volume of $V_b = 148 \text{ cm}^3$. The compression is done rapidly so that the process is adiabatic. How much work is done in compressing the gas, and what is the final pressure and temperature of the gas?

Approach: This problem is similar to Example 19-7-A except that the current process is adiabatic. We can thus not assume that the temperature is constant.

Solution: During an adiabatic compression the pressure and volume are related in the following manner:

$$PV^\gamma = \text{constant} = P_a V_a^\gamma \qquad \Rightarrow \qquad P = \frac{P_a V_a^\gamma}{V^\gamma}$$

We can use this relation between P and V to calculate the work done during the adiabatic compression:

$$W = \int_{V_a}^{V_b} P\, dV = \int_{V_a}^{V_b} \frac{P_a V_a^\gamma}{V^\gamma}\, dV =$$

$$= P_a V_a^\gamma \left. \frac{V^{1-\gamma}}{(1-\gamma)} \right|_{V_a}^{V_b} = \frac{P_a V_a^\gamma}{(1-\gamma)} \left(V_b^{1-\gamma} - V_a^{1-\gamma} \right)$$

For a monatomic ideal gas, the adiabatic gas constant γ is 5/3:

$$W = \frac{P_a V_a^{5/3}}{(1-5/3)}\left(V_b^{1-5/3} - V_a^{1-5/3}\right) = \frac{3}{2}\left(1.01\times10^5\right)\left(7.40\times10^{-4}\right)^{5/3}\left[\left(1.48\times10^{-4}\right)^{-2/3} - \left(7.40\times10^{-4}\right)^{-2/3}\right]$$

The work done by the gas is -216 J; the work done by the outside force to compress the gas is +216 J. We conclude that more work is done during an adiabatic compression than during an isothermal compression (see Example 19-7-A). The work done during the adiabatic process increases the temperature and thus the pressure, increasing the work that needs to be done. We can confirm this by calculating the final pressure and temperature of the gas.

The final pressure can be obtained by using the relationship between pressure and volume for an adiabatic process:

$$P_b = \frac{P_a V_a^{\gamma}}{V_b^{\gamma}} = \frac{\left(1.01\times10^5\right)\left(740\right)^{5/3}}{\left(148\right)^{5/3}} = 1.48\times10^6 \text{ N/m}^2$$

The final temperature of the gas can be obtained from the ideal gas law:

$$T_b = \frac{P_b V_b}{P_a V_a} T_a = \frac{\left(1.48\times10^6\right)\left(148\right)}{\left(1.01\times10^5\right)\left(740\right)}\left(300\right) = 879 \text{ K}$$

We see that, as expected, the final temperature and pressure obtained for an adiabatic process are larger than the final temperature and pressure obtained for an isothermal process.

Section 19-10. Heat Transfer: Conduction, Convection, Radiation

There are three mechanisms of heat transfer:
- **Conduction.** Conduction is the transfer of heat by the direct contact of atoms from a higher temperature substance with those of a lower temperature substance. The rate at which heat flows through a given cross section of material by conduction, dQ/dt, depends on the temperature gradient dT/dx, the cross-sectional area A through which the heat is transferred, and the **thermal conductivity k** of the material:

$$\frac{dQ}{dt} = -kA\frac{dT}{dx}$$

The minus sign indicates that heat flows in a direction opposite the direction of the gradient. If the gradient of the temperature through a uniform piece of material is approximately constant, we can rewrite dQ/dt as

$$\frac{dQ}{dt} = kA\frac{T_H - T_L}{l}$$

Materials with high values of thermal conductivity are called **conductors** and those with low values of conductivity are called **insulators**.
- **Convection.** Convection is the transfer of heat in fluids by the macroscopic motion of the fluid. If the motion of the fluid is driven by the density difference due to a temperature difference within the fluid, the convection is called **natural convection**. This is the phenomenon that most people are referring to when they say "heat rises." What they should say is that higher temperature air has a lower density than lower temperature air resulting in a buoyant force on the higher temperature air, forcing it upward. If a fan or pump is used to circulate the fluid, the convection is called **forced convection**.
- **Radiation.** Radiation is the transfer of energy by emission of electromagnetic radiation. All surfaces radiate electromagnetic radiation. The rate at which an object radiates energy depends on the absolute temperature T of the surface, the surface area A, and the emissivity e of the object. The rate at which energy leaves the surface of an object by radiation is

$$\frac{dQ}{dt} = e\sigma A T^4$$

where σ is the **Stefan-Boltzmann constant** ($\sigma = 5.67 \times 10^{-8}$ W/m²·K⁴.) This relationship is called the **Stefan-Boltzmann equation**.

All objects absorb electromagnetic radiation from the surroundings. If an object at temperature T_1 is in an environment at temperature T_2, then the net rate of heat flow from the object is

$$\frac{dQ}{dt} = e\sigma A\left(T_1^4 - T_2^4\right)$$

Example 19-10-A. Rate of heat transfer. An arrangement of two layers of materials separates a system at temperature T_1 and a system at temperature T_2, as shown in the diagram. The two layers have equal thickness d and a cross-sectional area A. The first layer has a conductivity k_1; the second layer has a conductivity k_2. What is the rate of heat transfer from the system at temperature T_1 to the system at temperature T_2?

Approach: Since energy is conserved, the flow of heat through the first layer must be equal to the flow of heat through the second layer. This requirement allows us to determine the temperature at the interface between the two layers and the flow of heat through the layers.

Solution: Let T_m be the temperature at the interface between the two layers. The rate of heat transfer from the system at temperature T_1 to the interface through the first layer is equal to

$$\frac{dQ_1}{dt} = k_1 A\frac{T_1 - T_m}{d}$$

The rate of heat transfer from interface to the body at temperature T_2 through the second layer is equal to

$$\frac{dQ_2}{dt} = k_2 A\frac{T_m - T_2}{d}$$

Since $dQ_1/dt = dQ_2/dt$ we conclude that

$$k_1 A\frac{T_1 - T_m}{d} = k_2 A\frac{T_m - T_2}{d} \quad\rightarrow$$

$$T_m = \frac{k_1 T_1 + k_2 T_2}{k_1 + k_2}$$

The rate of heat flow is equal to

$$\frac{dQ}{dt} = k_1 A\frac{T_1 - \dfrac{k_1 T_1 + k_2 T_2}{k_1 + k_2}}{d} = \frac{k_1 A}{d}\left\{\frac{k_2 T_1 - k_2 T_2}{k_1 + k_2}\right\} = \frac{A}{d}\frac{k_1 k_2}{k_1 + k_2}\left(T_1 - T_2\right) =$$

$$= \frac{2k_1 k_2}{k_1 + k_2} A\frac{\left(T_1 - T_2\right)}{2d} = k_{eff} A\frac{\left(T_1 - T_2\right)}{2d}$$

The effective conductivity of the two-layer system is $2k_1 k_2/(k_1 + k_2)$.

Example 19-10-B. Heat loss of the human body. Assume a human body has a surface temperature of 98.6° F and an emissivity of 0.8. Estimate the rate of heat emitted by the bare human body in the form of radiation.

Approach: To solve this problem we apply the Stefan-Boltzmann equation. We need to make sure we use the temperature in K. We will assume that the surface area of the human body is 1 m^2.

Solution: The rate at which energy will be radiated from the human body is equal to

$$\frac{dQ}{dt} = e\sigma AT^4 = \left(0.8\right)\left(5.67\times10^{-8}\right)\left(1\right)\left[\tfrac{5}{9}\left(98.6 - 32\right) + 273\right]^4 = 400\,\text{W}$$

Practice Quiz

1. An amount of work W is done by an ideal gas on its environment. During this process, the temperature of the ideal gas remains constant. How much heat was added to the ideal gas during this process?
   a) $Q = 0$
   b) $Q = W$
   c) $Q = -W$
   d) Need more information to determine the amount of heat added

2. What happens to the adiabatic gas constant as the value of the molar specific heat at constant volume is increased without bound?
   a) The adiabatic gas constant remains unchanged.
   b) The adiabatic gas constant increases unbounded.
   c) The adiabatic gas constant decreases to zero.
   d) The adiabatic gas constant decreases to one.

3. Usually, in home construction, the insulation in the ceiling of a home is made much thicker than the insulation in the walls of a home. Why?
   a) Ceilings have more area than walls.
   b) Insulation has a higher conductivity when laid horizontally than standing vertically.
   c) The ceiling is usually warmer than the walls because of convection in the house.
   d) Ceiling insulation is easier to install.

4. During which type of process on an ideal gas will the pressure change by the greatest factor for a given change in the volume?
   a) Isothermic
   b) Adiabatic
   c) Isobaric
   d) All processes will have the same pressure change.

5. A system consisting of a fixed amount of gas starts at pressure P_1 and volume V_1 and ends up at pressure P_2 and volume V_2 after some thermodynamic process. Which of the following quantities do not depend on the path taken on a pressure–volume diagram during the process?
   a) Work done on the environment by the system during the process.
   b) Work done by the environment on the system during the process.
   c) The heat added to the system during the process.
   d) The final temperature of the system.

6. You are in a room at normal room temperature. You touch two different objects that have been in the room for quite some time. One object is made of wood, the other is made of metal. The wood feels warmer than the metal. Why is that?
   a) Metal is always at a lower temperature than wood.
   b) The metal reflects the radiation from the surroundings more than wood.
   c) Metal has a higher thermal conductivity than wood.
   d) More work is done on the wood in touching it because it is softer than the metal and compresses further, raising its internal energy more than the metal.

7. An object is under conditions in which it is cooling mostly by conduction. When the object is at 50°C and is being cooled by conduction to a 20°C body, it takes 2 minutes to cool 1°C. Approximately how long would it take the body to cool 1°C if it started at 35°C and was cooled by conduction to the 20°C body?
   a) 0.5 minutes
   b) 1 minute
   c) 2 minutes
   d) 4 minutes

8. During an isochoric process on an ideal gas, what doesn't change?
   a) The temperature of the gas.
   b) The pressure of the gas.
   c) The internal energy of the gas.
   d) The volume of the gas.

9. By what mechanism is energy transferred from the Sun to the Earth?
   a) Conduction
   b) Convection
   c) Radiation
   d) Injection

10. A gas goes from an initial state with pressure P_1 and volume V_1 and reaches a final state with pressure P_2 and volume V_2. Which of the processes shown in the pressure–volume diagrams requires the most heat transfer to the gas during this change in pressure and volume?

a)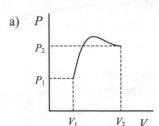

b)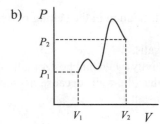

c)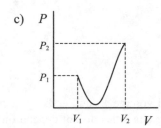

d)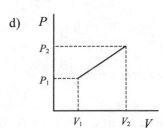

11. Warm water with a mass of 1.67 kg and a temperature of 85.2°C is added to cold water with a mass of 1.17 kg and a temperature 12.4°C. What is the temperature of the water when it reaches thermal equilibrium?

12. If heat is added at a rate of 8.56 kJ/s to a mixture of 0.500 kg of ice at -20°C and 0.678 kg of water at +30°C, how long does it take to boil away all of the water? Assume no heat is lost to the environment.

13. A process occurs such that the pressure is related to the volume by

$$P = 1.65 \times 10^5 \, \text{N} / \text{m}^8 \, V^2$$

while the volume changes from 2.00 m³ to 1.00 m³. How much work is done by the gas during this process?

14. Assume the Earth is at a uniform surface temperature of 50°F and the emissivity of the Earth is 1. What amount of energy would the Earth radiate into space in one year? (The emissivity of the Earth is not 1, and the presence of greenhouse gases in the atmosphere decreases the emissivity.)

15. Five moles of a monatomic ideal gas are in a container of fixed size. The pressure is initially 1.00×10^5 N/m² and the temperature is 300 K. 4.87 kJ of heat is added to the gas. What are the final pressure and temperature of the gas?

Responses to Select End-of-Chapter Questions

1. When a jar of orange juice is vigorously shaken, the work done on it goes into heating the juice (increasing the kinetic energy of the molecules), mixing the components of the juice (liquid and pulp), and dissolving air in the juice (froth).

7. When water at 100°C comes in contact with the skin, energy is transferred to the skin and the water begins to cool. When steam at 100°C comes in contact with the skin, energy is transferred to the skin and the steam begins to condense to water at 100°C. Steam burns are often more severe than water burns due to the energy given off by the steam as it condenses, before it begins to cool.

13. In an isothermal process, the temperature, and therefore the internal energy, of the ideal gas is constant. From the first law of thermodynamics, we know that if the change in the internal energy is zero, then the heat added to the system is equal to the work done by the system. Therefore, 3700 J of heat must have been added to the system.

19. When a gas is heated at constant volume, all of the energy added goes into increasing the internal energy, since no work is done. When a gas is heated at constant pressure, some of the added energy is used for the work needed to expand the gas, and less is available for increasing the internal energy. It takes more energy to raise the temperature of a gas by a given amount at constant pressure than at constant volume.

25. On a sunny day, the land heats faster than the water. The air over the land is also heated and it rises due to a decrease in density. The cooler air over the water is then pulled in to replace the rising air, creating an onshore or sea breeze.

31. Water has a greater thermal conductivity than air. Water at 22°C will feel cooler than air at the same temperature because the rate of heat transfer away from the body will be greater.

37. The temperature of the air around cities near oceans is moderated by the presence of large bodies of water which act like a heat reservoir. Water has a high heat capacity. It will absorb energy in the summer with only a small temperature increase, and radiate energy in the winter, with a small temperature decrease.

Solutions to Select End-of-Chapter Problems

1. The kcal is the heat needed to raise 1 kg of water by 1°C. Use this relation to find the change in the temperature.

$$\left(8700\,\mathrm{J}\right)\left(\frac{1\,\mathrm{kcal}}{4186\,\mathrm{J}}\right)\frac{\left(1\,\mathrm{kg}\right)\left(1\,\mathrm{C}^{\circ}\right)}{1\,\mathrm{kcal}}\left(\frac{1}{3.0\,\mathrm{kg}}\right) = 0.69\,\mathrm{C}^{\circ}$$

Thus the final temperature is $10.0°C + 0.69°C = \boxed{10.7°C}$.

7. The heat absorbed can be calculated from Eq. 19-2. Note that 1 L of water has a mass of 1 kg.

$$Q = mc\Delta T = \left[\left(18\,\mathrm{L}\right)\left(\frac{1\times10^{-3}\,\mathrm{m}^3}{1\,\mathrm{L}}\right)\left(\frac{1.0\times10^3\,\mathrm{kg}}{1\,\mathrm{m}^3}\right)\right]\left(4186\,\mathrm{J/kg\cdot C^\circ}\right)\left(95°C - 15°C\right) = \boxed{6.0\times10^6\,\mathrm{J}}$$

13. The heat gained by the glass thermometer must be equal to the heat lost by the water.

$$m_{\mathrm{glass}}c_{\mathrm{glass}}\left(T_{\mathrm{eq}} - T_{i\,\mathrm{glass}}\right) = m_{\mathrm{H_2O}}c_{\mathrm{H_2O}}\left(T_{i\,\mathrm{H_2O}} - T_{\mathrm{eq}}\right)$$

$$\left(31.5\,\mathrm{g}\right)\left(0.20\,\mathrm{cal/g\cdot C^\circ}\right)\left(39.2°C - 23.6°C\right) = \left(135\,\mathrm{g}\right)\left(1.00\,\mathrm{cal/g\cdot C^\circ}\right)\left(T_{i\,\mathrm{H_2O}} - 39.2°C\right)$$

$$T_{i\,\mathrm{H_2O}} = \boxed{39.9°C}$$

19. Assume that the heat from the person is only used to evaporate the water. Also, we use the heat of vaporization at room temperature (585 kcal/kg), since the person's temperature is closer to room temperature than 100°C.

$$Q = mL_{vap} \quad \rightarrow \quad m = \frac{Q}{L_{vap}} = \frac{180 \text{ kcal}}{585 \text{ kcal/kg}} = 0.308 \text{ kg} \approx \boxed{0.31 \text{ kg}} = 310 \text{ mL}$$

25. The kinetic energy of the bullet is assumed to warm the bullet and melt it.

$$\tfrac{1}{2} mv^2 = Q = mc_{Pb}\left(T_{melt} - T_i\right) + mL_{fusion} \quad \rightarrow$$

$$v = \sqrt{2\left[c_{Pb}\left(T_{melt} - T_i\right) + L_{fusion}\right]} = \sqrt{2\left[\left(130 \text{ J/kg} \cdot \text{C}°\right)\left(327°\text{C} - 20°\text{C}\right) + \left(0.25 \times 10^5 \text{ J/kg}\right)\right]}$$

$$= \boxed{360 \text{ m/s}}$$

31. (a) No work is done during the first step, since the volume is constant. The work in the second step is given by $W = P\Delta V$.

$$W = P\Delta V = \left(1.4 \text{ atm}\right)\left(\frac{1.01 \times 10^5 \text{ Pa}}{1 \text{ atm}}\right)\left(9.3 \text{ L} - 5.9 \text{ L}\right)\frac{1 \times 10^{-3} \text{ m}^3}{1 \text{ L}} = \boxed{480 \text{ J}}$$

(b) Since there is no overall change in temperature, $\Delta U = \boxed{0 \text{ J}}$.

(c) The heat flow can be found from the first law of thermodynamics.

$$\Delta U = Q - W \quad \rightarrow \quad Q = \Delta U + W = 0 + 3.5 \times 10^2 \text{ J} = \boxed{480 \text{ J} \left(\text{into the gas}\right)}$$

37. The work done by an ideal gas during an isothermal volume change is given by Eq. 19-8.

$$W = nRT \ln\frac{V_2}{V_1} = P_1V \ln\frac{V_2}{V_1} = \left(1.013 \times 10^5 \text{ Pa}\right)\left(3.50 \times 10^{-3} \text{ m}^3\right)\ln\frac{1.80 \text{ L}}{3.50 \text{ L}} = -236 \text{ J}$$

The work done by an external agent is the opposite of the work done by the gas, $\boxed{236 \text{ J}}$.

43. If there are no heat losses or mass losses, then the heating occurs at constant volume, and so Eq. 19-10a applies, $Q = nC_V\Delta T$. For an ideal diatomic gas, $C_V = (5/2)R$.

$$Q = nC_V\Delta T = \frac{5}{2}nR\Delta T = \frac{5}{2}\frac{P_0V_0}{T_0}\Delta T \quad \rightarrow$$

$$\Delta T = \frac{2QT}{5PV} = \frac{2\left(1.86 \times 10^6 \text{ J}\right)\left(293 \text{ K}\right)}{5\left(1.013 \times 10^5 \text{ Pa}\right)\left(3.5 \text{ m}\right)\left(4.6 \text{ m}\right)\left(3.0 \text{ m}\right)} = 44.55 \text{ K} \approx \boxed{45°\text{C}}$$

49. (a) The change in internal energy is given by Eq. 19-12.

$$\Delta E_{int} = nC_V\Delta T = n\left(\tfrac{5}{2}R\right)\Delta T = \tfrac{5}{2}\left(2.00 \text{ mol}\right)\left(8.314 \frac{\text{J}}{\text{mol} \cdot \text{K}}\right)\left(150 \text{ K}\right) = 6236 \text{ J} \approx \boxed{6240 \text{ J}} \text{ or}$$

$$\Delta E_{int} = nC_V\Delta T = \left(2.00 \text{ mol}\right)\left(4.96 \frac{\text{cal}}{\text{mol} \cdot \text{K}}\right)\left(4.186 \frac{\text{J}}{\text{cal}}\right)\left(150 \text{ K}\right) = 6229 \text{ J} \approx \boxed{6230 \text{ J}}$$

(b) The work is done at constant pressure.

$$W = P\Delta V = nR\Delta T = \left(2.00 \text{ mol}\right)\left(8.314 \frac{\text{J}}{\text{mol} \cdot \text{K}}\right)\left(150 \text{ K}\right) = 2494 \text{ J} \approx \boxed{2490 \text{ J}}$$

(c) The heat is added at constant pressure, and so Eq. 19-10b applies.

$$Q = nC_p \Delta T = n\left(\tfrac{7}{2}R\right)\Delta T = \tfrac{7}{2}\left(2.00\,\text{mol}\right)\left(8.314\,\frac{\text{J}}{\text{mol}\cdot\text{K}}\right)\left(150\,\text{K}\right) = \boxed{8730\,\text{J}} \quad\text{or}$$

$$\Delta E_{int} = Q - W \quad\rightarrow\quad Q = \Delta E_{int} + W = 6229\,\text{J} + 2494\,\text{J} = \boxed{8720\,\text{J}}$$

55. (a) To plot the graph accurately, data points must be calculated. V_1 is found from the ideal gas equation, and P_2 and V_2 are found from the fact that the first expansion is adiabatic.

$$P_1 V_1 = nRT_1 \quad\rightarrow\quad V_1 = \frac{nRT_1}{P_1} = \frac{\left(1.00\,\text{mol}\right)\left(8.314\,\dfrac{\text{J}}{\text{mol}\cdot\text{K}}\right)\left(588\,\text{K}\right)}{1.013\times10^5\,\text{Pa}} = 48.26\times10^{-3}\,\text{m}^3 = 48.26\,\text{L}$$

$$P_2 V_2 = nRT_2 \;,\; P_2 V_2^{\gamma} = P_1 V_1^{\gamma} \quad\rightarrow\quad V_2^{\gamma-1} = \frac{P_1 V_1^{\gamma}}{nRT_2} \quad\rightarrow$$

$$V_2 = \left(\frac{P_1 V_1^{5/3}}{nRT_2}\right)^{3/2} = \left(\frac{\left(1.013\times10^5\,\text{Pa}\right)\left(48.26\times10^{-3}\,\text{m}^3\right)^{5/3}}{\left(1.00\,\text{mol}\right)\left(8.314\,\dfrac{\text{J}}{\text{mol}\cdot\text{K}}\right)\left(389\,\text{K}\right)}\right)^{3/2} = 89.68\times10^{-3}\,\text{m}^3 = 89.68\,\text{L}$$

$$P_2 = \frac{nRT_2}{V_2} = \frac{\left(1.00\,\text{mol}\right)\left(8.314\,\dfrac{\text{J}}{\text{mol}\cdot\text{K}}\right)\left(389\,\text{K}\right)}{89.68\times10^{-3}\,\text{m}^3} = 3.606\times10^4\,\text{Pa}\left(\frac{1\,\text{atm}}{1.013\times10^5\,\text{Pa}}\right) = 0.356\,\text{atm}$$

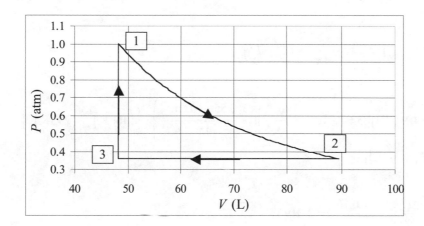

(b) Both the pressure and the volume are known at point 3, the lower left corner of the graph.

$$P_3 = P_2\;,\; V_3 = V_1 \quad\rightarrow\quad P_3 V_3 = nRT_3 \quad\rightarrow$$

$$T_3 = \frac{P_3 V_3}{nR} = \frac{P_2 V_1}{nR} = \frac{\left(3.606\times10^4\,\text{Pa}\right)\left(48.26\times10^{-3}\,\text{m}^3\right)}{\left(1.00\,\text{mol}\right)\left(8.314\,\dfrac{\text{J}}{\text{mol}\cdot\text{K}}\right)} = \boxed{209\,\text{K}}$$

(c) For the adiabatic process, state 1 to state 2:

$$\Delta E_{int} = \tfrac{3}{2}nR\Delta T = \tfrac{3}{2}\left(1.00\,\text{mol}\right)\left(8.314\,\frac{\text{J}}{\text{mol}\cdot\text{K}}\right)\left(389\,\text{K} - 588\,\text{K}\right) = -2482\,\text{J} \approx \boxed{-2480\,\text{J}}$$

$$Q = \boxed{0}\;\left(\text{adiabatic}\right)\;;\; W = Q - \Delta E_{int} = \boxed{2480\,\text{J}}$$

For the constant pressure process, state 2 to state 3:

$$\Delta E_{int} = \tfrac{3}{2} nR\Delta T = \tfrac{3}{2}(1.00\,\text{mol})\left(8.314\,\frac{J}{\text{mol}\cdot\text{K}}\right)(209\,\text{K} - 389\,\text{K}) = -2244.7\,\text{J} \approx \boxed{-2240\,\text{J}}$$

$$W = P\Delta V = (3.606\times10^4\,\text{Pa})(48.26\times10^{-3}\,\text{m}^3 - 89.68\times10^{-3}\,\text{m}^3) = -1494\,\text{J} \approx \boxed{-1490\,\text{J}}$$

$$Q = W + \Delta E_{int} = -1494\,\text{J} - 2244.7\,\text{J} = -3739\,\text{J} \approx \boxed{-3740\,\text{J}}$$

For the constant volume process, state 3 to state 1:

$$\Delta E_{int} = \tfrac{3}{2} nR\Delta T = \tfrac{3}{2}(1.00\,\text{mol})\left(8.314\,\frac{J}{\text{mol}\cdot\text{K}}\right)(588\,\text{K} - 209\,\text{K}) = 4727\,\text{J} \approx \boxed{4730\,\text{J}}$$

$$W = P\Delta V = 0 \; ; \; Q = W + \Delta E_{int} = \boxed{4730\,\text{J}}$$

(*d*) For the complete cycle, by definition $\Delta E_{int} = \boxed{0}$. If the values from above are added, we get ΔE_{int} = 4727 J - 2482 J - 2245 J = 0. Add the separate values for the work done and the heat added.

$$W = 2482\,\text{J} - 1494\,\text{J} = 998\,\text{J} \approx \boxed{990\,\text{J}} \; ; \; Q = -3739\,\text{J} + 4727\,\text{J} = 998\,\text{J} \approx \boxed{990\,\text{J}}$$

Notice that the first law of thermodynamics is satisfied.

61. (*a*) The rate of heat transfer due to radiation is given by Eq. 19-17. We assume that each teapot is a sphere that holds 0.55 L. The radius and then the surface area can be found from that.

$$V = \tfrac{4}{3}\pi r^3 \rightarrow r = \left(\frac{3V}{4\pi}\right)^{1/3} \rightarrow S.A. = 4\pi r^2 = 4\pi\left(\frac{3V}{4\pi}\right)^{2/3}$$

$$\frac{\Delta Q}{\Delta t} = \varepsilon\sigma A\left(T_1^4 - T_2^4\right) = 4\pi\varepsilon\sigma\left(\frac{3V}{4\pi}\right)^{2/3}\left(T_1^4 - T_2^4\right)$$

$$\left(\frac{\Delta Q}{\Delta t}\right)_{ceramic} = 4\pi(0.70)\left(5.67\times10^{-8}\,\frac{W}{\text{m}^2\cdot\text{K}^4}\right)\left(\frac{3(0.55\times10^{-3}\,\text{m}^2)}{4\pi}\right)^{2/3}\left[(368\,\text{K})^4 - (293\,\text{K})^4\right]$$

$$= 14.13\,\text{W} \approx \boxed{14\,\text{W}}$$

$$\left(\frac{\Delta Q}{\Delta t}\right)_{shiny} = \left(\frac{\Delta Q}{\Delta t}\right)_{ceramic}\left(\frac{0.10}{0.70}\right) = 2.019\,\text{W} \approx \boxed{2.0\,\text{W}}$$

(*b*) We assume that the heat capacity comes primarily from the water in the teapots, and ignore the heat capacity of the teapots themselves. We apply Eq. 19-2, along with the results from part (*a*). The mass is that of 0.55 L of water, which would be 0.55 kg.

$$\Delta Q = mc\Delta T \rightarrow \Delta T = \frac{1}{mc}\left(\frac{\Delta Q}{\Delta t}\right)_{radiation}\Delta t_{elapsed}$$

$$(\Delta T)_{ceramic} = \frac{14.13\,\text{W}}{(0.55\,\text{kg})\left(4186\,\dfrac{J}{\text{kg}\cdot\text{C}^\circ}\right)}(1800\,\text{s}) = \boxed{11\,\text{C}^\circ}$$

$$(\Delta T)_{shiny} = \frac{1}{7}(\Delta T)_{ceramic} = \boxed{1.6\,\text{C}^\circ}$$

67. (a) Choose a cylindrical shell of length l, radius R and thickness dR. Apply Eq. 19-16b, modified for the radial geometry. See the figure. Note that dQ/dt must be a constant, so that all of the heat energy that enters a shell also exits that shell.

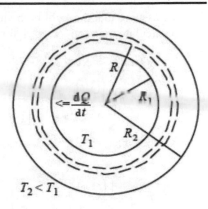

$$\frac{dQ}{dt} = -kA\frac{dT}{dR} = -k2\pi Rl\frac{dT}{dR} \quad \rightarrow$$

$$\frac{dR}{R} = -\frac{2\pi kl}{dQ/dt}dT \quad \rightarrow \quad \int_{R_1}^{R_2}\frac{dR}{R} = -\frac{2\pi kl}{dQ/dt}\int_{T_1}^{T_2}dT \quad \rightarrow$$

$$\ln\frac{R_2}{R_1} = \frac{2\pi kl}{dQ/dt}(T_1 - T_2) \quad \rightarrow \quad \boxed{\frac{dQ}{dt} = \frac{2\pi k(T_1 - T_2)l}{\ln(R_2/R_1)}}$$

(b) For still water, the initial heat flow outward from the water is described by Eq. 19-2.

$$\Delta Q = mc\Delta T \quad \rightarrow \quad \frac{\Delta Q}{\Delta t} = mc\frac{\Delta T}{\Delta t} = \frac{2\pi k(T_1 - T_2)l}{\ln(R_2/R_1)} \quad \rightarrow$$

$$\frac{\Delta T}{\Delta t} = \frac{2\pi k(T_1 - T_2)l}{mc\ln(R_2/R_1)} = \frac{2\pi k(T_1 - T_2)l}{\rho_{H_2O}\pi R_1^2 lc\ln(R_2/R_1)} = \frac{2k(T_1 - T_2)}{\rho_{H_2O}R_1^2 c\ln(R_2/R_1)}$$

$$= \frac{2\left(40\dfrac{J}{s\cdot m\cdot C^\circ}\right)(71^\circ C - 18^\circ C)}{(1.0\times 10^3\,kg/m^3)(0.033m)^2\left(4186\dfrac{J}{kg\cdot C^\circ}\right)\ln\left(\dfrac{4.0}{3.3}\right)} = 4.835\,C^\circ/s \approx \boxed{4.8\,C^\circ/s}$$

Note that this will lower the value of T_1, which would lower the rate of temperature change as time elapses.

(c) The water at the entrance is losing temperature at a rate of $4.835^\circ C/s$. In one second, it will have traveled 8 cm, and so the temperature drop per cm is $\left(4.835\dfrac{C^\circ}{s}\right)\left(\dfrac{1s}{8cm}\right) = \boxed{0.60\,C^\circ/cm}$.

73. The temperature rise can be calculated from Eq. 19-2.

$$Q = mc\Delta T \quad \rightarrow \quad \Delta T = \frac{Q}{mc} = \frac{(0.80)(200\,kcal/h)(0.5h)}{(70\,kg)(0.83\,kcal/kg\cdot C^\circ)} = 1.38\,C^\circ \approx \boxed{1\,C^\circ}$$

79. (a) The energy required to raise the temperature of the water is given by Eq. 19-2.

$$Q = mc\Delta T \quad \rightarrow$$

$$\frac{Q}{\Delta t} = mc\frac{\Delta T}{\Delta t} = (0.250\,kg)\left(4186\dfrac{J}{kg\cdot C^\circ}\right)\left(\dfrac{80\,C^\circ}{105\,s}\right) = 797\,W \approx \boxed{800\,W\,(2\text{ sig. fig.})}$$

(b) After 105 s, the water is at 100°C. So for the remaining 15 s the energy input will boil the water. Use the heat of vaporization.

$$Q = mL_v \quad \rightarrow \quad m = \frac{Q}{L_v} = \frac{(\text{Power})\Delta t}{L_v} = \frac{(797\,W)(15\,s)}{2260\,J/g} = \boxed{5.3\,g}$$

85. We assume that the light bulb emits energy by radiation, and so Eq. 19-18 applies. Use the data for the 75-W bulb to calculate the product $e\sigma A$ for the bulb, and then calculate the temperature of the 150-W bulb.

$$\left(Q/t\right)_{60\,W} = e\sigma A\left(T_{60\,W}^4 - T_{room}^4\right) \rightarrow$$

$$e\sigma A = \frac{\left(Q/t\right)_{60\,W}}{\left(T_{60\,W}^4 - T_{room}^4\right)} = \frac{(0.90)(75\,W)}{\left[(273+75)K\right]^4 - \left[(273+18)K\right]^4} = 9.006 \times 10^{-9}\ W/K^4$$

$$\left(Q/t\right)_{150\,W} = e\sigma A\left(T_{150\,W}^4 - T_{room}^4\right) \rightarrow$$

$$T_{150\,W} = \left[\frac{\left(Q/t\right)_{150\,W}}{e\sigma A} + T_{room}^4\right]^{1/4} = \left[\frac{(0.90)(150\,W)}{\left(9.006 \times 10^{-9}\ W/K^4\right)} + (291\,K)^4\right]^{1/4}$$

$$= 386\ K = 113°C \approx \boxed{110°C}$$

91. The work is given by $W = \int_{V_1}^{V_2} PdV$. The pressure is found from the van der Walls expression, Eq. 18-9. The temperature is a constant. Here is the analytic integration and subsequent evaluation.

$$W = \int_{V_1}^{V_2} PdV = \int_{V_1}^{V_2}\left(\frac{RT}{\dfrac{V}{n} - b} - \frac{a}{\left(\dfrac{V}{n}\right)^2}\right)dV = \int_{V_1}^{V_2}\left(\frac{nRT}{V - bn} - \frac{an^2}{V^2}\right)dV = \left[nRT\ln\left(V - bn\right) + \frac{an^2}{V}\right]_{V_1}^{V_2}$$

$$= \left[nRT\ln\left(V_2 - bn\right) + \frac{an^2}{V_2}\right] - \left[nRT\ln\left(V_1 - bn\right) + \frac{an^2}{V_1}\right] = nRT\ln\frac{\left(V_2 - bn\right)}{\left(V_1 - bn\right)} + an^2\left(\frac{1}{V_2} - \frac{1}{V_1}\right)$$

$$= (1.0\,mol)(8.314\,J/mol\cdot K)(373\,K)\ln\frac{\left[1.00\,m^3 - \left(3.0 \times 10^{-5}\ m^3/mol\right)(1.0\,mol)\right]}{\left[0.50\,m^3 - \left(3.0 \times 10^{-5}\ m^3/mol\right)(1.0\,mol)\right]}$$

$$+ \left(0.55\,N\cdot m^4/mol^2\right)(1.0\,mol)^2\left(\frac{1}{1.00\,m^3} - \frac{1}{0.50\,m^3}\right)$$

$$= 3101.122\ln\frac{\left[1.0 - \left(3.0 \times 10^{-5}\right)\right]}{\left[0.5 - \left(3.0 \times 10^{-5}\right)\right]} - 0.55 = 2149\,J$$

An acceptable answer from the numeric integration would need to be in the range of $2100 - 2200$ J.

To do the numeric integration, we partition the volume range. For each volume value, the pressure is calculated using Eq. 18-9. The pressure is assumed constant over each segment of the volume partition, and then the work for that segment is the (constant) pressure times the small change in volume. Even with a relatively crude partition size of $0.1\ m^3$, good agreement is found. The result from the numeric integration is 2158 J, which rounds to $\boxed{2200\,J}$.

Chapter 20: Second Law of Thermodynamics

Chapter Overview and Objectives

In this chapter, the second law of thermodynamics is introduced and its applications to physical systems are discussed. The principles of operation of heat engines, refrigerators, and heat pumps, and the method used to determine their efficiencies and coefficients of performance are described. The concepts of both macroscopic and microscopic entropy and their equivalence are discussed.

After completing this chapter you should:
- Know the various equivalent forms of the second law of thermodynamics.
- Know what heat engines, refrigerators, and heat pumps are.
- Know the definition of the efficiency of heat engines.
- Know the definition of the coefficient of performance of refrigerators and heat pumps.
- Know that the efficiency of a Carnot cycle heat engine is the maximum possible theoretical efficiency.
- Know how to calculate the efficiency of a Carnot cycle heat engine.
- Know the definition of a change in macroscopic entropy and how to calculate this change.
- Know that entropy measures the disorder of a system.
- Know the distinction between a macrostate and a microstate.
- Know how entropy can be related to the distribution of microstates belonging to macrostates.

Summary of Equations

Efficiency of a heat engine:
$$e = \frac{W}{Q_H}$$
(Section 20-2)

Efficiency of heat engine following a Carnot cycle:
$$e = 1 - \frac{T_L}{T_H}$$
(Section 20-3)

Coefficient of performance of a refrigerator:
$$CP = \frac{Q_L}{W}$$
(Section 20-4)

Coefficient of performance of a Carnot cycle refrigerator:
$$CP_{Carnot} = \frac{T_L}{T_H - T_L}$$
(Section 20-4)

Coefficient of performance of a heat pump:
$$CP = \frac{Q_H}{W}$$
(Section 20-4)

Definition of the change in the entropy of a system:
$$dS = \frac{dQ}{T}$$
(Section 20-5)

Change in the entropy for a finite amount of heat:
$$\Delta S = \int_a^b \frac{dQ}{T}$$
(Section 20-5)

Statistical definition of entropy:
$$S = k \ln W$$
(Section 20-9)

Definition of the thermodynamic temperature scale: $\quad T = \left(273.16\,K\right)\left(\dfrac{Q}{Q_{TP}}\right)$ (Section 20-10)

Chapter Summary

Section 20-1. The Second Law of Thermodynamics—Introduction

The **second law of thermodynamics** states that heat does not flow spontaneously from a cold object to a hot object, but only from a hot object to a cold object.

A **heat engine** is a mechanical device used to transform internal energy into mechanical energy. Its efficiency is limited by the second law of thermodynamics.

Section 20-2. Heat Engines

A simple model of a heat engine is a device that allows heat to be transferred from a high temperature reservoir to a low temperature reservoir, while transforming some of the heat into work. The heat engine satisfies the first law of thermodynamics in that the heat that leaves the high temperature reservoir is equal to the heat that enters the low temperature reservoir plus the work done by the heat engine. The temperatures of the two reservoirs involved in the heat transfer are called the operating temperatures of the heat engine.

The efficiency e of a heat engine is the ratio of the work done by the engine and the amount of heat taken from the high temperature reservoir:

$$e = \frac{W}{Q_H} = 1 - \frac{Q_L}{Q_H}$$

where W is the work done by the engine and Q_H is the heat extracted from the high temperature reservoir. It is found that Q_L cannot be reduced to zero, and this implies that no heat engine can have an efficiency of 1. This statement is called the **Kelvin-Planck statement of the second law of thermodynamics**. Another form of the second law of thermodynamics states: **no device can exist that transforms a given amount of heat completely into work.**

Example 20-2-A. The efficiency of a heat engine. A heat engine is needed that will output 240 horsepower. At what rate does heat need to be put into the engine if the engine efficiency is 0.323?

Approach: This problem provides us with the power requirement of the heat engine and its efficiency e. Using the definition of the efficiency of the heat engine, we can determine the heat that needs to be extracted from the hot reservoir. Note: 1 hp = 746 W.

Solutions: The efficiency of the heat engine is:

$$e = \frac{W}{Q_H}$$

During a one-second time interval, the work done is $W = 240 \times 746 = 1.79 \times 10^5$ J. The heat extracted from the hot reservoir during this time interval is

$$Q_H = \frac{W}{e} = \frac{1.79 \times 10^5}{0.323} = 5.54 \times 10^5 \text{ J}$$

The rate of heat flow from the hot reservoir is thus 5.54×10^5 J/s.

Section 20-3. Reversible and Irreversible Processes; Carnot Engine

A **reversible process** is a process in which the system is always in thermodynamic equilibrium. This means that the temperature must be constant throughout the system. It is called reversible because there is never a flow of heat from the higher temperature reservoir to the lower temperature reservoir, so that we could reverse the process without requiring heat to flow from the lower temperature reservoir to the higher temperature reservoir. A process that is not reversible is said to be an **irreversible process**. If we were to reverse the direction of an irreversible process, we would observe heat flowing from lower temperatures to higher temperatures. Any real process will create finite temperature differences in a system, and is irreversible.

A **Carnot cycle** is a thermodynamic cycle that is reversible. In a Carnot cycle, the system undergoes an isothermal change in volume at a temperature T_H. This is followed by an adiabatic expansion, an isothermal compression at temperature T_L, and finally an adiabatic compression to bring the system back to its original state. The efficiency of a Carnot cycle is

$$e = 1 - \frac{T_L}{T_H}$$

This is the maximum possible efficiency of a heat engine. The efficiency of any irreversible heat engine is always less than the efficiency of a Carnot cycle.

Example 20-3-A. Optimizing the efficiency of a heat engine. A heat engine loses its waste heat at a temperature of 30°C. What is the minimal possible temperature that heat must be added to the engine so that the efficiency of the engine is 82%?

Approach: Since no heat engine is more efficient than a Carnot engine, we can use the efficiency of the Carnot engine to define the minimum temperature of the high temperature reservoir.

Solution: The maximum efficiency of a heat engine is the Carnot efficiency e:

$$e = 1 - \frac{T_L}{T_H}$$

where the temperatures are absolute temperatures. The temperature of the cold reservoir is 30°C. Using the efficiency e we can determine the temperature of the hot reservoir:

$$T_H = \frac{T_L}{1-e} = \frac{30+273}{1-0.82} = \frac{303}{0.18} = 1.7 \times 10^3 \text{ K}$$

To achieve the efficiency of 82%, the minimum temperature of the hot reservoir must be 1.7×10^3 K.

Example 20-3-B. The combustion engine. The temperature of combustion in a car engine is $T_H = 4200$ K. The gas leaving the engine is still very hot, and it is possible to use this hot exhaust gas to add heat to an auxiliary engine to get additional work out of the energy put into the car. Suppose the exhaust temperature is $T_e = 600$ K and we could construct an auxiliary engine that can exhaust its waste heat at $T_L = 400$ K. Compare the efficiency of a single Carnot engine acting between a temperature of $T_H = 4200$ K and a temperature of $T_L = 400$ K with the total efficiency of two Carnot engines with the first engine operating between temperatures of $T_H = 4200$ K and $T_e = 600$ K and the second engine operating between temperatures of $T_e = 600$ K and $T_L = 400$ K, utilizing the waste heat from the first Carnot engine. The total efficiency of the pair of engines will be the total work done divided by the heat into the first engine.

Approach: The efficiency of the single Carnot engine is easy to calculate. To determine the efficiency of the double Carnot engine we need to calculate the efficiencies of the individual engines to determine the total work done. The heat added to the cold reservoir of engine 1 is equal to the heat extracted from the hot reservoir of engine 2. The total work done is the sum of the work done by engine 1 and engine 2.

Solution: The efficiency of the single Carnot engine is

$$e_{single} = 1 - \frac{T_L}{T_H}$$

Now consider the double engine system. The efficiency of the first engine is

$$e_1 = 1 - \frac{T_e}{T_H}$$

The work done by engine 1 is equal to

$$W_1 = e_1 Q_H = \left(1 - \frac{T_e}{T_H}\right) Q_H$$

The heat added to the cold reservoir of engine 1 is equal to

$$Q_e = (1 - e_1) Q_H = \left(\frac{T_e}{T_H}\right) Q_H$$

The heat added to the cold reservoir of engine 1 is equal to the heat extracted from the hot reservoir of engine 2. The efficiency of engine 2 is equal to

$$e_2 = 1 - \frac{T_L}{T_e}$$

The work done by this engine is equal to

$$W_2 = e_2 Q_e = \left(1 - \frac{T_L}{T_e}\right)\left(\frac{T_e}{T_H}\right) Q_H = \left(\frac{T_e}{T_H} - \frac{T_L}{T_H}\right) Q_H$$

The work done by the double engine is equal to the sum of the work done by engine 1 and engine 2:

$$W_{double} = W_1 + W_2 = \left(1 - \frac{T_e}{T_H}\right) Q_H + \left(\frac{T_e}{T_H} - \frac{T_L}{T_H}\right) Q_H = \left(1 - \frac{T_L}{T_H}\right) Q_H$$

The efficiency of the double engine can now be determined:

$$e_{double} = \frac{W_{double}}{Q_H} = \left(1 - \frac{T_L}{T_H}\right)$$

We conclude that the efficiency of the single engine is the same as the efficiency of the double engine, independent of the value of the intermediate temperature T_e. We should also have expected this from knowing that any two reversible engines acting between the same two temperatures have the same efficiency. Whether these are one or two component engines does not affect its reversibility. For the values provided in the problem, the efficiency is 90.5%.

Section 20-4. Refrigerators, Air Conditioners, and Heat Pumps

If we run a thermodynamic cycle in the opposite direction, we do net work on the system. In this case, heat is removed from the low temperature reservoir and added to the high temperature reservoir. If the system is used for its ability to remove heat from the low temperature reservoir we call the device a **refrigerator.** If the system is used for its ability to add heat to the high temperature reservoir we call the device a **heat pump.**

The **coefficient of performance**, CP, of the refrigerator is defined as:

$$CP = \frac{Q_L}{W}$$

where Q_L is the heat removed from the low temperature reservoir and W is the work done on the system by external forces. The coefficient of performance of a Carnot cycle refrigerator is

$$CP_{Carnot} = \frac{T_L}{T_H - T_L}$$

where T_L is the temperature of the low temperature reservoir and T_H is the temperature of the high temperature reservoir of the Carnot cycle.

The coefficient of performance of a heat pump is defined as the heat that is added to the high temperature side of the system Q_H divided by the work W done by external forces on the system during each thermodynamic cycle:

$$CP = \frac{Q_H}{W}$$

The coefficient of performance of a Carnot cycle heat pump is the maximum coefficient of performance for a given high and low temperature and is given by

$$CP_{Carnot} = \frac{T_H}{T_H - T_C}$$

Example 20-4-A. Heating a house with a heat pump. A heat pump is used to warm a house in winter. The desired temperature inside the house is $T_H = 68°$ F, which can be taken to be the high temperature of the heat pump. (In practice, the high temperature will be somewhat greater than the desired temperature.) Calculate the amount of work that must be done on a Carnot cycle heat pump to add 1.00×10^5 J to the house if the outside air temperature is $T_{L1} = 40°$ F and if it is $T_{L2} = -20°$ F.

Approach: In this problem we examine the coefficient of performance of a heat pump. The calculation of this coefficient and the work that needs to be done to operate the heat pump is straightforward, and we will be able to observe that the work required to add a fixed amount of heat to the inside of the house depends sensitively on the outside temperature. However, no matter what the outside temperature is, the heat added to the inside of the house is more than the work that needs to be done. Heat pumps are an attractive option to heat a house!

Solution: The coefficient of performance of a heat pump is the ratio of the heat added to the hot reservoir and the work done on the pump. A Carnot cycle heat pump has a coefficient of performance given by

$$CP = \frac{T_H}{T_H - T_L}$$

where the temperatures are absolute temperatures. We thus need to convert the temperatures from °F to Kelvin:

$$T_H = 68°F = \frac{5}{9}\left[68 - 32\right]°C = 20°C = 293 \text{ K}$$

$$T_{L1} = 40°F = \frac{5}{9}\left[40 - 32\right]°C = 4°C = 277 \text{ K}$$

$$T_{L2} = -20°F = \frac{5}{9}\left[-20 - 32\right]°C = -29°C = 244 \text{ K}$$

The coefficient of performance of the heat pump at the higher outdoor temperature is equal to

$$CP = \frac{T_H}{T_H - T_{L1}} = \frac{293}{293 - 277} = 18.3$$

The work required to operate the heat pump in this condition is:

$$W = \frac{Q_H}{CP} = \frac{1.00 \times 10^5}{18.3} = 5.5 \times 10^3 \text{ J}$$

The coefficient of performance of the heat pump at the lower outdoor temperature is equal to

$$CP = \frac{T_H}{T_H - T_{L2}} = \frac{293}{293 - 244} = 6.0$$

The work required to operate the heat pump in this condition is:

$$W = \frac{Q_H}{CP} = \frac{1.00 \times 10^5}{6.0} = 1.7 \times 10^4 \text{ J}$$

We observe that the coefficient of performance is reduced when the outside temperature is reduced, and more work is required to operate the heat pump at lower temperatures. This is one of the reasons why a traditional heating system may be required to serve as a backup when a heat pump is installed, especially in cold climates.

Section 20-5. Entropy

The **change in entropy** dS of a system with temperature T to which an infinitesimal amount of heat dQ is added is defined as

$$dS = \frac{dQ}{T}$$

If a finite amount of heat is added to the system, its temperature may change, and we need to integrate dQ/T in order to determine the change in entropy ΔS:

$$\Delta S = \int_a^b \frac{dQ}{T}$$

Note that T is only constant if the process is an isothermal process.

Example 20-5-A. Calculating ΔS. Calculate the entropy change when 1 kg of water at 100°C is changed to steam at 100°C.

Approach: Since the temperature is constant during this process, we only need to evaluate the heat added to the system when the water is changed to steam. The temperature used in our calculation needs to be expressed in Kelvin. The heat of vaporization of water is 2260 kJ/kg.

Solution: The change in the entropy of the system is

$$\Delta S = \int \frac{dQ}{T} = \frac{1}{T}\int dQ = \frac{mL}{T} = \frac{(1)(2260\times 10^3)}{373} = 6.1\times 10^3 \text{ J/K}$$

Section 20-6. Entropy and the Second Law of Thermodynamics

The second law of thermodynamics can be expressed in terms of entropy:

> *The entropy of an isolated system never decreases. It remains constant during reversible processes and it increases during irreversible processes.*

Example 20-6-A. Calculating ΔS - part II. A pot containing 10 kg of water at 20°C is placed on an electric element that is at a temperature 200°C. What is the change in entropy of the system as the water is warmed to 100°C? Ignore the change in entropy of the pot. Assume the range element remains at 200°C.

Approach: Our system contains water and the heating element. For each component we need to evaluate the integral of dQ/T. Assuming that the system is isolated, the heat required to heat the water is removed from the heating element. The magnitude of the entropy change of the water will be different from the magnitude of the entropy change of the heating element since the temperature change during the process is different for both components. The temperature used in our calculation should be the absolute temperature.

Solution: The change in entropy of each component is obtained by integrating $dS = dQ/T$. The heat extracted from the electric element can be found by determining the heat that is required to heat the water:

$$Q_{water} = mC\Delta T = (10)(4186)(373-293) = 3.35\times 10^6 \text{ J}$$

The change in entropy of the range element is

$$\Delta S_{range} = \int \frac{dQ}{T} = \frac{\int dQ}{T_{range}} = -\frac{3.35 \times 10^6}{473} = -7.08 \times 10^3 \text{ J / K}$$

The minus sign in this equation is due to the fact that heat is removed from the range; its energy decreases.
To calculate the change in the entropy of the water we need to evaluate the integral of dQ/T since the temperature of the water changes during the heating process. The change in entropy of the water is equal to

$$\Delta S_{water} = \int \frac{dQ}{T} = \int_{T_i}^{T_f} \frac{mC\,dT}{T} = mC \ln \frac{T_f}{T_i} = (10)(4186) \ln\left(\frac{373}{293}\right) = 1.01 \times 10^4 \text{ J/K}$$

The total change in the entropy of the system will be:

$$\Delta S = \Delta S_{range} + \Delta S_{water} = -7.08 \times 10^3 + 1.01 \times 10^4 = +3.0 \times 10^3 \text{ J/K}$$

Since the heating process is an irreversible process, as a consequence of the first law of thermodynamics, the entropy must increase ($\Delta S > 0$).

Section 20-7. Order to Disorder

Entropy is a measure of the disorder of the system. A state of low entropy is a state with a great amount of order; a state of high entropy is a state with a great amount of disorder. We can rewrite the second law of thermodynamics in terms of the order/disorder of a system: **natural processes tend to move systems from more ordered states to more disordered states.**

Section 20-8. Energy Availability; Heat Death

All real processes are irreversible thermodynamic processes. This means that the change in entropy is greater than zero for all real processes and that some energy that was available for transformation into some other form of energy has become internal energy at a lower temperature. This internal energy is unavailable to produce mechanical work. We conclude that in any natural process, some energy becomes unavailable to do useful work. The energy is **degraded** into internal or thermal energy.

Eventually, in any finite universe, all available energy will be transformed into internal energy at a uniform low temperature, and the entropy will have reached its maximum value. No further processes can occur. This is called the **heat death** of the universe.

Section 20-9. Statistical Interpretation of Entropy and the Second Law

The laws of classical physics are deterministic: if we completely know the exact position and velocity of all particles within a system and the forces that act between those particles, we can, in principle, determine the position and velocity of each particle at any future time. There are several reasons we are unable to apply this in practice:

- The number of particles (atoms) in any macroscopic system is too large to keep track of, let alone solve each of the equations of motion for each particle.
- The measured position and velocity of the particles in any system have limited accuracy. Small errors at one instant can result in very large errors in the predicted position and velocity at later times.

Specifying all the detailed information about a system, such as the position and velocity of each particle, determines a **microstate** of the system. Even for small amounts of gas, it is impossible to know the position and velocity of each particle at any given moment. However, it is possible to determine the temperature of the gas, which is equivalent to the average kinetic energy of the gas particles. The temperature, along with pressure and density, determines macrostates of the gas.

Even if we only have information about the macrostates of a system, we can turn to statistical physics to help us answer questions about the global evolution of the system. Statistical physics relies on the assumption that for any macrostate, it is equally probable to find the system in any of the microstates consistent with the given macrostate. The probability of finding a system in a given macrostate is thus equal to the number of microstates that produce a particular macrostate divided by the total number of possible microstates of the system.

It can be shown that the entropy change $dS = dQ/T$ is consistent with a definition of entropy of a given macrostate based on counting the number of corresponding microstates. A statistical definition of entropy of a given macrostate is

$$S = k \ln W$$

where k is Boltzmann's constant and W is the number of microstates corresponding to the macrostate.

The second law of thermodynamics can be rewritten as: **the most likely state to find a system in is that with the greatest entropy.** The second law of thermodynamics does not make a definite statement about what will happen, but only states what the most likely thing to happen is. For any macroscopic system, the probability to observe a decrease of its entropy is so small that we can ignore this possibility with a great deal of confidence, but we cannot rule it out.

Example 20-9-A. Entropy of tossing coins. Examine what happens if we toss three coins and observe whether they land as heads or tails. If we define our macrostates to be the number of coins that land as heads, we observe four possible microstates. Calculate the entropies for these macrostates.

Approach: We have to determine all possible results of the tossing coins to determine the possible microstates. We can determine the entropy of each macrostate by counting the number of microstates that are consistent with these macrostates.

Solution: We can make a table of all possible results of the coin toss. The first column represents the face showing on the first coin, the second column represents the face showing on the second coin, and the third column represents the face showing on the third coin.

H	H	H
H	H	T
H	T	H
H	T	T
T	H	H
T	H	T
T	T	H
T	T	T

This list exhausts all possibilities and represents the microstates of the system. Using this table we can determine the entropy of each of the four possible macrostates (number of heads = 0, 1, 2, 3):

- The number of microstates corresponding to the macrostate with three heads is 1, so

$$S_{3\ heads} = k \ln W = \left(1.38 \times 10^{-23}\right) \ln 1 = 0 \text{ J/K}$$

- The number of microstates corresponding to the macrostate with two heads is 3, so

$$S_{2\ heads} = k \ln W = \left(1.38 \times 10^{-23}\right) \ln 3 = 1.52 \times 10^{-23} \text{ J/K}$$

- The number of microstates corresponding to the macrostate with one heads is 3, so

$$S_{1\ head} = k \ln W = \left(1.38 \times 10^{-23}\right) \ln 3 = 1.52 \times 10^{-23} \text{ J/K}$$

- The number of microstates corresponding to the macrostate with no heads is 1, so

$$S_{no\ heads} = k \ln W = \left(1.38 \times 10^{-23}\right) \ln 1 = 0 \text{ J/K}$$

Section 20-10. Thermodynamic Temperature; Third Law of Thermodynamics

Temperature can be defined in terms of the amount of heat exchanged during the two isothermal processes of a Carnot cycle. The definition of the Kelvin or thermodynamic temperature scale is

$$T = \left(273.16 \text{ K}\right)\left(\frac{Q}{Q_{TP}}\right)$$

where Q is the heat exchanged with the heat reservoir maintained at a temperature T and Q_{TP} is the heat exchanged with the heat reservoir maintained at the triple point temperature of water. The triple point temperature of water is defined to be 273.16 K. This definition is in agreement with the definition using a constant volume ideal gas thermometer. It does not depend on the properties of a particular substance.

It is impossible to bring a system to absolute zero from any finite temperature above absolute zero in any finite number of cooling processes. This is a statement of the **third law of thermodynamics**.

Section 20-11. Thermal pollution, Global Warming, and Energy Resources

Since the second law of thermodynamics requires that any heat engine adds heat to its cold reservoir, it is impossible to operate any engine without adding heat to the environment. This heat is often referred to as **thermal pollution.** Other types of pollution, such as **air pollution** that contributes to **global warming**, may be reduced or eliminated by switching to other energy sources. However, thermal pollution is unavoidable.

Practice Quiz

1. You measure the efficiency of an engine to be different than a Carnot engine acting between the same two temperatures. What statement can you make about the efficiency and the engine cycle?
    a) The efficiency can be less than or greater than the Carnot efficiency and the cycle is a reversible cycle.
    b) The efficiency of the engine is less than the Carnot efficiency and the cycle can be reversible or irreversible.
    c) The efficiency of the engine is greater than the Carnot efficiency and the cycle is reversible.
    d) The efficiency of the engine is less than the Carnot efficiency and the cycle is irreversible.

2. The work done in proceeding around a thermodynamic cycle is the area enclosed by a cycle. The largest area for a given perimeter is that of a circle. Why isn't the highest efficiency cycle shaped like a circle on a pressure volume graph?
    a) A circular cycle is impossible.
    b) It isn't the same shape as a Carnot cycle.
    c) It could be if the cycle were followed reversibly.
    d) It requires negative pressure, which is impossible.

3. Heat pumps are used more often where winters are moderate than where winters are very cold. Why?
    a) The coefficient of performance decreases the further the outdoor temperature drops.
    b) The coefficient of performance increases the further the outdoor temperature drops.
    c) Fossil fuel for heating is cheap where the winters are cold.
    d) Heat pumps can't operate where the temperature is very cold.

4. Can you heat your home using the internal energy contained in ice?
    a) No, heat flows from hot to cold only.
    b) Yes, but only to heat your home from below the freezing temperature of water up to the freezing temperature of water.
    c) No, there is not enough internal energy in ice to heat anything.
    d) Yes, but it requires work.

5. Which of the following is closest to being a reversible process?
    a) A vase falls off a shelf and breaks upon hitting the floor.
    b) A cup of water at room temperature evaporates.
    c) An ice cube is melted by dropping it into a cup of very hot water.
    d) A car engine which operates with an efficiency of half that of a Carnot engine.

6. Which of the following changes would decrease the input work needed to remove a given amount of heat from the inside of a refrigerator using a Carnot cycle?
    a) Raise the exterior temperature.
    b) Lower the interior temperature of the refrigerator.
    c) Raise the interior temperature of the refrigerator.
    d) Change to an irreversible cycle refrigerator.

7. What is the heat death of the universe?
    a) Eventually, the universe will become so hot from waste heat that it will vaporize.
    b) Eventually, the entropy of the universe will maximize and no further irreversible thermodynamic changes will occur.
    c) One day the Sun will become a giant star and burn up everything.
    d) My brain will overheat from trying to learn thermodynamics and the universe will cease to exist as far as I am concerned.

8. Which of these statements is not a statement of the second law of thermodynamics?
    a) The net result of any physical process cannot be the conversion of heat completely into work.
    b) Heat cannot flow spontaneously from a low temperature system to a higher temperature system.
    c) Internal energy cannot be transferred from a lower temperature system to a higher temperature system.
    d) In any physical process, the change in entropy of the universe is never negative.

9. What is the probability of flipping a fair coin twice and getting identical results on both flips?
    a) 1
    b) 1/2
    c) 1/3
    d) 1/4

10. Flipping 1000 coins at one time, which of the following is the most probable outcome?
    a) All heads
    b) All tails
    c) 400 heads, 600 tails
    d) 472 heads, 528 tails

11. Determine the efficiency of a heat engine that lifts 200 kg a distance of 14.6 m when 5.67×10^4 J of heat are added to the engine.

12. Determine the rate at which heat must flow into a Carnot cycle engine acting between 2400 K and 860 K if the output power of the engine is 47.2 hp.

13. Calculate the total change in entropy of the water in a 50-g ice cube at $-10°$ C when it is warmed to the melting temperature, melted, and then warmed to $+20°$ C.

14. Calculate the total change in entropy of the universe if the ice cube in Question 13 received all of its heat from a source at 20° C.

15. Two dice are rolled simultaneously. Calculate the probability for each possible total of the spots (macrostate) showing on the upper surface. List the microstates that correspond to each macrostate.

Responses to Select End-of-Chapter Questions

1. Yes, mechanical energy can be transformed completely into heat or internal energy, as when an object moving over a surface is brought to rest by friction. All of the original mechanical energy is converted into heat. No, the reverse cannot happen (second law of thermodynamics) except in very special cases (reversible adiabatic expansion of an ideal gas). For example, in an explosion, a large amount of internal energy is converted into mechanical energy, but some internal energy is lost to heat or remains as internal energy of the explosion fragments.

7. The two main factors which keep real engines from Carnot efficiency are friction and heat loss to the environment.

13. 1 kg of liquid iron will have greater entropy, since it is less ordered than solid iron and its molecules have more thermal motion. In addition, heat must be added to solid iron to melt it; the addition of heat will increase the entropy of the iron.

19. No. Even if the powdered milk is added very slowly, it cannot be re-extracted from the water without very large investments of energy. This is not a reversible process.

Solutions to Select End-of-Chapter Problems

1. The efficiency of a heat engine is given by Eq. 20-1a. We also invoke energy conservation.

$$e = \frac{W}{Q_H} = \frac{W}{W + Q_L} = \frac{2600 \text{ J}}{2600 \text{ J} + 7800 \text{ J}} = 0.25 = \boxed{25\%}$$

7. (a) To find the efficiency, we need the heat input and the heat output. The heat input occurs at constant pressure, and the heat output occurs at constant volume.

$$e = 1 - \frac{Q_L}{Q_H} = 1 - \frac{nC_V(T_d - T_a)}{nC_P(T_c - T_b)} = 1 - \frac{(T_d - T_a)}{\gamma(T_c - T_b)}$$

So we need to express the temperatures in terms of the corresponding volume, and use the ideal gas law and the adiabatic relationship between pressure and volume to get those expressions. Note that $P_c = P_b$ and $V_d = V_a$.

$$PV = nRT \rightarrow T = \frac{PV}{nR} \rightarrow$$

$$e = 1 - \frac{\left(\frac{P_d V_d}{nR} - \frac{P_a V_a}{nR}\right)}{\gamma\left(\frac{P_c V_c}{nR} - \frac{P_b V_b}{nR}\right)} = 1 - \frac{(P_d V_d - P_a V_a)}{\gamma(P_c V_c - P_b V_b)} = 1 - \frac{V_a(P_d - P_a)}{\gamma P_b(V_c - V_b)} = 1 - \frac{\left(\frac{P_d}{P_b} - \frac{P_a}{P_b}\right)}{\gamma\left(\frac{V_c}{V_a} - \frac{V_b}{V_a}\right)}$$

$$= 1 - \frac{\left(\frac{P_d}{P_c} - \frac{P_a}{P_b}\right)}{\gamma\left(\frac{V_c}{V_a} - \frac{V_b}{V_a}\right)}$$

Use the adiabatic relationship between pressure and volume on the two adiabatic paths.

$$P_d V_d^\gamma = P_c V_c^\gamma \rightarrow \frac{P_d}{P_c} = \left(\frac{V_c}{V_d}\right)^\gamma = \left(\frac{V_c}{V_a}\right)^\gamma \; ; \; P_a V_a^\gamma = P_b V_b^\gamma \rightarrow \frac{P_a}{P_b} = \left(\frac{V_b}{V_a}\right)^\gamma$$

$$e = 1 - \frac{\left(\frac{P_d}{P_c} - \frac{P_a}{P_b}\right)}{\gamma\left(\frac{V_c}{V_a} - \frac{V_b}{V_a}\right)} = 1 - \frac{\left(\frac{V_c}{V_a}\right)^\gamma - \left(\frac{V_b}{V_a}\right)^\gamma}{\gamma\left[\left(\frac{V_a}{V_c}\right)^{-1} - \left(\frac{V_a}{V_b}\right)^{-1}\right]} = \boxed{1 - \frac{\left(\frac{V_a}{V_c}\right)^{-\gamma} - \left(\frac{V_a}{V_b}\right)^{-\gamma}}{\gamma\left[\left(\frac{V_a}{V_c}\right)^{-1} - \left(\frac{V_a}{V_b}\right)^{-1}\right]}}$$

(b) For a diatomic ideal gas, $\gamma = 7/5 = 1.4$.

$$e = 1 - \frac{\left(\dfrac{V_a}{V_c}\right)^{-\gamma} - \left(\dfrac{V_a}{V_b}\right)^{-\gamma}}{\gamma\left[\left(\dfrac{V_a}{V_c}\right)^{-1} - \left(\dfrac{V_a}{V_b}\right)^{-1}\right]} = 1 - \frac{(4.5)^{-1.4} - (16)^{-1.4}}{1.4\left[(4.5)^{-1} - (16)^{-1}\right]} = \boxed{0.55}$$

13. The maximum (or Carnot) efficiency is given by Eq. 20-3, with temperatures in Kelvin.

$$e = 1 - \frac{T_L}{T_H} = 1 - \frac{(330 + 273)\,K}{(665 + 273)\,K} = 0.354$$

Thus the total power generated can be found as follows.

Actual Power $= \left(\text{Total Power}\right)\left(\text{max. eff.}\right)\left(\text{operating eff.}\right) \rightarrow$

$$\text{Total Power} = \frac{\text{Actual Power}}{\left(\text{max. eff.}\right)\left(\text{operating eff.}\right)} = \frac{1.2\,\text{GW}}{(0.354)(0.65)} = 5.215\,\text{GW}$$

Exhaust Power = Total Power – Actual Power = 5.215 GW – 1.2 GW = 4.015 GW

$$= \left(4.015 \times 10^9\,\text{J/s}\right)\left(3600\,\text{s/h}\right) = \boxed{1.4 \times 10^{13}\,\text{J/h}}$$

19. (a) The pressures can be found from the ideal gas equation.

$$PV = nRT \rightarrow P = \frac{nRT}{V} \rightarrow$$

$$P_a = \frac{nRT_a}{V_a} = \frac{(0.50\,\text{mol})(8.314\,\text{J/mol}\cdot\text{K})(743\,\text{K})}{7.5 \times 10^{-3}\,\text{m}^3}$$

$$= 4.118 \times 10^5\,\text{Pa} \approx \boxed{4.1 \times 10^5\,\text{Pa}}$$

$$P_b = \frac{nRT_b}{V_b} = \frac{(0.50\,\text{mol})(8.314\,\text{J/mol}\cdot\text{K})(743\,\text{K})}{15.0 \times 10^{-3}\,\text{m}^3}$$

$$= 2.059 \times 10^5\,\text{Pa} \approx \boxed{2.1 \times 10^5\,\text{Pa}}$$

(b) The volumes can be found from combining the ideal gas law and the relationship between pressure and volume for an adiabatic process.

$$P_c V_c^{\gamma} = P_b V_b^{\gamma} \rightarrow \frac{nRT_c}{V_c} V_c^{\gamma} = \frac{nRT_b}{V_b} V_b^{\gamma} \rightarrow T_c V_c^{\gamma-1} = T_b V_b^{\gamma-1} \rightarrow$$

$$V_c = \left(\frac{T_b}{T_c}\right)^{\frac{1}{\gamma-1}} V_b = \left(\frac{743\,\text{K}}{533\,\text{K}}\right)^{2.5} (15.0\,\text{L}) = \boxed{34.4\,\text{L}}$$

$$V_d = \left(\frac{T_a}{T_d}\right)^{\frac{1}{\gamma-1}} V_a = \left(\frac{743\,\text{K}}{533\,\text{K}}\right)^{2.5} (7.5\,\text{L}) = 17.2\,\text{L} \approx \boxed{17\,\text{L}}$$

(c) The work done at a constant temperature is given by Eq. 19-8.

$$W = nRT \ln\left(\frac{V_b}{V_a}\right) = (0.50\,\text{mol})(8.314\,\text{J/mol}\cdot\text{K})(743\,\text{K}) \ln\left(\frac{15.0}{7.5}\right) = 2141\,\text{J} \approx \boxed{2100\,\text{J}}$$

Note that this is also the heat input during that process.

(d) Along process cd, there is no change in internal energy since the process is isothermic. Thus by the first law of thermodynamics, the heat exhausted is equal to the work done during that process.

$$Q = W = nRT \ln\left(\frac{V_d}{V_c}\right) = (0.50\,\text{mol})(8.314\,\text{J/mol·K})(533\,\text{K})\ln\left(\frac{17.2}{34.4}\right) = -1536\,\text{J} \approx \boxed{-1500\,\text{J}}$$

So 2100 J of heat was exhausted during process cd.

(e) From the first law of thermodynamics, for a closed cycle, the net work done is equal to the net heat input.

$$W_{\text{net}} = Q_{\text{net}} = 2141\,\text{J} - 1536\,\text{J} = 605\,\text{J} \approx \boxed{600\,\text{J}}$$

(f) $e = \dfrac{W}{Q_H} = \dfrac{605\,\text{J}}{2141\,\text{J}} = \boxed{0.28}$; $e = 1 - \dfrac{T_L}{T_H} = 1 - \dfrac{533\,\text{K}}{743\,\text{K}} = 0.28$

25. (a) The total rate of adding heat to the house by the heat pump must equal the rate of heat lost by conduction.

$$\frac{Q_L + W}{\Delta t} = (650\,\text{W/C}°)(T_{\text{in}} - T_{\text{out}})$$

Since the heat pump is ideal, we have the following.

$$1 - \frac{T_{\text{out}}}{T_{\text{in}}} = 1 - \frac{Q_L}{Q_H} = 1 - \frac{Q_L}{Q_L + W} = \frac{W}{Q_L + W} \;\rightarrow\; Q_L + W = W\frac{T_{\text{in}}}{T_{\text{in}} - T_{\text{out}}}$$

Combine these two expressions, and solve for T_{out}.

$$\frac{Q_L + W}{\Delta t} = (650\,\text{W/C}°)(T_{\text{in}} - T_{\text{out}}) = \frac{W}{\Delta t}\frac{T_{\text{in}}}{(T_{\text{in}} - T_{\text{out}})} \;\rightarrow\; (T_{\text{in}} - T_{\text{out}})^2 = \frac{W}{\Delta t}\frac{T_{\text{in}}}{(650\,\text{W/C}°)} \;\rightarrow$$

$$T_{\text{out}} = T_{\text{in}} - \sqrt{\frac{W}{\Delta t}\frac{T_{\text{in}}}{(650\,\text{W/C}°)}} = 295\,\text{K} - \sqrt{(1500\,\text{W})\frac{295\,\text{K}}{(650\,\text{W/C}°)}} = 269\,\text{K} = \boxed{-4°C}$$

(b) If the outside temperature is 8°C, then the rate of heat loss by conduction is found to be (650 W/°C)(14°C) = 9100 W. The heat pump must provide this much power to the house in order for the house to stay at a constant temperature. That total power is $(Q_L + W)/\Delta t$. Use this to solve for the rate at which the pump must do work.

$$(Q_L + W)/\Delta t = \frac{W}{\Delta t}\left(\frac{T_{\text{in}}}{T_{\text{in}} - T_{\text{out}}}\right) = 9100\,\text{W} \;\rightarrow$$

$$\frac{W}{\Delta t} = 9100\,\text{W}\left(\frac{T_{\text{in}} - T_{\text{out}}}{T_{\text{in}}}\right) = 9100\,\text{W}\left(\frac{14\,\text{K}}{295\,\text{K}}\right) = 432\,\text{W}$$

Since the maximum power the pump can provide is 1500 W, the pump must work 432 W/1500 W = 0.29 or $\boxed{29\%}$ of the time.

31. The coefficient of performance is the heat removed from the low-temperature area divided by the work done to remove the heat. In this case, the heat removed is the latent heat released by the freezing ice, and the work done is 1.2 kW times the elapsed time. The mass of water frozen is its density times its volume.

$$\text{COP} = \frac{Q_L}{W} = \frac{mL_f}{W} = \frac{\rho V L_f}{Pt} \;\rightarrow$$

$$V = \frac{(\text{COP})Pt}{\rho L_f} = \frac{(7.0)(1200\,\text{W})(3600\,\text{s})}{(1.0 \times 10^3\,\text{kg/m}^3)(3.33 \times 10^5\,\text{J/kg})} = 0.0908\,\text{m}^3 \approx \boxed{91\,\text{L}}$$

37. The same amount of heat that leaves the high temperature heat source enters the low temperature body of water.

$$\Delta S = \Delta S_1 + \Delta S_2 = -\frac{Q}{T_{high}} + \frac{Q}{T_{low}} = Q\left(\frac{1}{T_{low}} - \frac{1}{T_{high}}\right) \quad \rightarrow$$

$$\frac{\Delta S}{t} = \frac{Q}{t}\left(\frac{1}{T_{low}} - \frac{1}{T_{high}}\right) = \left(9.50\,\text{cal/s}\right)\left(\frac{4.186\,\text{J}}{1\,\text{cal}}\right)\left(\frac{1}{(22+273)\,\text{K}} - \frac{1}{(225+273)\,\text{K}}\right)$$

$$= \boxed{5.49\times10^{-2}\,\frac{\text{J/K}}{\text{s}}}$$

43. (*a*) To approximate, we use the average temperature of the water.

$$\Delta S = \int \frac{dQ}{T} \approx \frac{\Delta Q}{T_{avg}} = \frac{mc\Delta T_{water}}{T_{avg}} = \frac{\left(1.00\,\text{kg}\right)\left(4186\,\dfrac{\text{J}}{\text{kg}\cdot\text{K}}\right)\left(75\,\text{C}^\circ\right)}{\left(273+\tfrac{1}{2}75\right)\text{K}} = 1011\,\text{J/K} \approx \boxed{1010\,\text{J/K}}$$

(*b*) The heat input is given by $Q = mc\Delta T$, so $dQ = mcdT$.

$$\Delta S = \int \frac{dQ}{T} = \int_{T_1}^{T_2} \frac{mcdT}{T} = mc\ln\frac{T_2}{T_1} = \left(1.00\,\text{kg}\right)\left(4186\,\dfrac{\text{J}}{\text{kg}\cdot\text{K}}\right)\ln\left[\frac{(273+75)\text{K}}{(273)\text{K}}\right]$$

$$= 1016\,\text{J/K} \approx \boxed{1020\,\text{J/K}}$$

The approximation is only about 1% different.

(*c*) We assume that the temperature of the surroundings is constant at 75°C (the water was moved from a cold environment to a hot environment).

$$\Delta S = \int \frac{dQ}{T} = \frac{\Delta Q}{T} = \frac{-mc\Delta T_{water}}{T_{surroundings}}\frac{-\left(1.00\,\text{kg}\right)\left(4186\,\dfrac{\text{J}}{\text{kg}\cdot\text{K}}\right)\left(-75\,\text{C}^\circ\right)}{(273+75)\text{K}} = \boxed{-900\,\text{J/K}}\,(2\text{ sig. fig.})$$

If instead the heating of the water were done reversibly, the entropy of the surroundings would decrease by 1020 J/K. For a general non-reversible case, the entropy of the surroundings would decrease, but by less than 1020 J/K (as in the calculation here).

49. For a system with constant volume, the heat input is given by Eq. 19-10a, $Q = nC_V\Delta T$. At temperature T, an infinitesimal amount of heat would result in an infinitesimal temperature change, related by $dQ = nC_V dT$. Use this with the definition of entropy.

$$dS = \frac{dQ}{T} = \frac{nC_V dT}{T} \quad \rightarrow \quad \frac{dT}{dS} = \frac{T}{nC_V}$$

This is the exactly the definition of the slope of a process shown on a *T–S* graph, so $\boxed{\text{slope} = \dfrac{T}{nC_V}}$. A function with

this property would be $T = T_0 e^{S/nC_V}$.

55. From the table below, we see that there are a total of $2^6 = 64$ microstates.

Macrostate	Possible Microstates (H = heads, T = tails)						Number of microstates
6 heads, 0 tails	H H H H H H						1
5 heads, 1 tails	H H H H H T	H H H H T H	H H H T H H	H H T H H H	H T H H H H	T H H H H H	6
4 heads, 2 tails	H H H H T T	H H H T H T	H H T H H T	H T H H H T	T H H H H T		15
	H H H T T H	H H T H T H	H T H H T H	T H H H T H	H H T T H H		
	H T H T H H	T H H T H H	H T T H H H	T H T H H H	T T H H H H		
3 heads, 3 tails	H H H T T T	H H T H T T	H T H H T T	T H H H T T	H H T T H T		20
	H T H T H T	T H H T H T	H T T H H T	T H T H H T	T T H H H T		
	T T T H H H	T T H T H H	T H T T H H	H T T T H H	T T H H T H		
	T H T H T H	H T T H T H	T H H T T H	H T H T T H	H H T T T H		
2 heads, 4 tails	T T T T H H	T T T H T H	T T H T T H	T H T T T H	H T T T T H		15
	T T T H H T	T T H T H T	T H T T H T	H T T T H T	T T H H T T		
	T H T H T T	H T T H T T	T H H T T T	H T H T T T	H H T T T T		
1 heads, 5 tails	T T T T T H	T T T T H T	T T T H T T	T T H T T T	T H T T T T	H T T T T T	6
0 heads, 6 tails	T T T T T T						1

 (a) The probability of obtaining three heads and three tails is $20/64$ or $5/16$.
 (b) The probability of obtaining six heads is $1/64$.

61. We assume that the electrical energy comes from the 100% effective conversion of the gravitational potential energy of the water.

$$W = mgh \;\rightarrow$$

$$P = \frac{W}{t} = \frac{m}{t}gh = \rho\frac{V}{t}gh = \left(1.00\times10^3\,\text{kg/m}^3\right)\left(32\,\text{m}^3/\text{s}\right)\left(9.80\,\text{m/s}^2\right)\left(38\,\text{m}\right) = \boxed{1.2\times10^7\,\text{W}}$$

67. (a) The exhaust heating rate is found from the delivered power and the efficiency. Use the output energy with the relationship $Q = mc\Delta T = \rho V c\Delta T$ to calculate the volume of air that is heated.

$$e = W/Q_{\text{H}} = W/\left(Q_{\text{L}} + W\right) \;\rightarrow\; Q_{\text{L}} = W\left(1/e - 1\right) \;\rightarrow$$

$$Q_{\text{L}}/t = W/t\left(1/e - 1\right) = \left(9.2\times10^8\,\text{W}\right)\left(1/0.35 - 1\right) = 1.709\times10^9\,\text{W}$$

$$Q_{\text{L}} = mc\Delta T \;\rightarrow\; Q_{\text{L}}/t = \frac{mc\Delta T}{t} = \frac{\rho V c\Delta T}{t} \;\rightarrow\; V/t = \frac{\left(Q_{\text{L}}/t\right)}{\rho c\Delta T}$$

The change in air temperature is 7.0°C. The heated air is at a constant pressure of 1 atm.

$$V/t = \frac{\left(Q_{\text{L}}/t\right)t}{\rho c\Delta T} = \frac{\left(1.709\times10^9\,\text{W}\right)\left(8.64\times10^4\,\text{s/day}\right)}{\left(1.2\,\text{kg/m}^3\right)\left(1.0\times10^3\,\text{J/kg}\cdot\text{C}^\circ\right)\left(7.0\text{C}^\circ\right)}$$

$$= 1.757\times10^{10}\,\text{m}^3/\text{day}\left(\frac{10^{-9}\text{km}^3}{1\,\text{m}^3}\right) = 17.57\,\text{km}^3/\text{day} \approx \boxed{18\,\text{km}^3/\text{day}}$$

 (b) If the air is 200 m thick, find the area by dividing the volume by the thickness.

$$A = \frac{\text{Volume}}{\text{thickness}} = \frac{17.57\,\text{km}^3}{0.15\,\text{km}} = 117\,\text{km}^2 \approx \boxed{120\,\text{km}^2}$$

This would be a square of approximately 6 miles to a side. Thus the local climate for a few miles around the power plant might be heated significantly.

73. Take the energy transfer to use as the initial kinetic energy of the cars, because this energy becomes "unusable" after the collision – it is transferred to the environment.

$$\Delta S = \frac{Q}{T} = \frac{2\left(\frac{1}{2}mv_i^2\right)}{T} = \frac{\left(1100\,\text{kg}\right)\left[\left(75\,\text{km/h}\right)\left(\dfrac{1\,\text{m/s}}{3.6\,\text{km/h}}\right)\right]^2}{\left(15+273\right)\text{K}} = \boxed{1700\,\text{J/K}}$$

79. (*a*) For the Carnot cycle, two of the processes are reversible adiabats, which are constant entropy processes. The other two processes are isotherms, at the low and high temperatures. See the adjacent diagram.

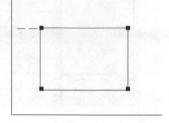

(*b*) The area underneath any path on the T-S diagram would be written as

$$\int T\, dS$$

This integral is the heat involved in the process.

$$\int T\, dS = \int T\frac{dQ}{T} = \int dQ = Q_{\text{net}}$$

For a closed cycle such as the Carnot cycle shown, since there is no internal energy change, the first law of thermodynamics says that

$$\int T\, dS = Q_{\text{net}} = \boxed{W_{\text{net}}},$$

the same as

$$\int P\, dV\,.$$

Quiz Answers

Chapter 1
1. c
2. a
3. a
4. c
5. c
6. b
7. b
8. b
9. b
10. a
11. 27.25
12. 0.2%
13. 5.7%
14. 100,000
15. B^2CD

Chapter 2
1. c
2. a
3. d
4. a
5. c
6. a
7. e
8. b
9. c
10. b
11. 32.1 m
12. 5.03 s
13. First car wins
14. 40 mph
15. 3.0 s

Chapter 3
1. d
2. d
3. b
4. c
5. d
6. a
7. a
8. a
9. c
10. c
11. 8.9, 4.6° N of E
12. 9.48, 88.7° S of W
13. 1.56 s, 1.57 m
14. 5.95×10^{-3} m/s^2
15. 59.1 km/hr, 68.4° W of N

Chapter 4
1. c
2. b
3. d
4. c
5. a
6. b
7. a
8. d
9. b
10. d
11. 98 m/s^2, 17° W of S
12. 560 m
13. 98.8 N
14. 149 N
15. 2.88×10^3 N

Chapter 5
1. b
2. d
3. c
4. d
5. c
6. d
7. b
8. a
9. a
10. a
11. 1.41×10^3 N, down the bank
12. 0.991 m/s^2
13. 13.3 kg
14. 19.2 s
15. 5.37 m/s^2 down

Chapter 6
1. c
2. d
3. b
4. a
5. b
6. c
7. d
8. d
9. c
10. a
11. 1.0×10^4 s
12. $r/(1 + 1/\sqrt{3})$
13. $d(M/m)^{1/3}$
14. 6.7×10^7 m
15. 1.20×10^{-5} N

Chapter 7
1. c
2. d
3. c
4. c
5. d
6. c
7. c
8. d
9. b
10. c
11. 7.0 N
12. 5.93 m/s
13. 246 J, 55°
14. 134 N
15. 1000 m

Chapter 8
1. We know there are non-conservative forces acting on the automobile because the automobile slows down if the engine does not continue to do work. Nonconservative forces that act on the car are the drag force of the air and the friction between the tires and the road.

2. The change in gravitational potential energy only depends on the initial and final height. It does not depend on the path taken.

3. There is not enough information to determine the displacement of the end of the spring. There are two different displacements with the same potential energy: one with the spring compressed and one with the spring stretched.

4. For a constant force, the potential energy increases linearly with position in the direction directly opposite to the direction of the force.

5. Energy conservation, in this context, means to conserve the energy in a useful form. We will learn more about

what that means in Chapter 20.

6. No. Power is proportional to speed, but the energy used is proportional to the time the power is being used, and is thus inversely proportional to speed v. The travel time of car B is half the travel time of the car A, and the total energy used by car A will be the same as the energy used by car B.

7. The equilibrium positions for a three-dimensional system will be those positions where the vector sum of the forces and the vector sum of the torques are equal to zero. The stability conditions are that a small displacement in any direction or rotation about any axis results in an increase in the potential energy of the system.

8. The more energy required to displace or rotate a system to a position in which it is no longer stable, the greater the stability of the system.

9. If there is greater output power than input power, the process does not conserve energy. All real machines have some nonconservative forces that are responsible for negative work done on the system. This means that for a real machine, the output power will always be less than the input power.

10. The force associated with this potential well have a linear dependence on x:
    $$F(x) = a - bx$$
    where a and b are constants.

11. 8.85 mm.

12. $U(x) = (1/2) kx^2 + (1/4) qx^4$

13. 5.48 cm above its initial position.

14. 11.0 N.

15. Equilibrium positions at
    $$x = 1 + \sqrt{(5/3)}$$
    and
    $$x = 1 - \sqrt{(5/3)}.$$

The first equilibrium is unstable and the second equilibrium is stable.

Chapter 9

1. The change in the momentum of the earth is equal in magnitude and opposite in direction to the change in momentum of the car.

2. The piece with less mass will fly out at the highest speed.

3. The two pieces must fly out in exactly opposite directions.

4. If the impulse is applied in the shortest time possible, the train will travel with the shortest time between stops, but the acceleration will be the most uncomfortable to the passengers. If the impulse is applied over a long time, the train will take longer to travel between stations, but the acceleration will be the most comfortable for the passengers.

5. In two and three dimensions, there are more than two unknown components of velocity after the collision. More information must be given to reduce the number of unknowns to the number of relationships between those unknowns. For example, in two dimensions, there are four unknown velocity components after the collision. There are only three relationships between the initial and final velocities: the conservation of momentum in the two perpendicular directions and the conservation of kinetic energy. One additional piece of information must be known to solve the problem.

6. Yes. A horseshoe and a donut are examples of such objects.

7. Many collisions occur in a short time interval with the collision force much larger in magnitude than the external

forces that the effect of the external forces during the collision can be ignored.

8. The padding on the surface extends the time that the collision occurs. For a given change in momentum, this implies a reduced force during the collision.

9. The collapse of the car implies that the driver moves for a greater amount of time before stopping than if the car remained intact and rigid. The same impulse is given to the driver regardless of what happens to the car, but the destructible car allows the impulse to occur over a greater time, reducing the force.

10. A bullet that bounces off of an object has a greater momentum change than a bullet that sticks in the object. This implies that the momentum change of the object struck by the bullet also has a greater momentum change when the bullet bounces off the object.

11. 170 N·s

12. 5.65×10^{-23} m/s

13. 15.0 m/s

14. 0.209 m from the end with the sphere attached

15. 11.8 m/s 29.7° N of W

Chapter 10

1. $x = R \cos \omega t$ and
    $y = R \sin \omega t$

2. No, unless $R = 0$ m.

3. Yes.

4. The angular velocity is directly into the face of the clock.

5. The velocity of the top of the wheel relative to the ground is twice the velocity of the center of mass of the wheel, relative to the ground.

6. The moment of inertia is undefined until the axis of rotation is defined.

7. _____

8. The argument ignores the fact that the slowing of the rotation is caused by an external torque. If there is an external torque, the angular momentum is not conserved.

9. For a given direction of rotation axis, the moment of inertia is minimum when the rotation axis passes through the center of mass of the object.

10. If the knob is in the center of the door, the moment arm is shorter than if the knob is at the edge of the door opposite the hinge. This reduces the torque applied to the door for a given magnitude of force.

11. 23.0 rad/s^2

12. 15.1 rad/s^2

13. $\hat{r}_{cm} = 0.5$ m $\hat{\mathbf{i}} + 0.33$ m $\hat{\mathbf{j}}$

    $I_{cm} = 5.20$ kg m^2

14. 57.5 rev/min

15. 329 rad/s

Chapter 11

1. d
2. b
3. a
4. d
5. a
6. d
7. c
8. d
9. c
10. b
11. $5\hat{\mathbf{i}} + \hat{\mathbf{j}} + 4\hat{\mathbf{k}}$
12. $7\hat{\mathbf{k}}$ Nm
13. 5.5×10^{-14} rad
14. 2.20×10^{-2} kg m^2/s
15. 0.057 rad/s

Chapter 12

1. d
2. d
3. a
4. d

5. c
6. b
7. b
8. c
9. c
10. d
11. 35 N
12. 45 kg
13. 13 kg
14. 9.7×10^{-4} m
15. 0.42

Chapter 13

1. b
2. c
3. d
4. b
5. c
6. d
7. d
8. c
9. d
10. d
11. 680 kg/m^3
12. 3.89×10^3 s
13. 1.56×10^5 N/m^2
14. 47.6 N
15. 13.0 m

Chapter 14

1. d
2. d
3. c
4. c
5. b
6. d
7. d
8. b
9. b
10. d
11. 1.74×10^4 N/m
    939 kg
12. 5.13×10^{-2} s
13. 161 s
14. $\pi/2, 0, -\pi/2$
15. 2.38 cm

Chapter 15

1. b
2. c
3. c
4. c
5. d
6. a

7. a
8. b
9. d
10. b
11. 9.4×10^{10} N/m^2
12. 2.38×10^f J
13. 3.2 cm, 5.2 cm, 1.1 Hz, 0.92 s, 5.7 cm/s
14. 109 Hz
15. 9.56 m/s

Chapter 16

1. b
2. b
3. d
4. a
5. b
6. c
7. d
8. c
9. d
10. d
11. 82 Hz, 164 Hz, 246 Hz
12. 90.4 dB
13. 2.2 m/s
14. 584 m/s
15. 2.1%

Chapter 17

1. a
2. d
3. b
4. c
5. c
6. c
7. c
8. a
9. c
10. b
11. 0.051 in
12. $-40°$ C
13. 8.67×106 N/m^2
14. 173 moles
15. 3.2×10^6

Chapter 18

1. c
2. a
3. c
4. d
5. d
6. b
7. d
8. d

9. a
10. d
11. 8.0×10^{-9} K
12. $\dfrac{2N}{\sqrt{\pi}} \dfrac{\sqrt{E}}{(kT)^{3/2}} e^{-E/kT} dE$
13. 20 g
14. 6.25×10^{18} m^{-3}
15. $a = 50.6$ N·m^4
    $b = 1.00 \times 10^{-2}$ m^3

6	1-5
	2-4
	3-3
	4-2
	5-1
7	1-6
	2-5
	3-4
	4-3
	5-2
	6-1
8	2-6
	3-5
	4-4
	5-3
	6-2
	7-1
9	3-6
	4-5
	5-4
	6-3
10	4-6
	5-5
	6-4
11	5-6
	6-5
12	6-6

Chapter 19

1. b
2. d
3. c
4. b
5. d
6. c
7. d
8. d
9. c
10. a
11. 55.2°
12. 135 s
13. -3.85×10^5 J
14. 5.88×10^{24} J
15. 1.26×10^5 N/m^2
    378 K

Chapter 20

1. b
2. b
3. a
4. d
5. b
6. c
7. b
8. c
9. b
10. d
11. 50.5%
12. 5.49×10^4 J/s
13. 1.9×10^{-2} kcal/K
14. 1.2×10^{-3} kcal/K
15.

macro	micro
2	1-1
3	1-2
	2-1
4	1-3
	2-2
	3-1
5	1-4
	2-3
	3-2
	4-1

NOTES

NOTES